Electrocardiography in Veterinary Medicine

J. P. Varshney

Electrocardiography in Veterinary Medicine

Springer

J. P. Varshney
Veterinary Medicine
Shri Surat Panjarapole Prerit Nandini Veterinary Hospital
Surat, Gujarat
India

ISBN 978-981-15-3698-4 ISBN 978-981-15-3699-1 (eBook)
https://doi.org/10.1007/978-981-15-3699-1

This Springer imprint is published by the registered company Springer Nature Singapore Pte Ltd.
The registered company address is: 152 Beach Road, #21-01/04 Gateway East, Singapore 189721, Singapore

Dedicated
To
My Parents and Teachers
Who
Taught Me The Philosophy of Endurance
And
Karma with Unflinching Faith in Almighty

Foreword

In India, during the last couple of decades or so the interest of practising veterinarians, especially the small animal clinicians and those associated with clinical teaching veterinary schools towards electrocardiography has increased considerably. This is due to the fact that the clinicians have realized that the manifestations of cardiac involvement are seen not only in diseases of cardiac origin but also in the ailments which do not originate in the heart. During these years, the focus on electrocardiography has mainly remained confined to short-term training workshop for small animal clinicians in big cities and/or veterinary educational institutes. Not much has been documented about the electrocardiographic studies undertaken on farm and companion animals and therefore, a need for a textbook on electrocardiography has been felt for quite some time.

The book *Electrocardiography in Veterinary Medicine* written by Dr. J.P. Varshney, Retired Principal Scientist, Indian Veterinary Research Institute, a well-known senior consultant with a private veterinary hospital, is a remarkable collection of data based on his rich clinical experience. The author has attempted to address the long-standing demand of veterinary practitioners, teachers and veterinary students by providing simple, useful and concise information on electrocardiography of various animal species including canines, ruminants, equines and some uncommon species like reptiles, chelonians, mongoose, etc. The book has support in the form of ample number of illustrations and electrocardiographic patterns.

I strongly believe that the book will be a useful guide to the veterinary clinicians, teachers and students in understanding the intricacies of clinical electrocardiography and help in the diagnosis of animals. I am confident that the veterinary professionals will appreciate and use this textbook for the purpose of study and clinical practice.

D. Shekhar Nauriyal, MVSc, PhD
President, Indian Society of Veterinary Medicine
Department of Medicine College of Veterinary Medicine and Animal Husbandry
Anand Agricultural University
Anand, Gujarat, India

Preface

Veterinary medicine is a branch of medicine concerned with diagnosis, and treatment (both curative and preventive) of diseases, disorders and injuries of animals. This specialty of medicine is unique because of its diversified nature of patients right from cattle, buffaloes, sheep, goats, yaks, mithun, camels, pigs, dogs, cats, horses, donkeys, poultry, pigeons, parrots, birds, chelonians (tortoise/turtles), reptiles, squirrel, rabbits, rats, mice, hamsters, guinea pigs, monkeys, mongoose, to wild animals. Each species of animal has variation in physiology, disease occurrence and drug pharmacology. In India, there has been a transition in practice of veterinary medicine largely from equine centred to bovine centred, and recently pet animal practice has occupied prominent place in urban areas. This change in practice of veterinary medicine during recent years has given birth to the concept of various specialties with an increasing demand of specialists who have a thorough knowledge of disease pathogenesis, multi-systemic effects of diseases, well conversant with modern diagnostic modalities and latest therapeutic interventions. The success of treatment depends solely on diagnosis. There is no short cut for diagnosis. During my professional career of 53 years as clinician, teacher and researcher, I always emphasized the importance of history taking, detailed systematic examination of each body system and confirming clinical suspicion with the help of modern diagnostic techniques to arrive at correct diagnosis. Omission of any vital organ during examination may mislead diagnosis.

In routine clinical practice, many of us take many things for granted and leave many organs unexamined when there is no direct history or pathognomonic symptom related to particular organ. Vital body organs are affected both by primary diseases of the organs as well as by diseases of other organs. Hence, there is no compromised approach in diagnosis. The heart, being a central vital organ, is affected by both cardiac and non-cardiac diseases and has a potential to influence the disease outcome. Disturbances of cardiac rate, rhythm and conduction are quite common in almost all species of animals. Cardiomyopathy is of great clinical significance in dogs and cats. Cursory auscultation of the heart, most common technique of evaluation of the heart, fails to differentiate the type of arrhythmia and to assess the size of the heart chambers particularly in dogs and cats. Next step examination of the heart by electrocardiography overcomes these lacunae. Proper and detailed examination of the heart in ailing animals adds to understanding of the disease in totality and thus is of great value in undertaking rational therapeutic measures and predicting clinical outcome of the case. Nowadays, diseases of cardiovascular system (arrhythmias and cardiomyopathies), both as primary and/or secondary, in canines and felines are being confronted in substantial proportions in routine practice. Arrhythmias are also common in other species of animals. It seems imperative for veterinary clinicians and veterinary students to get themselves acquainted with techniques employed in the diagnosis of cardiovascular diseases. Electrocardiography is an important non-invasive tool, one step ahead of auscultation, being employed in clinical examination of the heart in health and diseases of canines and felines in routine clinical practice and also in other species of animals in developed countries. Of late, teaching institutes in India have started focusing on cardiac evaluation techniques in their clinics. Still we have to go a long way to make cardiac evaluation an easy task in our routine clinical practice. Electrocardiography has now started attracting attention of veterinary medical specialists to detect electrical vectors, their magnitude and direction or diagnosing cardiac diseases as well as diseases that secondarily affect the heart, such as metabolic disorders. Nowadays, electrocardiography, an important diagnostic intervention in canine medicine, has a proven usefulness and utility. In India, electrocardiography has still not been exploited fully even in canine and feline clinical practice what to say of large animal practice. These days, a majority of practising field veterinarians and clinical scientists in our country are fully convinced that a systematic and comprehensive evaluation of cardiovascular system should be an integral part of clinical examination of critically ailing animals and a regular health checkup. Clinicians are interested in clinical electrocardiography. But a large number of clinicians are deterred from using an electrocardiograph because of lack of basic knowledge of the potentialities and limitations of the technique; non-availability of a simple and easily understandable handbook on electrocardiography that can assist them in interpreting the electrocardiograms. A number of textbooks and huge literature dealing with fundamental aspects of electrocardiography and experimental studies are available in advanced countries. However, a simple, methodical and point-oriented book suiting to the requirement of veterinary students, practising veterinarians as well as researchers is not available particularly in India and is the demand of the day.

Electrocardiographic examination in isolation has little value and therefore should be conducted in conjunction with basic physical and thoracic radiographical examinations. Advanced techniques viz. Holter electrocardiographic monitoring, echocardiography, and Doppler echocardiography are now also available and are becoming a very important non-invasive diagnostic technique in cardiology. Important information on these aspects has also been added in the book to improve diagnostic interpretations of electrocardiography.

Electrocardiography in Veterinary Medicine has been written to meet the requirement of a book that provides very useful practical information in the simple format on basic elements of electrocardiography with illustrations, interpretations of normal and abnormal wave forms and various electrocardiographic patterns in cardiac and non-cardiac diseases along with treatment approach. A consistent and pragmatic approach to interpretations of canine electrocardiograms will widen the understanding of the clinicians to assist them in interpretation of routine electrocardiograms. Separate sections on electrocardiography of felines (cats), ruminants (cattle, buffaloes, mithun, sheep and goats) and other animals (equines, reptiles, pigeon, rabbits, chelonians, mongoose, leopard, etc.) have provided important information on electrocardiographic indices and their abnormalities in these species to bridge the gap in practical aspects of electrocardiography in species other than canines.

It is hoped that this book will encourage clinicians and clinical scientists to make greater use of this valuable non-invasive diagnostic tool in cardiac evaluation in animals in health and diseases. Many time reading of electrocardiograms does not provide clear-cut information in accordance to set patterns of abnormality associated with specific disease. These situations should be taken as a challenge and the clues are to be searched for and interpreted in the light of clinical, radiographic, echocardiographic and other laboratory findings.

Surat, Gujarat, India J. P. Varshney

Acknowledgements

I am grateful to my better half Mrs. Jai PrabhaVarshney for her emotional and moral support, patience, sacrificing and stoic nature, encouragement and always caring inclination with broad smile. Without her support, this piece of work would not have seen the light of the day. My thanks are also due to my other family members Dr. Jai PratibhaVarshney (daughter) and Col.(Dr.) Atul Seth (son-in-law); Jai Prabhat Varshney (son) and Mrs. Ritu Gupta (daughter-in-law); Mrs. Jai Prabhanshi Varshney (daughter) and Ajit Varshney (son-in-law); and grandchildren (Agrima Seth, Vedaant Varshney, Devansh Varshney and Anvi Varshney) who always make me happy and proud with their love, respect and caring attitude.

I am also indebted to all my animal patients and their owners for their faith and confidence in me and cooperation in willingly undertaking electrocardiography and other investigations without which this text *Electrocardiography in Veterinary Medicine* would not have been possible. I also thank my students (Dr. Tarun Gaur, Dr. Pooja Dixit, Dr. Bendangla Chinkija, Dr. Subhmita Chaudhuri, Dr. Sarita,) who were of great support.

Information on mithun shared by Dr. Akhilesh Kumar, Scientist, National Research centre on Mithun, Dimapur, Nagaland, is duly acknowledged. I thank Dr. Neetu Saini, Associate Professor, Department of Medicine, Guru Angad Dev Veterinary and Animal Sciences University, Ludhiana, for sharing photographs of horse with electrocardiographic leads and an ECG strip.

I take this opportunity to thank Shri. Nayan N. Bharatia Managing Trustee and Board of Trustees, Nandini Veterinary Hospital, Surat, for providing facilities and encouragement.

J. P. Varshney

About the Book

The book *Electrocardiography in Veterinary Medicine* is a unique attempt to provide basic information and methodology for recording electrocardiogram in various species of animals viz. dogs, cats, cattle, buffaloes, sheep, goats, mithun, chelonians, snakes, avians, equines, rabbits, Indian gray mongoose and leopard at one place. The book is divided into four parts. Part I is devoted to canines (dogs) as it is the major animal species where the use of electrocardiography is more extensive, logical and fruitful. Part II is devoted to felines (cats), another species kept as pet, wherein electrocardiography can play a significant role in the diagnosis of arrhythmias. Part III deals with electrocardiographic procedures in ruminants and Part IV describes electrocardiography in other pet and exotic animals. The book is intended not only to provide basic information on electrocardiography, electrocardiographic physiology, generation of electrocardiogram, normal criteria for different species of animals, electrocardiograms in health and diseases, interpretations of abnormal electrocardiograms; cardiomyopathy, arrhythmias along with their treatment protocols but also deals with other evaluation approaches for the heart. The book will provide a very useful to-the-point information in the simplest form on basic elements of electrocardiography with illustrations, interpretations of abnormal wave forms and various electrocardiographic patterns in primary and secondary diseases of the heart along with treatment approach. A consistent and pragmatic approach to interpretations of electrocardiograms of dogs, cats, ruminants, tortoises, pigeons and other animals will widen the understanding of the clinicians to assist them in interpretation of routine electrocardiograms. Electrocardiographic patterns of different species of animals at a glance will evoke interest in furthering research on electrocardiography and evaluation of the heart in different species of animals in veterinary practice.

It is hoped that this book will encourage clinicians and clinical scientists to make greater use of this valuable non-invasive diagnostic tool in the diagnosis of heart diseases as well as general health examination.

Contents

About the Author

J. P. Varshney, BVSc, AH, MVSc, PhD has served in the Division of Veterinary Medicine, Indian Veterinary Research Institute, Izatnagar, Bareilly-243,122 (U.P.), National Research Centre on Equines, Hisar (Haryana), and Indian Grassland and Fodder Research Institute, Jhansi (U.P.), in various capacities for about 35 years conducting research on clinical diseases and managing health of different species of animals. Though, superannuated in February, 2007 (after 35 years of ICAR service) as Principal Scientist (Veterinary Medicine), he is still contributing to the cause of Veterinary Medicine as Senior Consultant (Medicine) at Nandini Veterinary Hospital, Surat (Gujarat). He is a clinical scientist of repute with a vast clinical experience of 53 years and teaching experience of 11 years at post graduate level. He has guided 9 MVSc and 3 PhD scholars. He had also been a visiting Faculty in the area of Animal Cardiology at Department of Medicine, Veterinary College, Dantiwada (11th to 22nd March, 2013) and Veterinary College, GADVASU, Ludhiana (11th to 18th March, 2019). He is the recipient of as many as 33 honours and awards including Dr. K.S. Nair Memorial Gold Medal—1990, Dr. C.G. Bhaskar Gold Medal—1996, Ram Lal Agrawal Gold Medal—1997 for contributions in Veterinary Medicine, Intas Polivet best clinical and research article award (6 times), Award of Merit—2001 and Award of Honour—2002 (IVRI, Izatnagar, for contributions in Research and Academics), Best Teacher Award 2003–2004 (IVRI, Izatnagar), PETCARE Award for Canine Excellence—2005, Vijay Rattan Award and Certificate of Excellence—2005 (India International friendship Society, New Delhi), Sarmaya-e-Hind—2007 Award and Citation for

Outstanding Achievement in Veterinary Medicine (2019) by Indian Society for Veterinary Medicine. In recognition of his contributions in Veterinary Homeopathy, he was appointed as a member of Consultative Committee of Experts in Veterinary Science and had been a member of Special Committee for Fundamental Research in Homoeopathy at Central Council for Research in Homeopathy, New Delhi (Ministry Of Ayush, Gov. of India). He has served as a Member, General Body of U.P. Agricultural Research Council, Lucknow. Dr. Varshney has 382 research and clinical publications in national/international scientific journals of repute, 2 chapters in book, 6 bulletins and monographs, 1 book, 23 review articles, 16 popular articles besides 150 presentations in different National and International Conferences and 10 webinars at his credit.

Abbreviations

<	Less than
>	More than
AI	Aortic Insufficiency
AMPK	Adenosine Monophosphate-Activated Protein Kinase
APC	Atrial Premature complex
AST	Aspartate dehydrogenase
AV block	Atrio-ventricular Block
AV junction	Atrio-ventricular junction
AV node	Atrio-ventricular node
aVF	Augmented Vector Left Foot
aVL	Augmented Vector Left Fore Limb
aVR	Augmented Vector Right Fore limb
B. gibsoni	*Babesia gibsoni*
BBB	Bundle Branch Block
BID	Bis in die (two times a day)
BPM	Beat Per Minute
Ca^{++}	Calcium
CHF	Congestive Heart failure
CK	Creatine Kinase
CK-MB	Creatine Kinase Myocardial Band
cm	Centimetre
CMVI	Chronic Mitral Valve Insufficiency
cTn-I	Cardiac Troponin-I
cTn-T	Cardiac Troponin-T
CVP	Central Venous Pressure
CW	Continuous Wave
D. immitis	*Dirofilaria immitis*
DCM	Dilated cardiomyopathy
DV	Dorso-Ventral
E. canis	*Ehrlichia canis*
E. coli	*Escherichia coli*
ECG/EKG	Electrocardiogram
HCF	Hypertrophic Cardiomyopathy
HR Variability	Heart Rate Variability

i.e.	That is
IL	Interleukin
IM	Intramuscular
IV	Intravenous
K^+	Potassium
Kg	Kilogram
LBBB	Left Bundle Branch Block
LVE	Left ventricular Enlargement
M	Myoglobin
Mg	Milligram
MIMI	Microscopic Intramural Myocardial Infarction
Mm	Millimetre
MR	Mitral Regurgitation
MRI	Magnetic Resonance Imaging
MS	Milli Second
MUGA	Multiple Gated Acquisition
mV	Milli Volt
Na^+	Sodium
NCX-1	Sodium-Calcium Exchanger-1
NT-pro BNP	N-Terminal Pro-B-type Natriuretic Peptide
NYHA	New York Heart Association
OD	Omne in die (Once a day)
OPD	Out Patient Department
PDE-III Inhibitor	Phosphodiesterase-III inhibitor
PE	Pericardial Effusion
PET	Positron Emission Tomography
PI	Pulmonary Insufficiency
PNS	Parasympathetic Nervous System
PO	per Os (orally)
PO_2	Partial Pressure of Oxygen
PW	Pulse Wave
QID	Quater in die (four times a day)
RBBB	Right Bundle Branch Block
RFCA	Radio-frequency Catheter Ablation
RMSF	Rocky Mountain Spotted Fever
RVE	Right Ventricular Enlargement
SA node	Sino-atrial node
SA	Sino-atrial Arrest
SB	Sino-atrial Block
SID	Semel in die (once a day)
SNS	Sympathetic Nervous System
ST-2	Suppression of Tumorigenicity-2
T. brucei	*Trypanosoma brucei*
T. cruzi	*Trypanosoma cruzi*
T. evansi	*Trypanosoma evansi*

TEE	Trans-Esophageal Echocardiography
TI	Tricuspid Insufficiency
TID	Ter in die (three times a day)
TNF	Tumour Necrosis Factor
VD	Ventro-dorsal
VHS	Vertebral Heart Score
VPC	Ventricular Premature Complex
VVTI	Vaso Vagal Tonus Index

Section I

Canine

Cardiac Evaluation Approaches

1

Canine heart is a four-chambered organ having two atria, two ventricles, mitral valve, tricuspid valve, aortic valve, pulmonary valve, arteries, arterioles, veins, and venules. The major function of the heart is to provide nutrients to all organs of the body and to remove the waste products of metabolism from the organs of the body through its pumping action. Malfunctioning of pumping action and/or electrical events in the heart adversely affects nutrient and oxygen supply to organs and removal of waste products from the body making the survival reasonably difficult. During recent years cardiac disorders in canines have assumed greater significance owing to its frequent occurrence and variable outcome. Cardiac functioning is affected not only in primary heart diseases but also in diseases of other organs. Without correct diagnosis of cardiac abnormalities, treatment may be futile with a fatal outcome. Nowadays diagnosis of cardiac abnormalities is facilitated with the aid of modern diagnostic technology. Though clinical examination plays a significant role in the diagnosis of heart diseases, clinical significance of murmurs and/or arrhythmias is baffling without further investigations. Sometimes non-cardiac ailments also manifest symptoms mimicking heart diseases. Clinical signs such as coughing, tiredness, weakness, dyspnea, and respiratory crackling are also evident in lung diseases and need differentiation whether these signs are due to cardiac or pulmonary origin. Before the advent of electrocardiography much reliance was paid to ancillary approach (analyzing history, clinical symptoms, and clinical examination) that lacks differentiating ability of various cardiac diseases. Nowadays it is possible to evaluate animals at risk of cardiac diseases even when clinical manifestations are not apparent. Recent advances in cardiology during last few decades, most notably in the areas of diagnostic imaging and biomarkers, have considerably improved our diagnostic skill in differentiating various heart diseases of which diagnosis was unimaginable during the early twentieth century. A bird eye view of different systematic approaches for evaluating heart is presented below.

J. P. Varshney, *Electrocardiography in Veterinary Medicine*,
https://doi.org/10.1007/978-981-15-3699-1_1

1.1 History and Physical Examination

Despite the technical ability of electrocardiography and echocardiography, the history and physical examination remain the most important and desirable diagnostic steps in making a correct diagnosis of heart diseases. Findings from these examinations prompt the clinician to undertake further specific examinations on the basis of clinical suspicion so as to make specific diagnosis. History of coughing particularly during night in dogs with cardiac involvement indicates pulmonary venous congestion and edema owing to left atrial enlargement affecting the bronchi adversely. Coughing episodes worsening with exercise/excitement or at night in a lying animal are strongly suggestive of cardiac cough. Reduced cardiac output in heart diseases leads to exercise intolerance, cyanosis, syncope, marked weakness, inappetence, and weight loss. These manifestations are commonly narrated by the owners in the history.

Physical examination of cardiovascular system is comprised of inspection (looking for abducted elbow, jugular distension or pulse, edema of ventral abdomen or limbs, tachypnea/dyspnea), vital indices (temperature, pulse, respiration, mucus membrane color, capillary refill time), palpation (ascertain position of apex beat, presence or absence of precordial thrill, femoral pulse, ascites, hepatomegaly), and auscultation (rate, rhythm, intensity of heart sounds, murmurs, or abnormal respiratory sounds).

Auscultation is facilitated by the stethoscope. It was invented by a French physician Laennec Rene Theophile Hyacinthe in 1816 in France (Laennec 1819). Since its invention the stethoscope has become almost universal insignia of a physician or a veterinarian. Auscultation of chest/thorax (Figs. 1.1 and 1.2) is a very important part of cardiac examination to provide first-hand information on heart rate and its rhythm; presence of muffled heart sound; absence of heart sound; heart murmurs and gallop sounds (S3 or S4); systolic clicks and split heart sound (S1 or S2); as well as presence or absence of crackles and rales in the lungs. Normal heart sound

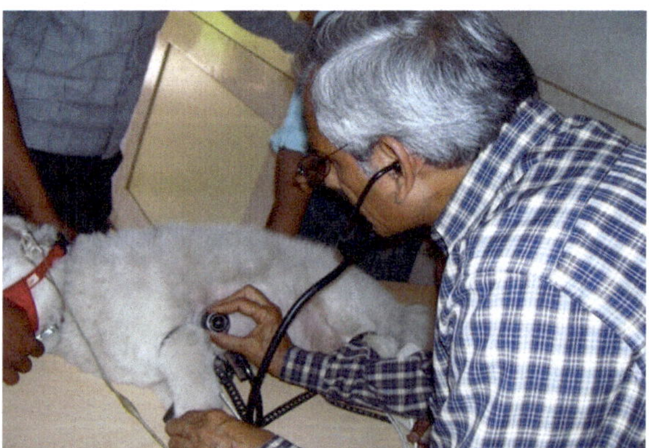

Fig. 1.1 Clinical examination of the dog showing auscultation of the heart

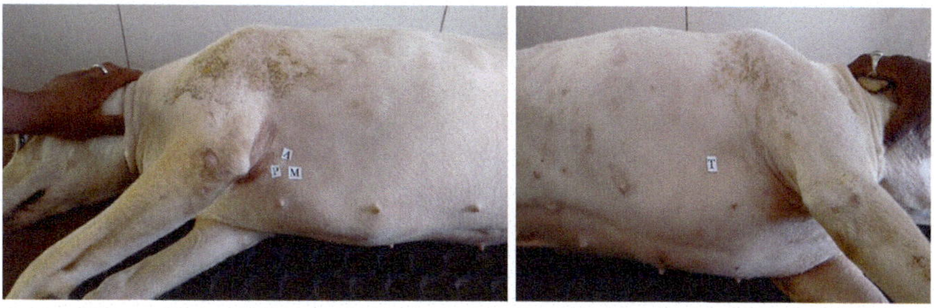

Fig. 1.2 Site for auscultation of different valves. On left side of the chest pulmonary valve (P) is auscultated at third intercostal space near the elbow. Aortic valve (A) is auscultated slightly dorsal and caudal to pulmonary valve at fourth intercostal space on left side. Slightly caudal and ventral to aortic valve in fifth intercostal space; mitral valve (M) is auscultated. On the right side of the chest at third/fourth intercostal space tricuspid valve (T) is auscultated. The site for auscultation of the valves have been marked as P, A, and M on left side and T on the right chest

is basically a three-character sound, viz., LUB-DUB-PAUSE. The first heart sound (S1-LUB) is associated with the closure of mitral and tricuspid valves. The second heart sound (S2-DUB) is associated with the closure of aortic and pulmonic valves. Heart is to be auscultated at all the valve areas, i.e., mitral valve at fifth intercostal space around costochondral junction on the left side (opposite to elbow point in standing dog), aortic valve at fourth intercostal space dorsal to mitral valve on the left side (at the level of the point of shoulder), pulmonic valve at third intercostal space at sternal border (axilla) on the left side, and tricuspid valve on third to fourth intercostal space at costochondral junction on the right side of the thorax. The intensity of normal heart sound increases in dogs with thin body, in young dogs, and in dogs with fever, anemia, or hyperthyroidism. In obese dogs the intensity of heart sound decreases. The intensity of heart sound also decreases in dogs with pleural or pericardial effusions. Though arrhythmias and murmurs are detected on heart auscultation, differentiation of arrhythmias or cardiomyopathy is not possible only on auscultation. Sometimes murmurs are physiological as observed in cases with anemia and/or hypoproteinemia. These physiological murmurs may be detected in young growing animals even in the absence of heart disease, and these are resolved as the animal grows. Heart murmurs are abnormal extra sound of relatively long duration and are generated due to turbulence within the heart owing to disturbance in blood flow. Detection of pathological murmurs may be indicative of valvular insufficiency, valvular stenosis, interatrial or interventricular septal defects, patent ductus arteriosus (PDA), or defect of great vessels. As per the timings of occurrence of the murmurs at the point of maximum intensity, these are divided into three classes as systolic (occurring during systole), diastolic (occurring during diastole), and continuous (occurring all the time). Systolic murmurs are generally soft and occur during early systole. Diastolic murmurs are of low frequency, and their occurrence is not very common. Continuous murmurs vary in intensity and are associated with patent ductus arteriosus (PDA). As per their intensity, murmurs have been graded into six grades as follows:

- Grade 1. Murmurs are very soft and localized. Generally detected on prolonged auscultation.
- Grade 2. Murmurs are soft and localized. Detected easily.
- Grade 3. Murmurs are moderately intense and detected at more than one place.
- Grade 4. Murmurs are moderately intense and detected at many places in left and right chest.
- Grade 5. Murmurs are loud over point of maximum intensity. Precordial thrill is also there.
- Grade 6. Murmurs are very loud associated with precordial thrill.

1.2 Blood Pressure Monitoring

Another important tool for cardiac evaluation is blood pressure. Blood pressure monitoring is a routine clinical determinant in humans during clinical examination. On the contrary, blood pressure monitoring has been given a little casual attention in dogs. Blood pressure monitoring may serve as a valuable major determinant of ventricular wall stress and myocardial oxygen consumption. The advantage of measuring blood pressure in canines and felines is being increasingly recognized owing to deleterious effect of hypertension on ocular, renal, cardiovascular, and cerebrovascular systems. There is a growing assumption that hypertension is quite common in dogs with renal and endocrinological disorders. Though direct blood pressure monitoring in dogs dates back to 1800, its cumbersomeness restricts it's clinical utility. For taking blood pressure in dogs, auscultative method (Fig. 1.3), ultrasonic Doppler sensing device (Fig. 1.4), oscillometric method, or electronic devices have been tried. Blood pressure has also been monitored in dogs using an aneroid sphygmomanometer with Velcro cuff of 5 cm × 22 cm size (Fig. 1.3). The dogs are restrained in right lateral recumbency, and Velcro cuff is applied over left hind limb

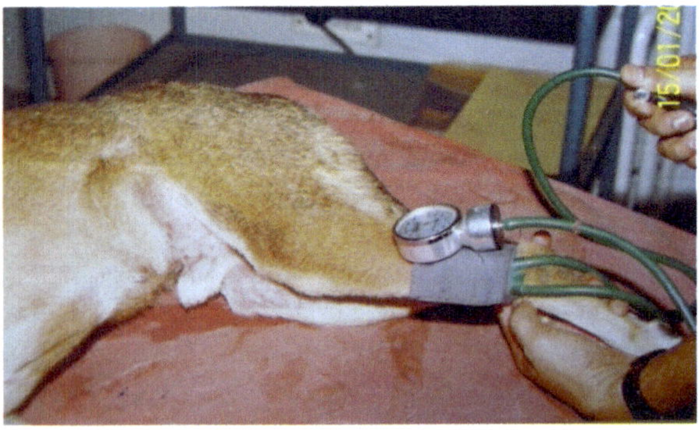

Fig. 1.3 Blood pressure measurement by indirect technique using a small Velcro cuff (5 × 22 cm size) tied over left hind leg on cranial tibial artery at distal medial aspect of tibia and aneroid sphygmomanometer. The hand bulb coupled to aneroid pressure gauge calibrated in mm of mercury (Hg) is used to inflate and vary the pressure in the cuff. The systolic and diastolic arterial pressures are indicated by the appearance of palpable beat with cuff deflation or appearance of throbbing of manometer or sound and ceasing of throbbing or sound, respectively

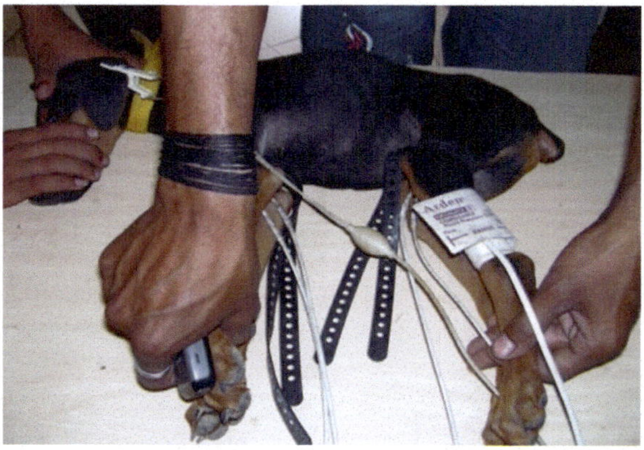

Fig. 1.4 Blood pressure measurement by indirect technique using ultrasonic Doppler sensing device

on cranial tibial artery at the distal medial aspect of tibia (Fig. 1.3). Mean systolic (102.5–162.5 mm Hg), diastolic (44.5–76.0 mm Hg), and arterial blood pressure (70–102.5 mm Hg) is variable in healthy dogs. Hypertension (>160 mm Hg systolic, >100 mm Hg diastolic) is generally seen secondary to diseases like diabetes mellitus, left ventricular enlargement, and chronic renal failure. Hypotension (fall in blood pressure) is observed in severe gastroenteritis, hypothermia, and shock. It is in fitness of things that blood pressure measuring is adopted in routine clinical practice.

1.3 Thoracic Radiography

Chest/thorax radiography continues to play a significant role in the assessment of cardiovascular diseases or assessing congestive heart failure. For better results lateral (Fig. 1.5) and dorso-ventral (Fig. 1.6)/ventro-dorsal (Fig. 1.7) views are taken with high kilo voltage peak (kVp) and low milliamperes (mAs). When evaluating cardiac size and shape in dog chest, chest conformation should always be considered. It is always better to take radiograph at the peak of inspiration.

Ventricles occupy approximately three intercostal spaces in normal thorax in dogs. Location of the heart extends from third to sixth ribs and heart touches nearly diaphragm. Right atrium and right ventricle are on the cranial border; and left atrium and left ventricles are on caudal border of heart silhouette. Both atria, pulmonary arteries, aorta, and vena cava (cranial and caudal) are situated dorsally in the right lateral radiographs (Figs. 1.8, 1.9, and 1.10).

Vertebral heart score (VHS)—VHS is calculated on the lateral thoracic radiographs to assess cardiac size. Long axis of cardiac silhouette is measured from the carina of the main bronchus to the apex of the heart and short axis at the widest part of the heart. These axis (long and short) measurements are transferred to the vertebrae starting from cranial edge of T4 and the number of vertebrae fall under each axis are counted. The number of vertebrae falling under the both axes (Fig. 1.11) is counted. A VHS > 10.5 is generally denotes cardiomegaly. It is a rough estimate of the size of the heart and cannot be taken as sole criteria of heart enlargement.

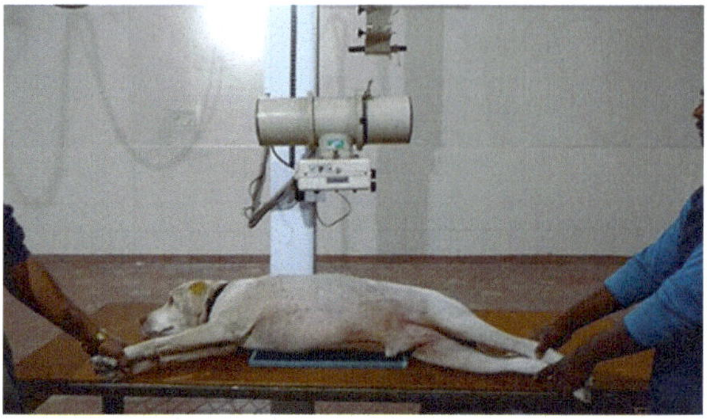

Fig. 1.5 Right lateral recumbent position for taking lateral view of chest radiograph

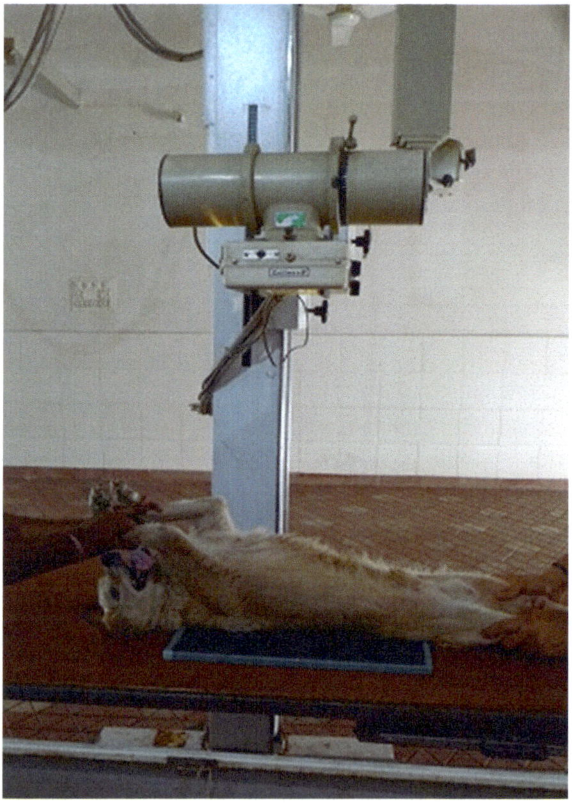

Fig. 1.6 Dorsal recumbent position for taking ventro-dorsal view of chest radiograph

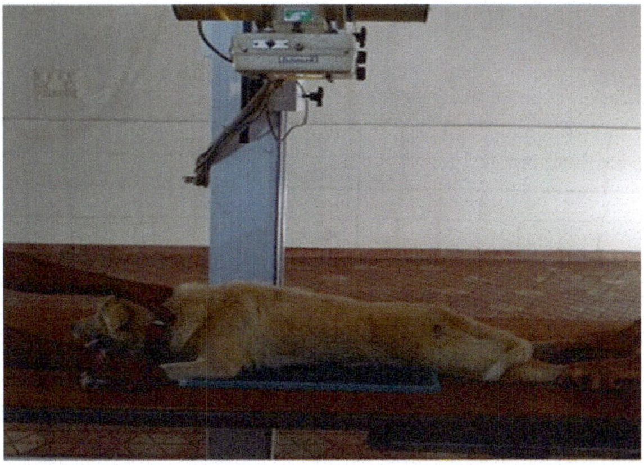

Fig. 1.7 Sternal recumbent position for taking dorso-ventral view of chest radiograph

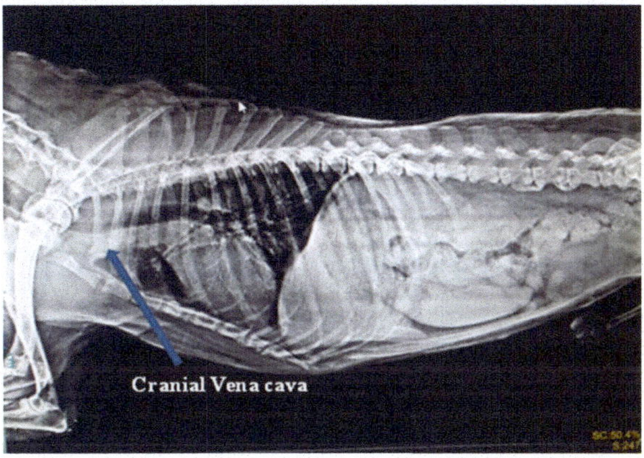

Fig. 1.8 Right lateral radiograph of a 18-year-old male Pomeranian showing distended cranial vena cava

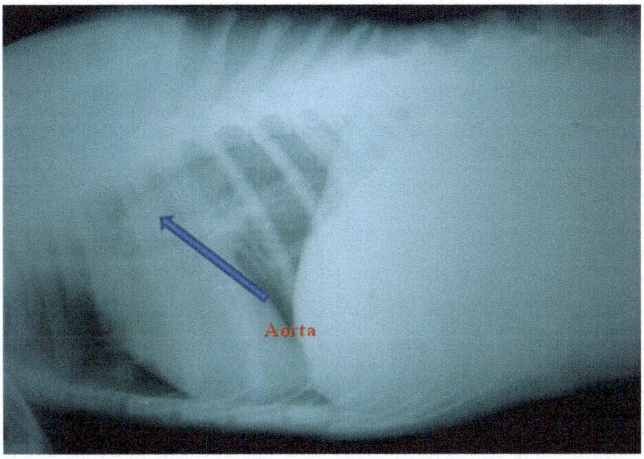

Fig. 1.9 Right lateral radiograph of 2-year-old male nondescript dog showing aorta, elongated shape of the heart touching the sternum and diaphragm

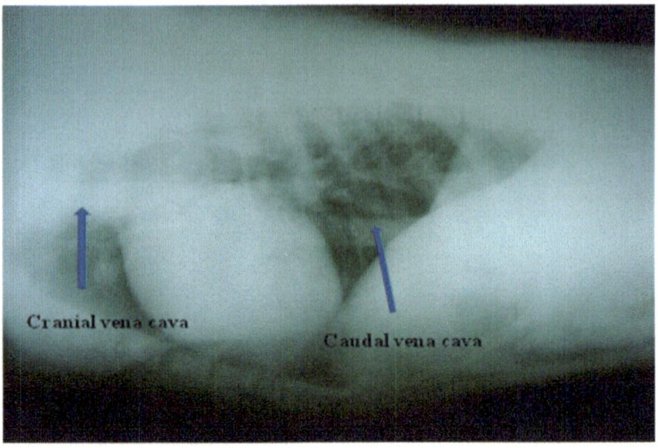

Fig. 1.10 Right lateral radiograph of 6-year-old female German Shepherd showing cranial and caudal vena cava and globoid heart touching the diaphragm

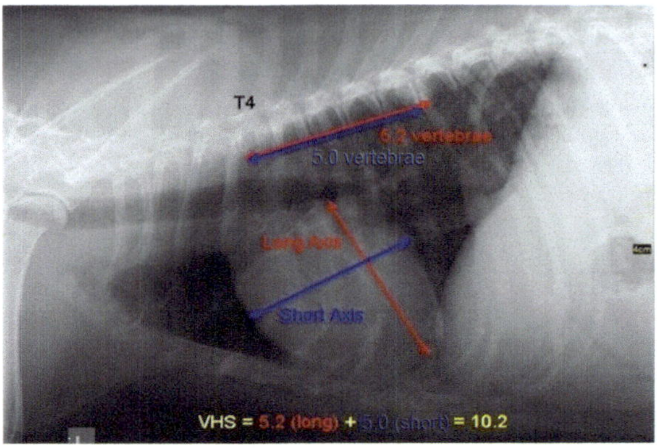

Fig. 1.11 Right lateral radiograph of a Labrador dog showing measurements of VHS

Chamber enlargement pattern and enlargement of great vessels can also be detected by radiography. A bulge in the 2–3 o'clock position of cardiac silhouette is concurrent with left atrial enlargement. Rounding and bulging in the 2–5 o'clock position is suggestive of left ventricle enlargement. Bulge in 9–11 o'clock position suggests right atrium and a reverse D configuration is suggestive of right ventricle enlargement. Chest radiographs give valuable information, not only about heart size but also about the status of the pulmonary vasculature (Fig. 1.12) and changes in the lungs that help differentiate left-sided congestive heart failure (pulmonary edema) from primary pulmonary disease. The presence of an alveolar pattern (characterized by the presence of air bronchograms), pulmonary venous distension (the artery and vein that sit on either side of the bronchus are normally of equal size), and pulmonary edema dorsal hilar region is suggestive of congestive left heart failure (Fig. 1.13a). Different radiological patterns to help in differentiating lung changes of cardiac or pulmonary origin have been shown in Fig. 1.13b–g. Globoid shape of the heart (Fig. 1.14) is considered to be indicative of pericardial effusion or cardiomyopathy. Deviation of the heart toward right (Fig. 1.15) or left (Fig. 1.16) can also be visualized in ventro-dorsal/dorso-ventral radiographs.

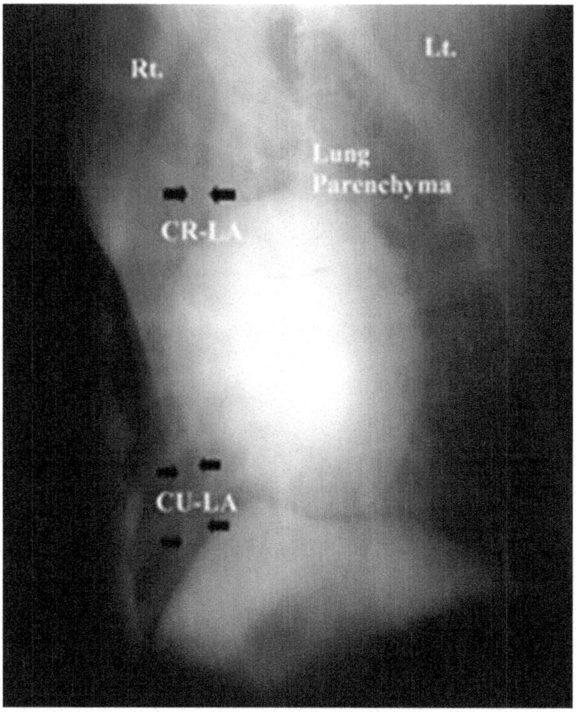

Fig. 1.12 Ventro-dorsal radiograph of a dog showing enlargement of cranial and caudal lobar arteries in a dog diagnosed with dirofilariasis

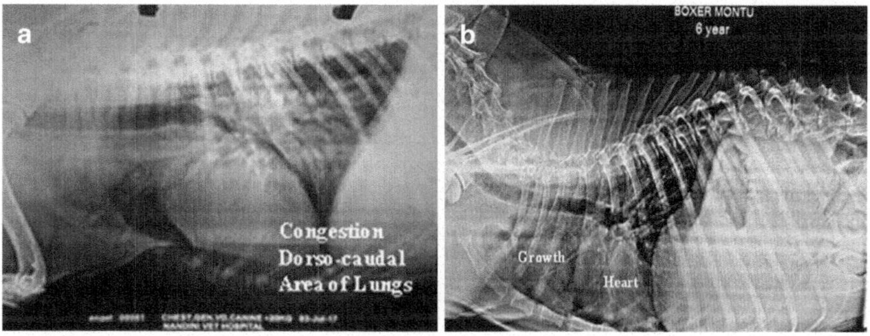

Fig. 1.13 (**a**–**g**) Right lateral radiograph of thorax of dogs showing different radiological patterns to assist in differentiating lung changes of cardiac or pulmonary origin. (**a**) Edema in dorsal hilar region of lung suggesting lung edema of cardiac origin due to left heart failure. (**b**) Changed radio-opacity in the cranio-ventral area of the lung with irregular margins suggesting growth. (**c**) Increased soft tissue opacity in the cranio-ventral area of the lung touching heart and inflamed bronchi suggesting pneumonia. (**d**) Changed radio-opacity of whole lung area suggesting pleural effusion. (**e**) Conspicuous bronchial pattern showing diffuse thickening of airways wall giving the appearance of thick lines and rings suggesting bronchitis. (**f**) Miliary opacities in lungs suggestive of Fungal pneumonia. (**g**) Cotton wool appearance in lungs of a Pomeranian dog having osteolytic change in proximal extremity of right ulna suggesting metastasis in lungs

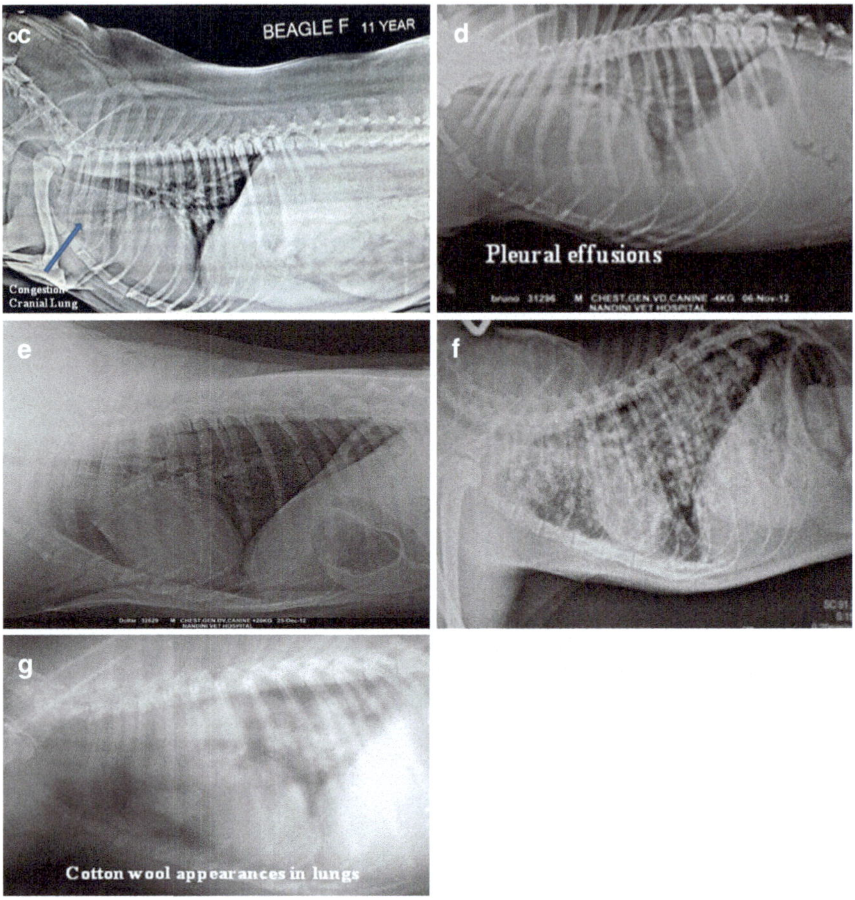

Fig. 1.13 (continued)

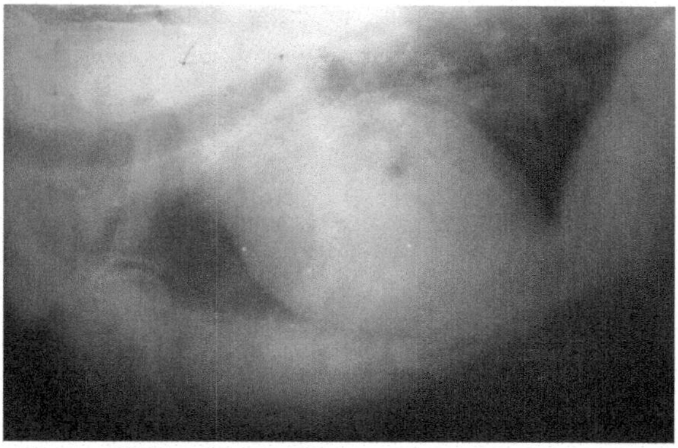

Fig. 1.14 Right lateral radiograph of a dog showing globoid heart suggestive of pericardial effusion/cardiomyopathy

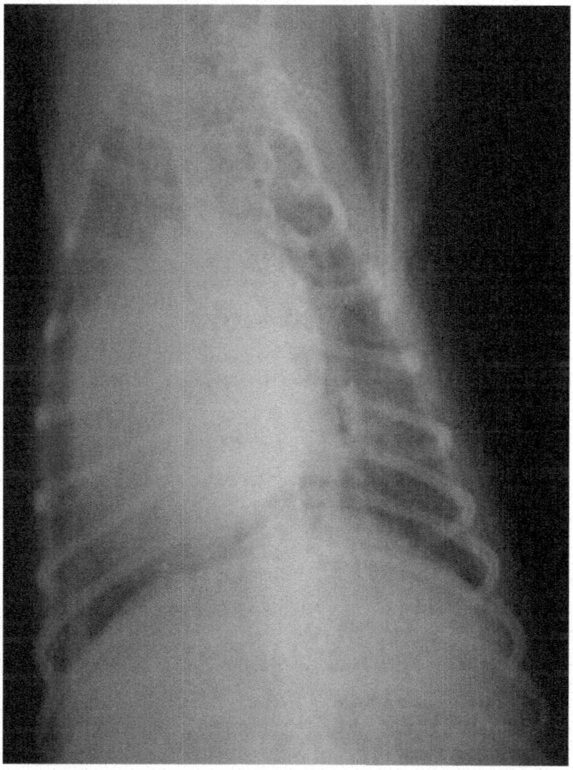

Fig. 1.15 Ventro-dorsal radiograph of a Pomeranian dog showing right deviation/enlargement

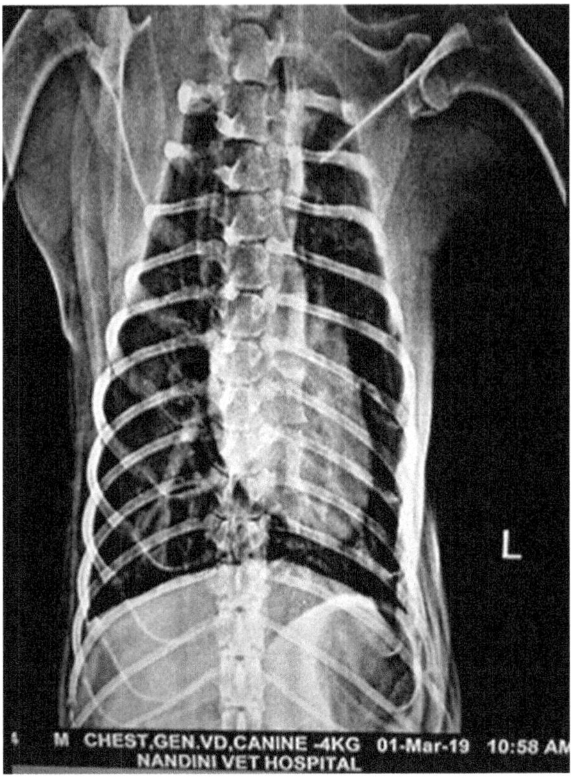

Fig. 1.16 Ventro-dorsal radiograph of an Apso dog showing left deviation/enlargement

1.3.1 Summary of Radiographic Findings in Heart Diseases

1. Right atrium enlargement is associated with bulging at 9–11 o'clock position (VD view), whereas left atrium enlargement is associated with bulging at 2–3 o'clock position (VD view) and bulging at dorso-caudal area of heart (lateral view).
2. Right ventricle enlargement is associated with rounding of heart at 6–11 o'clock position, reverse D shape (VD view), and elevated trachea, whereas left ventricle enlargement is associated with rounded cardiac apex, elongated cardiac shape, or nearness of the heart to chest wall (VD view) and rounded left heart (lateral view).
3. Biventricular enlargement is associated with rounded heart with increased sternal contact.
4. Left-sided congestive heart failure shows edema in dorsal perihilar region (lateral view). While dilated cardiomyopathy is associated with globoid heart and pulmonary venous congestion (lateral view).
5. Pericardial effusion is associated with globoid heart (lateral view). Hepatomegaly and ascites are observed in cases of pericardial effusion associated with congestive heart failure.

1.4 Electrocardiography

Electrocardiography (Fig. 1.17) is a noninvasive diagnostic tool commonly used to record the electrical activity of the heart in man and animals. A normal electrocardiogram does not rule out cardiac abnormalities or other cardiac changes. Similarly all changes in electrocardiogram are not necessarily indicative of heart diseases. An abnormal ECG may suggest side of the heart affected and disturbance of rhythm or rate. Electrocardiography is being used for routine health checkup; cardiac monitoring during anesthesia and surgery; evaluation of trauma; evaluation of heart size, shape, rhythm, and rate; evaluation of electrolyte disorders; routine presurgical examination; and preventive health checkup of geriatric dogs, cats, and horses. ECG cannot be taken as a last tool in itself for the diagnosis of cardiac diseases. It opens up avenues for further confirmatory investigations. It is an essential tool for the diagnosis and treatment of cardiac arrhythmias. Since treatment of arrhythmias is specific, error in diagnosis may result in fatalities. ECG is no panacea as it has its own limitations. It cannot detect mechanical status of the heart, pathology of valves, coronary arteries, endocardium, and pericardium, and predict prognosis always. Therefore, ECG should always be considered as a part of clinical findings, and its results should be interpreted in conjunction with history and other laboratory investigations.

Clinical indications for taking electrocardiogram include arrhythmias (tachycardia or bradycardia or irregular heart beat), shock, sudden onset of dyspnea, syncope or seizures, cardiac murmurs, increased area of heart auscultation, renal disease, endocrinopathies (Addison's disease, Cushing syndrome, thyroid dysfunctions), systemic diseases (pyometra, pancreatitis, uremia, neoplasms), acid-base, and/or ionic imbalances. Electrocardiography does have limitations in evaluating heart and serves as a rough guide for evaluating the heart and should, therefore, be interpreted in conjunction with clinical state of the ailing dog. Its interpretation in isolation may be misleading. It reflects functional status of the heart only and not of mechanical status. Diagnosis of diseases of valves (mitral, tricuspid, pulmonary, and aortic), coronary arteries, endocardium, or pericardium is not feasible by

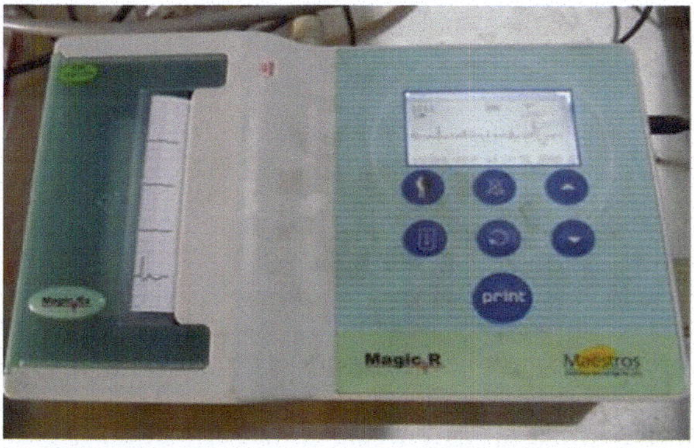

Fig. 1.17 Automatic six-lead electrocardiographic machine with display of six leads

electrocardiography. Significant cardiac disease may sometimes produce only minor changes or no change at all in electrocardiogram. Therefore a complete workup starting with clinical history, clinical examination, electrocardiography, echocardiography, chest radiography, and other diagnostic procedures, as per requirement, needs to be followed. Body conformation and breeds of the dog may alter mean accepted measurements of electrocardiographic indices. Hence standard measurements for different breeds have to be worked out.

- *Continuous Ambulatory ECG (Holter Monitoring)*—It is an improved version of electrocardiography that provides a continuous 24 h information of cardiac electrical activity during normal daily activities, strenuous exercise, or sleep. This technique is valuable for detecting intermittent arrhythmias and quantifying their rate, assessing efficacy of antiarrhythmic drugs, and screening for subclinical cardiomyopathy. This technique is valuable in patients with syncope wherein routine resting electrocardiogram and laboratory investigations fail to arrive at any conclusion.
- *Post Exercise Electrocardiography*—This technique is employed when resting electrocardiogram is normal in cases suspected with cardiac disease, arrhythmias, and syncope or showing exercise intolerance. The animal is given vigorous exercise and then electrocardiogram is recorded. T wave changes or S-T segment abnormalities, not detected in electrocardiogram taken at rest, may become apparent after post exercise electrocardiography. Such abnormalities detected in post exercise electrocardiogram may indicate underlying myocardial disease.
- *Event Recorders*—Event recorders or loop recorders are important diagnostic tool for identifying arrhythmias as a cause of syncope. These tools are very small ECG recorders easily worn by small dogs. These devices record a continuous 5 min loop of electrocardiogram. After attaching a recorder, the animal is sent home. When there is an event (syncope, seizures, or collapse), owner pushes a button to activate memory feature. ECG is preserved in memory for a programmed time. That recorded ECG can be transmitted trans telephonically to the expert for interpretation.
- *Atropine Response ECG*—Dogs with symptomatic bradyarrhythmia are given atropine (0.05 mg/kg IM or SC), and electrocardiogram is recorded 30 min later. The role of vagal tone in bradyarrhythmia can be distinguished using atropine response electrocardiogram.

1.5 Echocardiography

Echocardiography (Fig. 1.18), also known as cardiac sonography or cardiac ultrasound, is another important noninvasive diagnostic modality for imaging the heart and its surrounding structures. It provides important information about size of the heart chambers, thickness of the wall, wall motion, valve configuration and motion, the proximal great vessels, and detection of the pericardial and pleural effusions. Identification of mass lesion within and adjacent to the heart is also possible by echocardiographic examination. Like other diagnostic modalities its interpretation

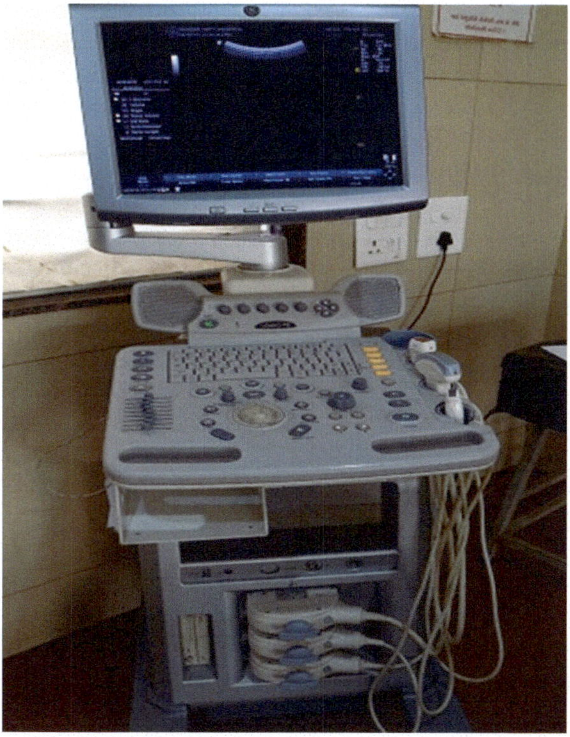

Fig. 1.18 Echocardiographic machine

should be viewed within the context of a thorough history, clinical picture, complete cardiovascular examination, and other tests.

M-mode, two-dimensional (2-D, real-time), and Doppler echocardiography is used in clinical practice. The basic echocardiographic examination includes M-mode measurements and all standard 2-D imaging planes from both sides of the chest. Doppler examination adds information about blood flow patterns. Tranquilization facilitates echocardiographic examination. M-mode echocardiography (oldest form of echocardiography) provides one-dimensional view (depth) into the heart with clear images of cardiac borders, and measurements of cardiac dimensions (diastolic and systolic thickness of left ventricular wall and interventricular septum, ventricular chamber dimensions; chamber volume and ejection fraction are calculated) and motions (mitral valve) are accurately recorded. M-mode can be used to measure left ventricular indices, mitral valve measurements, and systolic time intervals. 2-D echocardiography provides information of the width and depth of the tissues. Doppler echocardiography adds information not only about blood flow pattern but also about its velocity, detection, and quantification of valvular insufficiency, valvular stenosis, and cardiac shunts.

Transesophageal echocardiography (TEE) has been used in human medicine for many years; its use in veterinary medicine is being explored. This technique gives clear picture of heart structures those at or above the AV junction. The technique requires heavy sedation and structures are imaged through the esophageal wall with specialized transducer mounted on a flexible steerable endoscope tip.

1.6 Angiocardiography

There are two types of angiocardiography, viz., nonselective and selective. The former one is of diagnostic value in feline cardiomyopathy, feline dirofilariasis, vasculature, severe pulmonic or sub aortic stenosis, PDA and tetralogy of Fallot. While the latter one is used to evaluate specific areas of the heart or great vessels by placing cardiac catheters. Angiography provides information on the path of blood flow and anatomic abnormalities. This technique has now been superseded by Doppler echocardiography.

1.7 Pneumopericardiography

This technique is being used for delineating the causes of pericardial effusions.

1.8 Endomyocardial Biopsy

This is a biopsy technique for the heart. Special bioptome is passed into right ventricle via a jugular vein to obtain a small sample of endocardium and adjacent myocardium to evaluate myocardial metabolic abnormalities. This technique is of more academic interest rather than of clinical use.

1.9 Central Venous Pressure (CVP) Measurement

Measuring CVP is valuable in differentiating elevated right heart filling pressure (pericardial disease or right heart failure) from other causes of pleural or peritoneal effusions. To measure CVP, a catheter is passed via jugular vein that goes up to right atrium and is connected to manometer. Changes in the height of the fluid with the heart beat suggests either tricuspid insufficiency or that the catheter is in right ventricle. The technique is not used in routine clinical practice.

1.10 Pulmonary Capillary Wedge Pressure

This technique is used to monitor left heart filling pressure. It is obtained by passing an end-hole balloon-tipped catheter through right side of the heart and into the main pulmonary artery. This invasive technique is able to differentiate a cardiogenic or noncardiogenic pulmonary edema. It is of academic interest.

1.11 Nuclear Imaging

To evaluate cardiac function, radionuclide method is being used to provide noninvasive assessment of cardiac output, ejection fraction, myocardial blood flow, and metabolism. At present nuclear imaging technique is being used in research institutes for furthering the knowledge.

1.12 Cardiac Magnetic Imaging

MRI is imaging technique that takes the advantage of the property of certain atomic nuclei to vibrate or resonate when exposed to burst of magnetic energy. When the hydrogen nuclei resonate in response to changes in a magnetic field, they emit radiofrequency energy. The MRI machine detects this energy and converts it to a 3-D image. Difference in blood flow emit different amount of energy in different shades of gray, and MRI offers a potential means of detecting areas of cardiac tissue that have poor blood flow (coronary artery disease) or that has been damaged (heart attacks).

1.13 PET (Positron Emission Tomography)

It is a type of nuclear imaging that can evaluate heart function. PET scans can be used to look for coronary artery disease by examining how blood flows through the heart; it can evaluate damage to heart tissue after a heart attack.

1.14 MUGA (Multiple Gated Acquisition) Scan

It is also called radionuclide angiography (RNA). It is a test that is used to evaluate heart function by measuring how much blood is pumped out of the ventricles of the heart with each heartbeat (ejection fraction). A small amount of a safe radioactive tracer solution is introduced into a vein. This substance attaches to red blood cells, which are visualized by a special camera and computer as they travel through the heart, and the ejection fraction is calculated based on the computer-generated images.

1.15 Cardiac Catheterization

Catheter is passed into different area of heart and vasculature via jugular vein, carotid artery, or femoral vessels. Measurement of pressures, cardiac output, and blood oxygen concentrations can be obtained from specific areas. It is still considered gold standard for estimating severity of heart defects. These techniques have now been superseded by Doppler echocardiography.

1.16 Cardiac Biomarkers

Cardiac biomarkers are comparatively a new diagnostic modality gaining great prominence in human medicine these days, and their use in canine cardiology is also being increasingly investigated. While a number of cardiac biomarkers have been used to assess the myocardial status in humans, only two (cTn-I and NT pro BNP) are of practical importance in canine cardiology. Cardiac Troponin-I (cTn-I) is a fundamental component of cardiac muscle that is released in response to myocardial insult. It is one of the most important biomarkers used in human medicine where myocardial infarction with subsequent muscle damage is the predominant cause of cardiac disease. In human medicine, the cardiac biomarkers, cardiac troponin T (cTn-T) and I (cTn-I), and the cardiac isoenzyme of creatine kinase (CK-MB) are used extensively to diagnose and provide valuable prognostic information in patients with ischemic, traumatic, and septic myocardial injury and necrosis. Heart failure is the final stage of cardiac disease, which occurs due to structural and functional cardiac dysfunction. Chronic mitral valve insufficiency (CMVI) is the most common cause of heart failure in small breeds of dogs. Although many diagnostic imaging technologies are currently in use, these are unable to detect the early stages of heart failure. Cardiac troponin-I has been found to be a very sensitive serum biomarker of physical or metabolic myocardial injury, myocardial ischemia, or necrosis in humans with a cardiac specificity of 100%. Four to six hours after acute myocardial cell injury, the cardiac troponin concentration in blood increases in a biphasic manner. Diagnostic sensitivity of electrocardiography (ECG) or echocardiography to diagnose minor myocardial injury is very poor. However, early diagnosis of myocardial injury may be important from a therapeutic and prognostic perspective. cTn-I is 100% specific for the heart, i.e., it is never expressed by skeletal muscle. Circulating levels of cTn-I in healthy dogs are normally very low (<0.07 ng/ml) (Sleeper et al. 2001). Nowadays both qualitative (Fig. 1.19) and quantitative estimation methods for cTn-I are available. Elevated serum concentrations of cTn-I is seen in dogs with congestive heart failure (CHF), hypertrophic cardiomyopathy (HCM), myocardial inflammation, and gastric dilatation/volvulus, heat stroke, ehrlichiosis, babesiosis, liver cirrhosis, renal failure, respiratory distress, and fungal pneumonia. I have observed elevated levels of cTn-I levels in dogs with cardiac diseases, viz., right heart failure due to dirofilariasis (1.8–4.0 ng/ml),

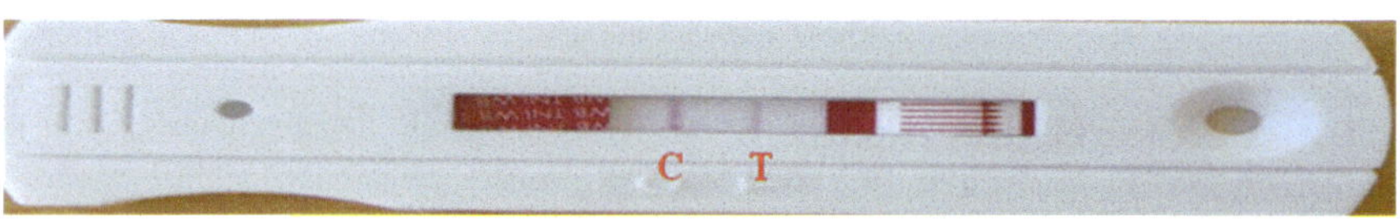

Fig. 1.19 Cardiac Troponin-I test kit for the qualitative estimation of cTn-I level to detect cardiac insult. Appearance of purple colored line, in test region 'T' and pinkish purple line in the control region "C" indicates that the sample is reactive to Troponin-I. Difference in the intensity of color between "C" and "T" is due to variations in cTn-I concentrations. The test can detect cTn-I in serum/plasma/whole blood as low as 0.3 ng/ml

left side congestive heart failure (0.23–8.97 ng/ml), atrial fibrillation (0.26 ng/ml), ventricular premature complexes (0.26–0.28 ng/ml), and left heart deviation (0.24 ng/ml), and also in dogs with non-cardiac diseases affecting the heart secondarily, viz., hepatic cirrhosis (0.20–0.40 ng/ml), tracheal narrowing (0.15–0.368 ng/ml), renal failure grade 5 (0.20–1.2 ng/ml), and fungal pneumonia (0.12–0.26 ng/ml), at the hospital during routine clinical practice. Recently increased levels of cTn-I in cases of canine monocytic ehrlichiosis caused by *E. canis* (Varshney et al. 2015), babesiosis caused by *B. gibsoni* (Varshney 2017a), heat stroke (Varshney 2017b), and viper bites (Varshney and Monapara 2019), suggesting myocardial damage, have been reported. The levels of cTn-I in these diseases varied widely (0.04–3.36 ng/ml in cases of ehrlichiosis; 0.5–14.6 ng/ml in cases of babesiosis; 1.4–6.0 ng/ml in cases of heat stroke; 0.1–5.8 ng in cases of viper bites). The increased level of cTn-I in cardiac and non-cardiac diseases indicates cardiac insult, and variability in levels of cTn-I may be associated with variable duration of cardiac insult. Acute or chronic cardiac insult induces release of cTn-I in the circulation. Cardiac insult in various diseases seems related to systemic inflammatory response, hypoxia, hyperthermia, or severe anemia. It seems that cTn I is released in reversible and/or irreversible cardiac damage. While cTn-I elevations indicate myocardial injury, they do not give any information as to the mechanism of injury. Cardiac Troponin-I level can rise with very small amounts of myocardial cell damage.

While myocardial damage releasing troponin-I does occur in many diseases in canine patients, it is usually released much later when the disease process is advanced. A more useful marker would be one that is released early in the disease process and rises in proportion to the extent of disease. Studies have shown that NTpro-BNP is a cardiac biomarker that behaves in this way in both dogs and cats and can provide useful information in the process of diagnosis. BNP was originally isolated from pig brain. NT pro-BNP is found in ventricle and is released in to circulation upon cardiac injury. NT Pro BNP is being used to detect the dogs at the risk of heart disease or heart failure in the early stage when symptoms are not much pronounced except dyspnea. I have observed that the clinically healthy dogs having no clinical, radiographic, electrocardiographic, or echocardiographic evidence of arrhythmia, heart enlargement, vulvular heart disease, or heart failure had very low levels (4.1386–16.5543 pmol/l) of NT-pro BNP. It seems that NT-pro BNP is a good indicator of differentiating severe panting or dyspnea of cardiac or pulmonary origin. A number of other biochemical markers of cardiac injury such as aspartate dehydrogenase (AST), creatine kinase (CK), lactate dehydrogenase (LDH), and myoglobin (M) have also been considered in human medicine. While being reasonably sensitive indicators of cardiac cell necrosis, these markers suffer from lack of specificity, as their circulating levels are also influenced by hepatic insult, skeletal muscle injury, or renal disease. Recent studies have revealed that, many pro-inflammatory cytokines are involved in the pathophysiology of heart failure. Cytokines (interleukin IL-1, IL-6, TNF-α, osteopontin, cardioprotrophin-1) and ST-2 (suppression of tumorigenicity 2, a member of interleukin-1 receptor family) seem to be a strong predictor of cardiovascular outcomes in both chronic and acute heart failures. Another type of cardiac biomarker, under investigation, is the expression of the sodium-calcium exchanger-1 (NCX-1) gene. It is a crucial gene in cardiac excitation-contraction coupling. It has a potential to become as an alternative cardiac biomarker in dogs.

New diagnostic modalities (cardiac biomarkers) add value to diagnostic database and have potential to indicate cardiac muscle damage. But do not indicate the cause of cardiac muscle damage and type of abnormalities in the heart. Despite an early detection of cardiac muscle damage by cardiac biomarkers, none can replace the basic approach starting from examination of history, thorough clinical examination, electrocardiographic and echocardiographic examinations, and chest radiography to biochemical investigations including biomarkers to arrive at a correct diagnosis of heart ailments with forecasting a probable outcome.

References

Laennec RTH (1819) De l'auscultation mediate ou traite du diagnostic des maladies des poumon et du coeur. Brosson and Chaude, Paris

Sleeper MM, Cliffoer CA, Laster LL (2001) Cardiac troponin I in the normal dog and cat. J Vet Intern Med 15:501–503

Varshney JP (2017a) Cardiac evaluation in anaemic dogs with clinical babesiosis caused by Babesia gibsoni. Indian J Vet Med 37:14–16

Varshney JP (2017b) Evaluation of myocardial damage in dogs with heat stroke. J Anim Res 7:9–981

Varshney JP, Monapara HD (2019) Evaluation of cardiac toxicity in dogs bitten by viper snake. Vet Pract 20:191–194

Varshney JP, Deshmukh VV, Chaudhary PS (2015) Evaluation of myocardial injury in acute canine monocytic ehrlichiosis. Intas Polivet 16:340–344

Further Reading

Anjos DS, Cintra CA, Roch AJR, Junior DP (2015) Cardiac biomarkers—an ally in the prognosis of heart disorders in small animals. Rev Investig Med Vet 14:38–45

Babuin L, Jaffe AS (2005) Biomarker of choice for detection of cardiac injury. Can Med Assoc J 173:1191–1202

Bers DM (2002) Cardiac excitation-contraction coupling. Nature 415:198–205

Cummins B, Cummins P (1987) Cardiac specific troponin I release in canine experimental myocardial infarction: development of a sensitive enzyme-linked immunoassay. J Mol Cell Cardiol 19:999–1010

Gaur T (2000) A clinico-epidemiological study on blood pressure in dogs. M.V.Sc. thesis submitted to Deemed University, Indian Veterinary Research Institute, Izatnagar

Gaur T, Varshney JP (2002) Non-invasive measurement of blood pressure and prevalence of blood pressure anomalies in dogs. Indian J Anim Sci 72:68–69

Gaur T, Varshney JP (2003) Non-invasive measurement of arterial blood pressure in conscious dogs—an experimental study. J Canine Dev Res 3:39–42

Gaur T, Varshney JP (2003) A comparative study of invasive (direct) and non- invasive (indirect) measurements of arterial blood pressure in conscious dogs. J Canine Dev Res 3:51–53

Hedayat M, Mahmoudi MJ, Rose NR, Rezaei N (2010) Proinflammatory cytokines in heart failure: double edged swords. Heart Fail Rev 15:543–562

Jongasaki M, Tachibana I, Luchner A (2000) Augmented cardiotrophin-I in experimental congestive heart failure. Circulation 101:14–17

Minoretti C, Falcone C, Calcagnino M, Emanuele E, Buzi MP, Coen E (2006) Prognostic significance of plasma osteopontin levels in patients with chronic stable angina. Eur Heart J 27:802–807

Moon HS, Choi E, Hyun C (2008) The cardiac sodium–calcium exchanger gene (NCX-1) is a potential canine cardiac biomarker of chronic mitral valvular insufficiency. J Vet Intern Med 22:1360–1365

Myer W, Bonagura JD (2000) Cardiovascular radiography. In: Birchard SJ, Sherding RG (eds) Saunders manual of small animal practice, 2nd edn. W.B. Saunders Company, Philadelphia, PA, pp 450–458

Nelson RW, Couto CG (1998) Small animal internal medicine (part 1), 2nd edn. C.G. Mosby, St. Louis

Neumayr G, Gaenzer H, Pfiste R (2001) Plasma levels of cardiac troponin I after prolonged strenuous exercise. Am J Cardiol 87:369–371

Rifai N, Douglas PS, O'Toole M (1999) Cardiac troponin T and I, echocardiographic wall motion analyses, and ejection fractions in athletes participating in the Hawaii Ironman Triathlon. Am J Cardiol 83:1085–1089

Schober K (2005) Biochemical markers of cardiovascular disease. In: Ettimger SJ, Feldman EC (eds) Textbook of veterinary medicine, vol 2. Elsevier, St. Louis, pp 940–948

Weinberg EO, Shimpo M, De Keulenae GW, MacGillivray C, Tominaga S, Solomon SD (2002) Expression and regulation of ST2, an interleukin-1 receptor family member, in cardiomyocytes and myocardial infarction. Circulation 106:2961–2966

Electrocardiography: Its Uses and Limitations

2

2.1 Electrocardiograph, Electrocardiography and Electrocardiogram

Electrocardiograph is a device/machine used to diagnose heart diseases by measuring electrical potentials on the surface of the body. Electrocardiography is the process of producing an electrocardiogram using an electrocardiographic machine. ECG stands for electrocardiogram, or electrocardiograph. In some countries, "EKG" is abbreviated for electrocardiogram. An electrocardiogram is a graphic record of the voltage produced by cardiac muscle cells during atrial and ventricular depolarization and repolarization plotted against time. In other words it is a graphic representation of electrical activities going on in the heart during depolarization and repolarization. It can provide useful information even about congenital heart diseases. A normal electrocardiogram does not rule out cardiac malformation or other cardiac changes. An abnormal ECG may suggest side of the heart affected or about some other abnormalities related to rate and rhythm of the heart.

2.2 Landmarks in the Development of Electrocardiography

- In 1616, Harvey discovered that blood circulation is due to heartbeat.
- Later it was recognized that heartbeat was an electrical process.
- In 1887, Augustus Desire Waller recorded the electrical changes of the human heart for the first time using the capillary electrometer.
- In 1895, Einthoven coined the terms P, Q, R, S and T for the electrocardiographic deflections.
- Around 1901, Einthoven invented string galvanometer-based 3-lead EKG machine for accurate tracings of cardiac electrical activity.
- In 1908, EKG was introduced in the United States.
- In 1909 Waller reported electrical activity of canine heart with Einthoven's string galvanometer. He used the term electrocardiogram for the first time.

J. P. Varshney, *Electrocardiography in Veterinary Medicine*,
https://doi.org/10.1007/978-981-15-3699-1_2

- Clinical application of electrocardiography in dogs was initiated by Norr in 1922.
- In 1924 Einthoven won Nobel Prize for his extraordinary contribution.
- During 1934–1938, Wilson invented the central terminal precordial leads.
- In 1954 American Heart Association standardized 12 lead EKG.
- Electrocardiogram of dogs received major attention in 1949, in decades of the 1960s and 1970s, and now again during recent years.

2.3 Application of Electrocardiography in Canine Medicine

- Differential diagnosis of arrhythmias.
- Cardiac monitoring during anesthesia and surgery.
- Evaluation of pets with gasping/panting/dyspnea/seizures/syncope.
- Evaluation for cardiac enlargement.
- Prognostic monitoring of hypothermic/hyperthermic dogs.
- Cardiac evaluation in diseases associated with multi organ failure.
- Evaluation of animals in acute emergencies and in intensive care units.
- Evaluation of geriatric pets.
- Evaluation of pets with peritoneal effusions.
- Evaluation of sports/snuffer animals.
- Routine health checkup.
- Evaluation of dogs with trauma/accidents.
- Evaluation of cardiac diseases with respect to anatomical changes particularly heart enlargements in dogs.
- Evaluation of electrolyte disorders.
- Assessing the efficacy of drugs (digitalis, quinidine) used for the treatment of heart diseases.
- Evaluation of disease progression.
- Database to differentiate nonspecific causes of weakness, fatigue, lethargy, syncope, collapse, seizure and fever.
- Evaluation of metabolic problems (eclampsia, ketoacidosis).
- Evaluation of patients with epilepsy.
- Routine presurgical examination.
- Routine medical checkup.

2.4 Utility of Electrocardiography

- It provides additional information to facilitate diagnosis and treatment of diseased conditions.
- It is noninvasive and quick tool.
- Electrocardiogram is immediately available.
- It is easy to read and results are immediately available.
- Evaluation of arrhythmic animals in a better way.
- Evaluation of cardiac compromised animals in a better way.
- ECG provides visual records of patient's cardiac status.

2.5 Limitations of Electrocardiogram

Despite a very useful technique, it has its limitation as with any other technique. It cannot provide definite diagnosis of congestive heart failure. Prognosis during anesthesia or surgical procedures is unpredictable. Heart chamber wall thickness cannot be measured. Diseases of heart valves, endocardium, or pericardium cannot be diagnosed. Obstructive blood flow diseases also remain undiagnosed. Wide variations among breeds and dogs of different body conformations make the interpretation of ECG difficult. Normal electrocardiogram may not always rule out cardiac disease.

2.6 The Electrocardiograph and Recording

2.6.1 Machine

The electrocardiograph is a sensitive voltage metering instrument capable of recording differences of electrical potential up to 1000th of a volt, i.e., 1 mV. The machine consists of a voltage meter and amplifier that is combined with a strip recording device (Fig. 2.1a, b) to give a permanent record. Electrocardiographic tracings are created by the movement of a heated stylus on heat-sensitive recording graph paper. The control panel of electrocardiograph depicts speed, sensitivity, lead selector, off/on switch, and filter (Fig. 2.1c). Electrocardiographic machine is able to run electrocardiograms at the paper speed of 25 or 50 mm/s. The sensitivity switch controls amplification. Normally this switch is positioned at 1, i.e., 1 mV input will cause the stylus to move 1 cm on the paper. Position 2 means 2 cm = 1 mV and 0.5 position means 0.5 cm = 1 mV. Filter is used to reduce disturbances. Stylus position is kept in the center by the position control switch in manual machines. Lead selector switch is used to select the leads or in automatic six channels machine there is no need to select the leads manually. Electrocardiographic machine by different companies may have different setup but basics are the same. It is always advisable to follow the instructions given in the operating manual of the machine for operation.

2.6.2 Electrodes

There are four wire electrodes in standard electrocardiographic cable (shown in bundle marked as 1—Fig. 2.2) and are labeled as right arm (RA), red colored; left arm (LA), yellow colored; right foot or leg (RL), black colored; and left leg or foot (LL), green colored. The other cable (marked as 2—Fig. 2.2) has V leads or exploring chest leads (consisting of six separate electrodes marked as V1, V2, V3, V4, V5, and V6). Alligator clips (Fig. 2.2), flat contact clips, or needles (Fig. 2.3) are used to attach to these electrodes so that these can be attached to the skin of the dog.

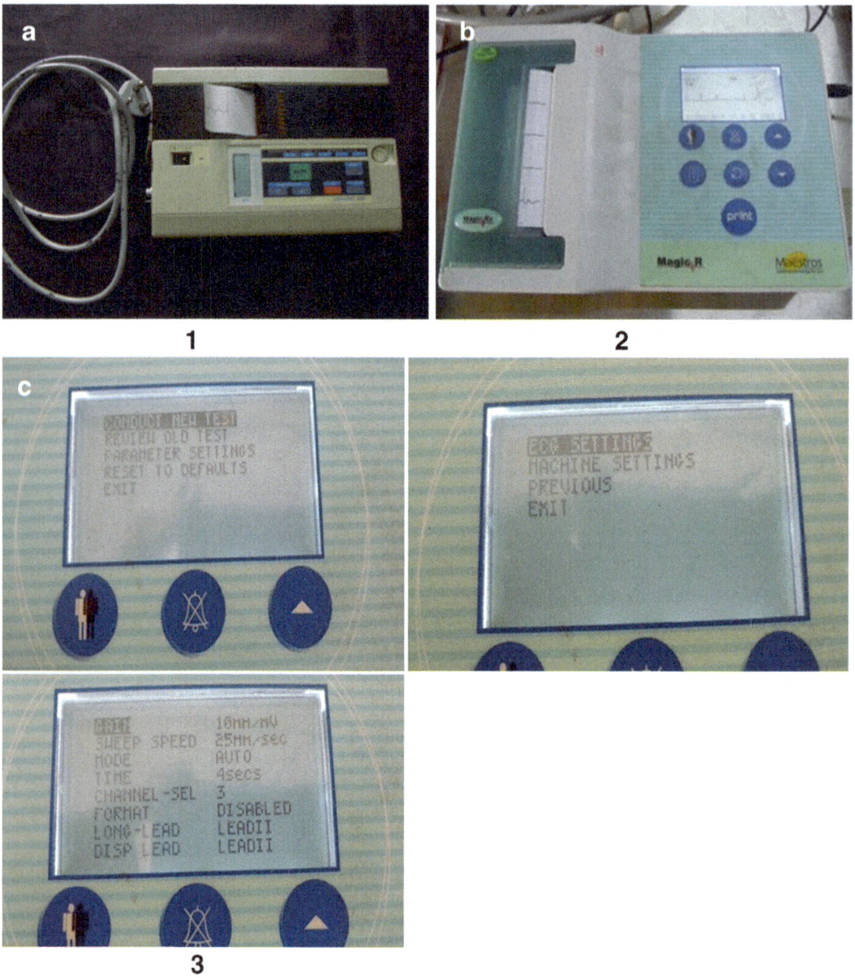

Fig. 2.1 (**a**) BPL 408 Single lead display ECG machine. (**b**) Magic RX (Maestros Mediline System Limited) automatic six lead recording ECG machine. (**c**) (1, 2, 3) Control panel of Magic Rx ECG machine for setting ECG parameters

2.6.3 Lead Systems

Each different angle or pair of electrodes is termed as lead. There are various leads used to evaluate heart. Four lead systems are given below. Each lead has direction and polarity and records components of depolarization and repolarization process aligned to it.

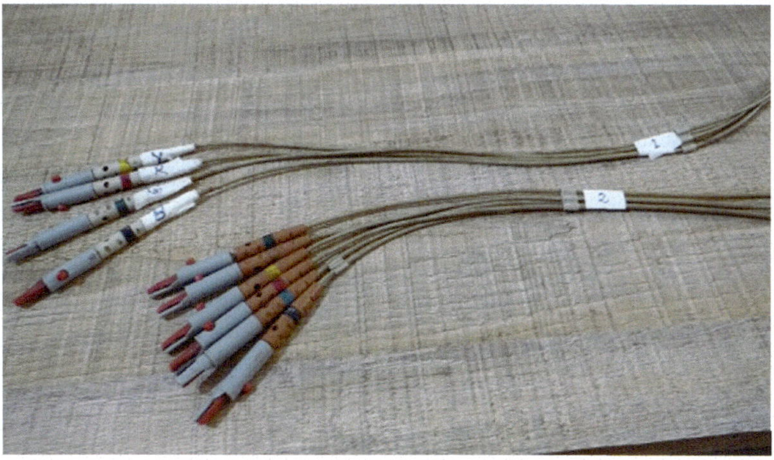

Fig. 2.2 Cable 1 has four electrodes of standard limb lead. Yellow, left arm electrode; red, right arm electrode; green, left leg electrode; black, right leg electrode. Cable 2 has six chest electrodes (V1, V2, V3, V4, V5, and V6). Four electrodes (V1, V2, V3, and V4) can be used in canines as CV5RL, CV6LL, CV6LU, and V10, respectively

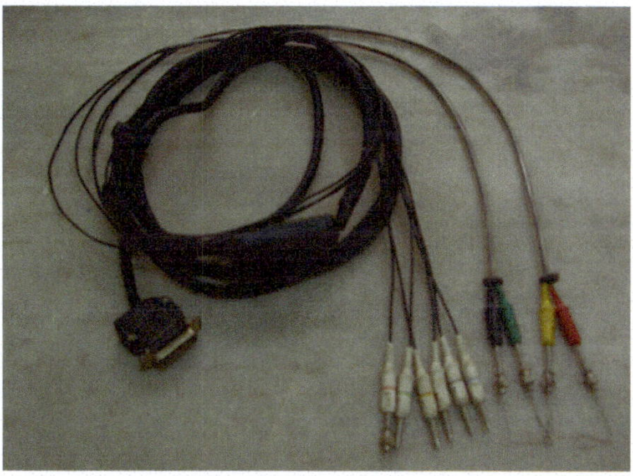

Fig. 2.3 Cable showing needle electrodes

1. Bipolar standard lead system—This consists of three leads designated as lead I, II, and III. These leads are placed as follows and compares potential difference of two limbs. It evaluates activation of the heart in the frontal plane.

   (a) Lead I—Lead I is formed by a pair of electrode applied on right forelimb and left forelimb and compares the electric potential of right forelimb (−) with that of left forelimb (+).

   (b) Lead II—Lead II is formed by a pair of electrode applied on right forelimb and left hind limb and compares the electric potential of right forelimb (−) with that of left hind limb (+).

(c) Lead III—Lead III is formed by a pair of electrode applied on left forelimb and left hind limb and compares the electric potential of left forelimb (+) with that of left hind limb (+).

2. Augmented unipolar limb lead system—The augmented unipolar limb leads (aVR, aVL, and aVF) are the other leads which provide additional information about cardiac vector. An augmented lead compares the electrical potential at the reference limb to the sum of electrical activity at the other two limbs. The word "a" stands for augmented, "V" stands for vector and words R, L, and F stand for right forelimb, left forelimb, and frontal limb (represents left hind limb), respectively. Therefore there are three leads in this system called aVR (augmented vector lead of the right arm or right forelimb), aVL (augmented vector lead of left arm or left forelimb), and aVF (augmented vector lead of frontal plane, i.e., left hind limb).

   (a) aVR—Electrical potential of right forelimb (+) is compared to the sum of electrical potential of left forelimb and left hind limb (−).
   (b) aVL—Electrical potential of left forelimb (+) is compared to the sum of electrical potential of right forelimb and left hind limb (−).
   (c) aVF—Electrical potential of left hind limb (+) is compared to sum of electrical potential of right and left forelimbs (−).

3. Bailey's hex-axial lead system—This is most widely used lead system in canine and feline electrocardiography. It combines both bipolar standard and augmented unipolar lead systems.
4. Special lead system—There are two other popular lead systems, viz., chest lead system and base-apex lead system:

   (a) Unipolar precordial chest leads—In dogs four chest leads are used. These leads view the heart from the transverse plane and provide important information for the diagnosis of heart enlargement (right or left), bundle branch blocks, ischemic heart changes, and arrhythmias.

      - rV_2 (CV_5 RL)
      - V_2 (CV_6 LL)
      - V_4 (CV_6 LU)
      - V_{10}

   (b) Base-apex bipolar monitoring lead—Recording is done on lead I. This lead system is widely used in cattle, buffaloes, sheep, goat, horses, and donkeys as electrocardiographic complexes (P, Q, R, S, and T) are more prominent in lead I. Details of this lead system are given under Sect. 3 Chap. 18. "Electrocardiography of Ruminants."

In dogs and cats most of the electrocardiographic investigations are based on evaluation of bipolar standard leads particularly lead II. Augmented unipolar leads are used to determine mean electrical axis or the position of the heart and confirming the changes observed in standard leads. The unipolar precordial chest leads are not well established in canine electrocardiography. V_{10} lead is most frequently used among the precordial chest leads to confirm changes in right heart.

2.6.4 Electrocardiographic Paper

Electrocardiographic paper is a graph paper lined into boxes (Fig. 2.4). There are two types of boxes, i.e., large and small. In electrocardiographic paper every fifth horizontal and vertical line is darker than others. Horizontal lines are 1 mm apart that represents 0.1 mV when ECG is set to the sensitivity of 1 (i.e., 1.0 cm or 10 mm = 1 mV), or 0.05 mV when sensitivity is set at 2, or 0.2 mV when sensitivity is set at 0.5. The vertical lines are also 1 mm apart and represents the time interval of 0.04 s when recording is done at the speed of 25 mm/s or 0.02 s when recording is done at the speed of 50 mm/s. Magnitude of electrocardiographic complexes is expressed in mV and duration in seconds.

2.6.5 Time and Speed

ECG machines record changes in electrical activity by drawing a trace on the moving paper strip. All ECG machines run at 25 mm/s (Fig. 2.5) or 50 mm/s (Fig. 2.6) speed and use paper with standard sized squares.

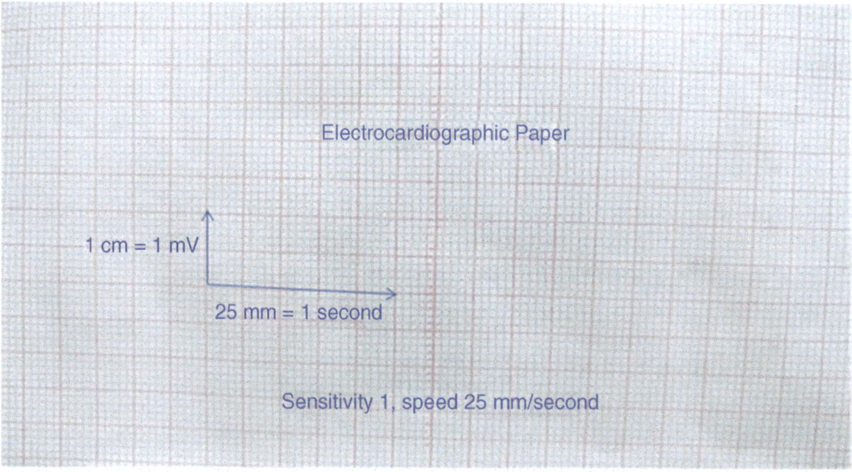

Fig. 2.4 Electrocardiographic paper and grid lines. At the paper speed of 25 mm/s, the interval between two vertical lines is 0.04 s and between two heavily drawn vertical lines is 0.2 s. At the normal setting of 1, distance between two horizontal lines indicates 0.1 mV

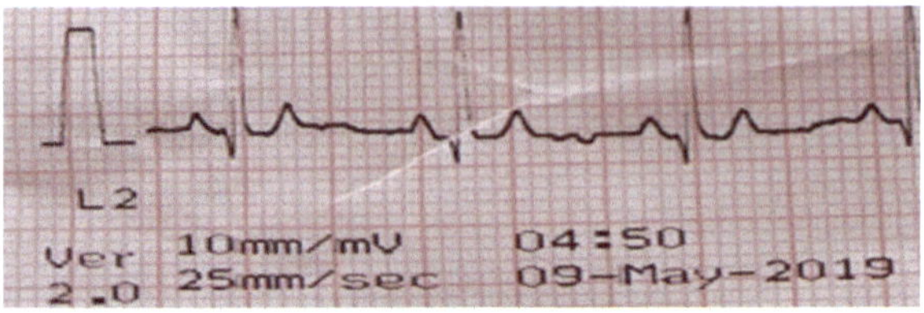

Fig. 2.5 Electrocardiogram of an adult Labrador taken at the speed of 25 mm/s. The distance between two vertical lines denotes 0.04 s

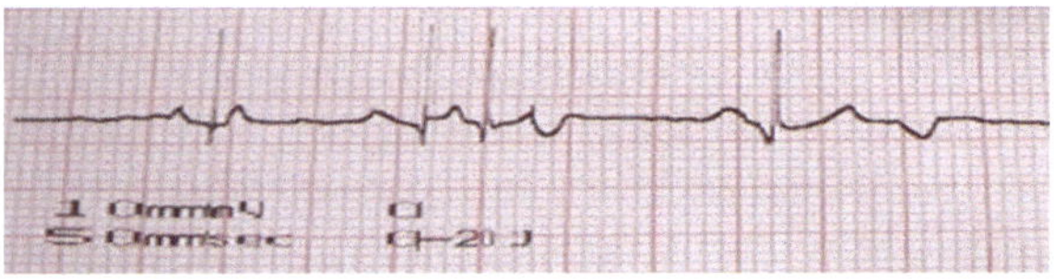

Fig. 2.6 Electrocardiogram of the same Labrador (as above) taken at the speed of 50 mm/s. The distance between two vertical lines denotes 0.02 s

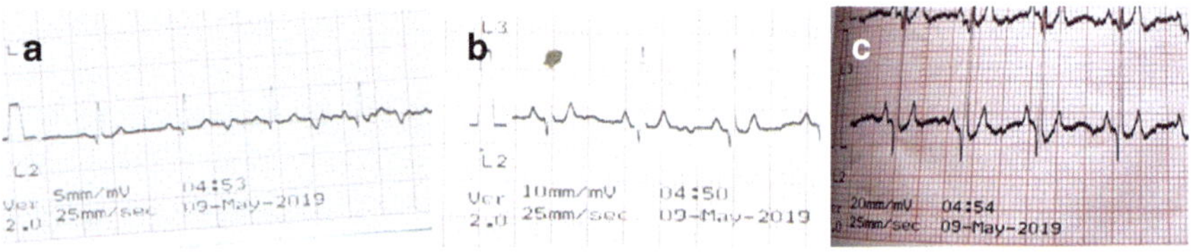

Fig. 2.7 Electrocardiogram of an adult Labrador (same as above) taken at different sensitivities. (**a**) At 0.5 (0.5 cm = 1 mV), (**b**) at 1.0 (1.0 cm = 1 mV), and (**c**) at 2.0 (2.0 cm = 1 mV). As sensitivity increases complexes also increase in size

2.6.6 Sensitivity

Sensitivity usually used in recording is 10 mm or 1 cm = 1 mV. If the complexes are too large, it can be reduced to 5.0 mm or 0.5 cm = 1 mV. If the complexes are too small, it can be increased to 20 mm or 2.0 cm = 1 mV (Fig. 2.7).

2.7 Position and Restrain of Dogs

Dogs are positioned on the table covered with a foam mattress and nonconductive rubber sheet in right lateral recumbency as shown in Fig. 2.8. An attendant restrains the dog in the manner that his right arm rests over the neck and left arm over hind quarter of the dog and both legs (left and right) do not touch each other. The forelimbs are kept perpendicular to the long axis of the body. If the left thoracic limb

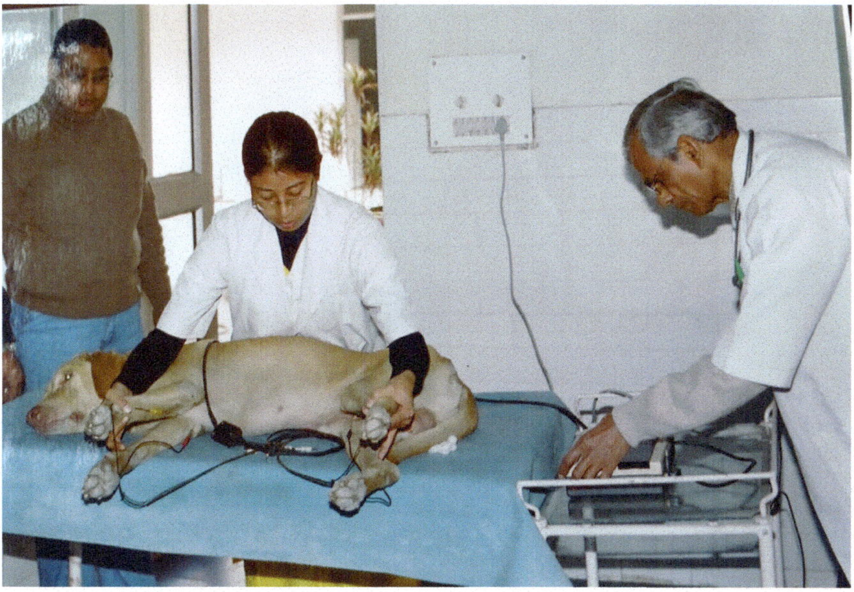

Fig. 2.8 The dog is positioned in right lateral recumbency for electrocardiography and is restrained as shown in the picture

is pulled caudally or back, the form of QRS in lead I appears like that of aVF, whereas, if that thoracic limb is pulled cranially, QRS in lead I appears like that recorded in aVR.

Generally no chemical or drug is used to restrain the dog. In most of the cases, the dog feels comfortable with the presence of the owner at the table. Uncooperative dogs may be calm down by the administration of diazepam (1 mg/kg orally 1 h before examination) or acetylpromazine (2 mg/kg orally 1 h before the examination or 0.5 mg/kg intramuscularly or intravenously 15 min before examination).

2.8 Placement of the Electrodes

Before putting the electrodes, both electrode and skin are moistened with electrocardiographic gel, paste, or alcohol. Alcohol works well. Electrodes are attached directly to the skin. In case nothing is available, water can be used to increase contact between electrode and skin in the unfavorable conditions. Placement of electrode is shown in Figs. 2.9 and 2.10.

- RA—Right forelimb clip or needle electrode is attached proximal to the olecranon on the caudal aspect of the right forelimb.
- LA—Left forelimb clip or needle electrode is attached proximal to the olecranon on the caudal aspect of the left forelimb.

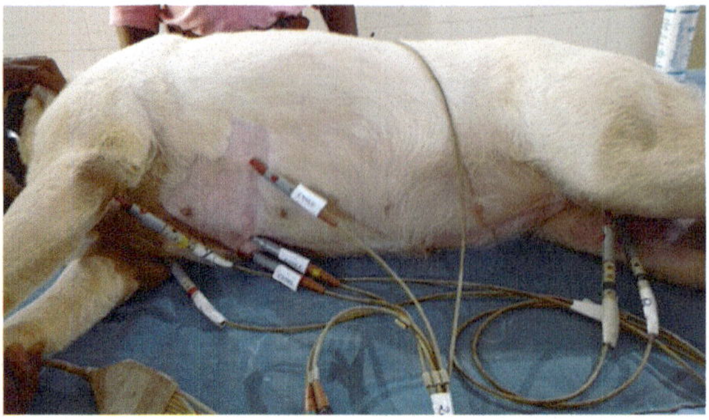

Fig. 2.9 Placement of electrodes. Attach left arm (yellow) electrode to the skin proximal to olecranon on the caudal aspect of the left forelimb, right arm (red) electrode to the skin proximal to olecranon on the caudal aspect of the right forelimb, left limb (green) electrode to the skin over patellar ligament on the anterior aspect of the left hind limb, right limb (black) electrode to the skin over patellar ligament on the anterior aspect of the right hind limb; CV5RL electrode at fifth right intercostal space near edge of sternum, CV6LL electrode at sixth left intercostal space near edge of sternum, CV6LU at sixth left intercostal space at costochondral junction, and V10 over dorsal spinous process of the seventh thoracic vertebra

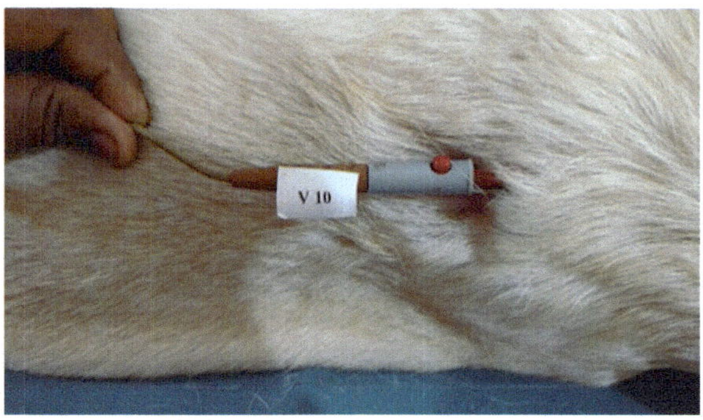

Fig. 2.10 Attachment of V10 electrode on dorsal spinous process of seventh thoracic vertebra

- RL—Right hind limb clip or needle electrode is attached over patellar ligament on the anterior aspect of the right hind limb.
- LL—Left hind limb clip or needle is attached over patellar ligament on the anterior aspect of the left hind leg.

Above four electrodes (RA, LA, RL, LL), placed on limbs, make six leads as lead I, lead II, lead III, lead aVR, lead aVL, and lead aVF. These six leads are frontal plane leads.

V leads or unipolar precordial chest leads are horizontal plane leads and are attached as follows:

- Electrode of lead CV_5RL (rV_2)—It is attached at fifth right intercostal space near edge of sternum.
- Electrode of lead CV_6LL (V_2)—It is attached at sixth left intercostal space near edge of sternum.
- Electrode of lead CV_6LU (V_4)—It is attached at sixth left intercostal space at costochondral junction.
- Electrode of lead V_{10}—It is attached over dorsal spinous process of seventh thoracic vertebra.

Base-Apex Bipolar monitoring lead. Electrodes are positioned as follows:

- LA electrode—It is applied over apex of the heart on left side just behind elbow.
- RA electrode—It is applied over spine of right scapula near the vertebra or in jugular furrow.
- RL electrode—It is applied over near hump.

Recording is done in lead I (details given under "Electrocardiography of Ruminants"). This lead system is generally not used in canines.

2.9 Special Features of Leads

- Standard limb leads examine the electrical activity of the heart in the frontal plane of the body.
- Chest leads determine the electrical activity of the heart in sagittal and horizontal planes of the body.
- V leads record the direction and voltage of electrical activity from the heart to the electrode position.

2.10 Recording of the Electrocardiogram

1. Turn power on.
2. Apply earth clip to some metal plate/sheath/object.
3. Set the sensitivity usually at position 1. It can also be set at position 0.5 or 2.
4. Set the paper speed at 25 mm/s. Paper speed can also be adjusted at 50 mm/s as per requirement.
5. Press print/start as per machine.
6. Display and printing of electrocardiogram will begin and will stop as soon all leads are recorded.
7. Turn power off.
8. Remove the lead electrodes from the dog.
9. Write case number, details of the dog, owners name, and date of the tracings.

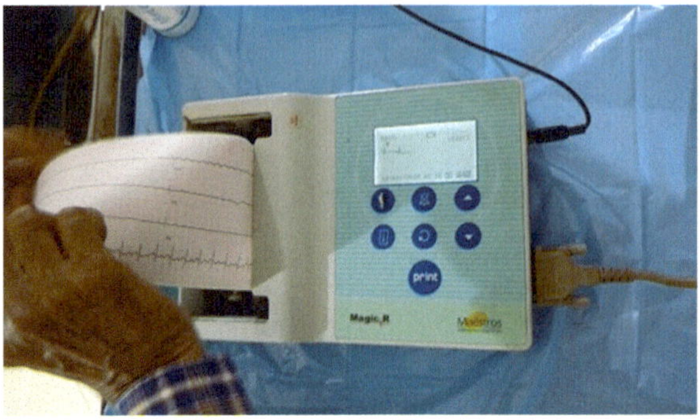

Fig. 2.11 Recording of an electrocardiogram

10. Electrocardiographic machines from different companies have different settings. Therefore machines should be operated as per their instructions given in their manual.
11. Recording is shown in Fig. 2.11.

2.11 Common Artifacts Observed During Electrocardiography

Artifacts are abnormalities in electrocardiogram. These abnormalities are not related to heart diseases in any way and arise from technical or mechanical errors or from malfunctioning during recording of an electrocardiogram. Sometimes these artifacts create confusion. Generally there are four types of artifacts, described below, which need to be managed for a good electrocardiogram.

1. *Electrical interference*: These defects are also called 60-cycle disturbances (Fig. 2.12). Regular sawtooth appearance of the baseline is due to improper grounding of the dog, electrocardiograph, or electrical source. This interference causes a lot of disturbances and ECG cannot be interpreted properly. Electrical interference can be corrected by taking following precautions.
   (a) By ensuring proper grounding of the power cord.
   (b) By ensuring secured attachment of all electrodes.
   (c) By ensuring liberal gel application on the skin.
   (d) By ensuring secured attachment of electrodes to the cable.
   (e) By ensuring that no fluorescent light is on in the room.
   (f) By ensuring that legs of the dog are held apart.
   (g) By ensuring no touching of the electrode clips with each other.
   (h) By ensuring no touch of leads by the attendant.
   (i) Metal table should be properly cushioned with foam mattress and rubber sheet.

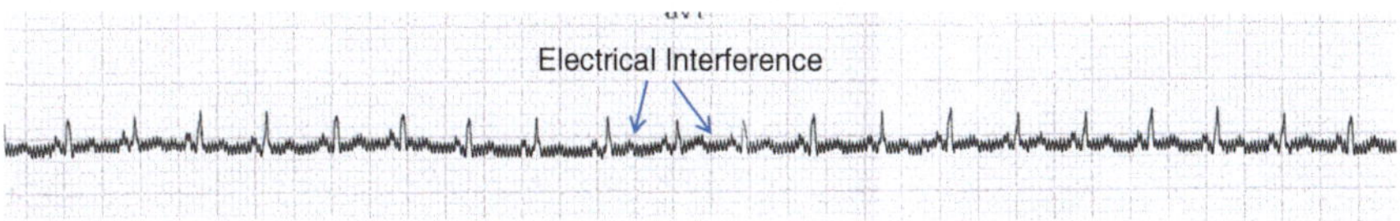

Fig. 2.12 Electrical interference or 60-cycle disturbance in electrocardiogram

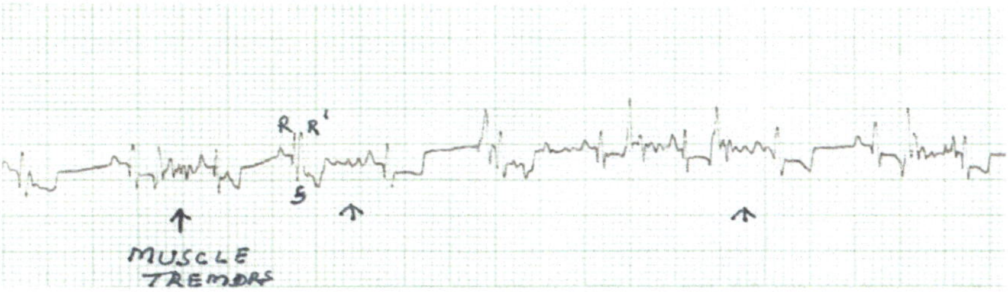

Fig. 2.13 Muscle tremors in electrocardiogram

(j) By making sure that electric chords are not touching the metal table.
(k) By preventing hair or skin contact with electrodes on the adjacent limb.
(l) By ensuring that the dog is not bearing any metal collar or halter.

2. *Muscle tremor*: Trembling of the body of the dog can also produce artifacts (Fig. 2.13). Muscle fasciculation, panting, shivering, or purring produces an irregular, rapid sawtooth vibrations of the baseline leading to difficulty in reading an electrocardiogram. These defects can be reduced as follows:
   (a) By ensuring comfortable positioning of the dog.
   (b) By keeping the owner with nervous dogs in the room during ECG recording.
   (c) By giving mild tranquilizer to uncooperative dogs to calm down. In such cases care should be taken in interpreting the electrocardiogram.
   (d) Muscle tremors are also reduced by placing hand over the chest wall during recording ECG.
   (e) Sometimes reducing sensitivity to 1/2 also serves the purpose.
3. *Wandering baseline*: Changes in the resistance between the electrode and skin of dog may cause wandering of the baseline (Fig. 2.14). Respiratory movements, coughing, or panting may cause wandering baseline. This can be overcome by:
   (a) Ensuring comfortable position of the dog. Dogs with heart failure are not comfortable in lateral recumbency. Therefore their ECG should be taken in the position the dog feels comfortable. There is not much effect on heart rate and rhythm.

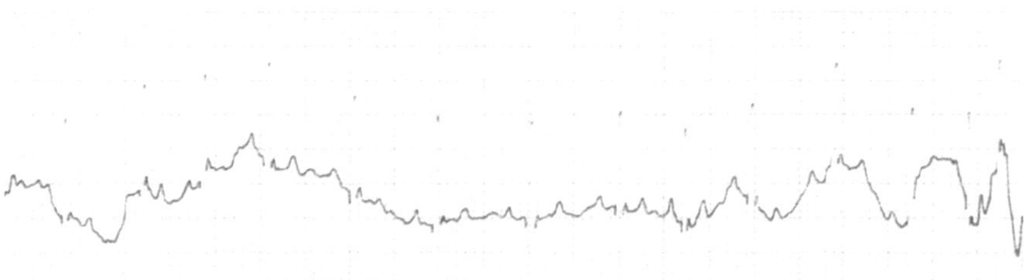

Fig. 2.14 Electrocardiogram showing wandering baseline

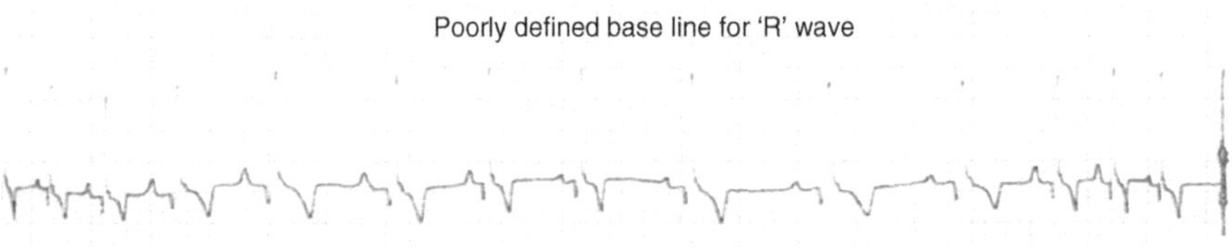

Fig. 2.15 Electrocardiogram of a dog taken at sensitivity of 1.0 and paper speed of 25 mm/s. Note the poorly defined "QRS" complex. The "R" wave is poorly defined

   (b) Closing dog's mouth for 3–4 s during recording.
   (c) Placing the electrode cable on the table with the dog.
   (d) Applying gentle pressure of the palm on chest of the dog.
4. *Poorly defined baseline*: Poorly defined baseline are seen for large complexes such as "R" wave (Fig. 2.15). The problem can be resolved by:
   (a) Cleaning the stylus in manual ECG machines.
   (b) Checking the stylus arm in manual ECG machines.
   (c) Increasing stylus heat in manual ECG machines.
   (d) If complexes are large, reducing the sensitivity.

Misplacement of electrodes: If electrodes are misplaced, complexes can be reversed in polarity or be mirror images resembling the aberrant conduction patterns. When LA is misplaced on right arm and RA on left arm, the amplitude of QRS increases in lead I, III, and aVL and decreases in lead II, aVR, and aVF. "T" wave shows change in polarity (Fig. 2.16b). Misplacement of hind limb electrode (LL on right hind limb and RL on left hind limb) increases amplitude in lead I and aVR and decreases in aVL (Fig. 2.16c). In case electrodes are totally misplaced as arm electrodes (LA and RA) on hind limbs and hind limb electrodes (LL and RL) on arms, the amplitude of QRS complex increases in lead I and aVR and decreases in lead III and aVL (Fig. 2.16d). In such situations electrode placement should be checked and need to be placed at right position.

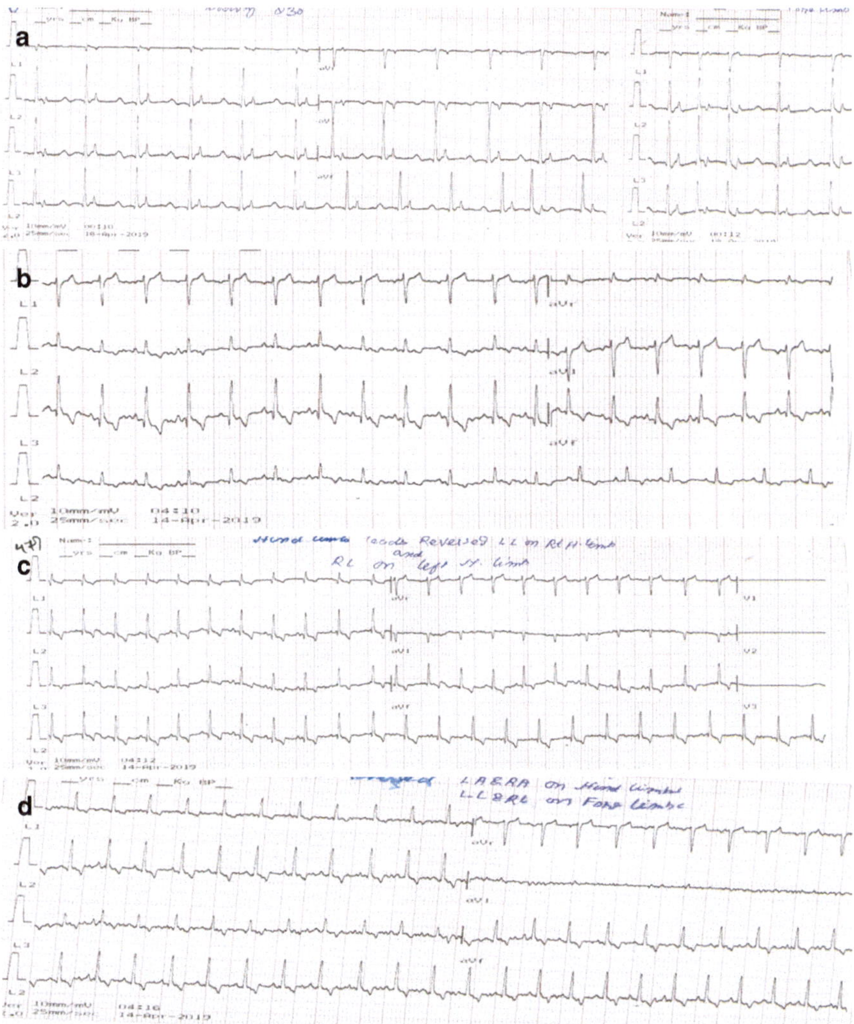

Fig. 2.16 (**a**) Limb electrodes (LA, RA, LL, and RL) placed in right place. (**b**) Arm electrodes misplaced. LA placed on right forelimb and RA placed on left forelimb. Note changes in the amplitude of complexes in different leads in comparison to (**a**). The amplitude of QRS has increased in lead I, III, and aVL and decreased in lead II, aVR, and aVF. The polarity of T wave has changed from –ve to + ve in lead aVL at two places. (**c**) Hind limb electrode misplaced. LL placed on right hind limb and RL placed on left hind limb. The amplitude of QRS increased in lead I and aVR and decreased in aVL when compared to (**a**). (**d**) Electrodes totally misplaced. Arm electrodes (LA and RA) were placed on hind limbs and hind limb electrodes (LL and RL) were placed on arms. The amplitude of QRS complex increased in lead I and aVR and decreased in lead III and aVL when compared to (**a**)

Further Reading

Bolton GR (1975) Handbook of canine electrocardiography. W.B. Saunders Company, Philadelphia, PA

Tilley LP (1985) Essentials of canine and feline electrocardiography, 2nd edn. Lea and Febiger, Philadelphia, PA

Tilley LP (1992) Essentials of canine and feline electrocardiography, 3rd edn. Lea and Febiger, Philadelphia, PA

Generation and Shape of Electrocardiogram

3

3.1 What Is an Electrocardiogram (ECG)?

An electrocardiogram is a graphic representation of the final total surface voltage and direction of electrical activity produced during depolarization and repolarization of cardiac cells plotted against time (Fig. 3.1). The potential difference occurs over the surface of the body owing to electrical activity in the heart. The electrical activity in the heart is associated with depolarization and repolarization of the myocardium. The amplitude of these potential differences between various points on the body surface is measured as millivolt (mV), and their duration is measured in seconds. ECG provides an information about heart rate; its rhythm; size of chambers in dogs and humans; conduction defects; myocardial diseases or ischemia; and electrolyte imbalances such as hypocalcemia, hypokalemia, hypercalcemia, or hyperkalemia. Information about some drug toxicities can also be obtained. Electrocardiographic parameters in different breeds of dog differ slightly.

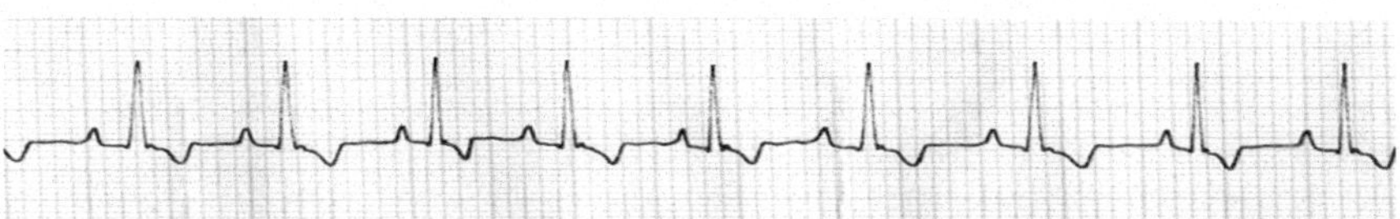

Fig. 3.1 Electrocardiogram is a graphic representation of the sum up surface voltage and direction of electrical activity produced during depolarization and repolarization of cardiac muscle cells plotted against time. On *X* axis of the graph (horizontal), time taken by the electrical activity is represented, and on *Y* axis of the graph (vertical), sum up surface voltage and direction of electrical activity is represented. In this figure the direction of electrical activity for P and R wave is positive (current flowing toward positive electrode) while for T wave is negative (current flowing away from positive electrode)

J. P. Varshney, *Electrocardiography in Veterinary Medicine*,
https://doi.org/10.1007/978-981-15-3699-1_3

3.2 Electrical Activity in Cardiac Cell

The inner and outer surfaces of cardiac muscle cell have different electrical charge. The inside of the cell is negatively charged, and outer cell surface is positively charged (Fig. 3.2). The contraction of K^+ inside the cell is higher, while concentration of sodium and calcium is higher outside the cell. Cardiac muscle cells can differentiate between sodium and potassium ions. The sodium ions are kept out, while

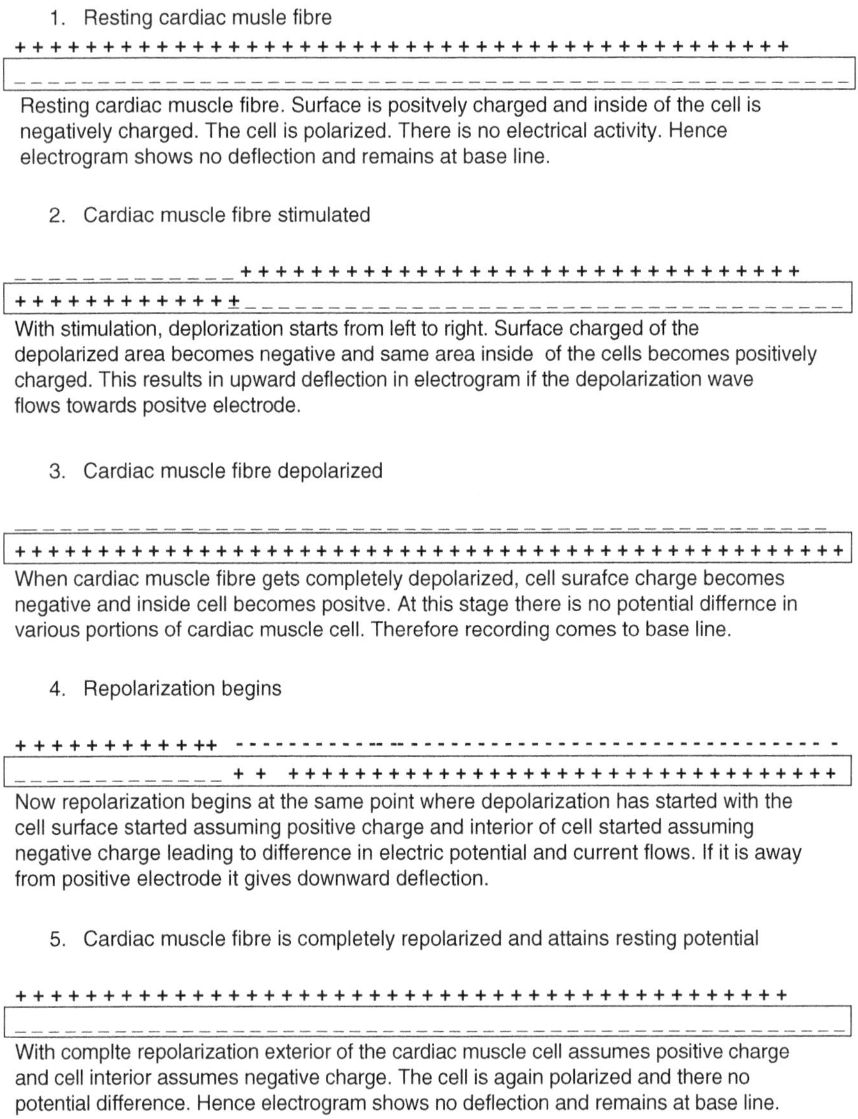

Fig. 3.2 Electrical activity in the single cardiac muscle fiber is illustrated in five steps

the potassium ions are retained within the cell. By this process (sodium and potassium pump), resting membrane potential in cardiac cell is maintained. On stimulation membrane potential of resting myocardial cell is reduced suddenly with a change in cell membrane permeability facilitating the entry of Na^+ ions into the cell and exit of K^+ ions from the cell. As a result the surface charge of the cardiac muscle cell starts becoming negative, and inside of the cell starts becoming positive leading to the event of depolarization. Electrocardiogram records this event as an upward deflection if depolarization flows toward positive electrode. On complete depolarization of the cardiac cell, potential difference between various portions of the cell surface is equalized, hence recording returns to baseline. Now repolarization begins with the exit of Na^+ from the cell and entry of K^+ into the cell, and outer surface starts getting positive charge leading to deflection of stylet as per current flow from positive to negative or vice versa.

3.3 Current Generation and Conduction System in the Heart

Canine and feline heart has four chambers, but from electrical point of view, it is considered as having only two units. The right and left atria are synchronized as one unit, and right and left ventricles are synchronized as other unit. The contraction of one unit (atrium) is followed by the contraction of the other unit (ventricles). The seat of generation of the current for each cardiac cycle is "sinoatrial" node (SA node) that is situated in the right atrium. Cardiac conduction system (Fig. 3.3) consists of sinoatrial node, the interatrial bundle, the atrioventricular node (AV node), the common bundle of His, the right and left bundle branches, and the Purkinje fibers. The SA node is a powerhouse from where depolarization is initiated and spread across the right and left atrium and to the AV node through internodal fibers. The right atrium depolarizes first followed by the depolarization of the left atrium. There is a difference of approximately 0.01 s between depolarization of both atrium. The speed of depolarization wave from SA node to the right to left atrium, to the bundle of His, to the bundle branches, and to the Purkinje fibers is variable. Depolarization wave is stopped for a fraction at all points except atrioventricular node (AV node). In ECG this delay is represented by a straight line, i.e., P-R segment. From the AV node current passes to the bundle of His to bundle branches to Purkinje fibers to ventricular myocardium. The bundle of His is divided into the right and left bundle branches. The right bundle branch passes down the right ventricle side. The left bundle branch passes down the left ventricular side of the septum. The left bundle branch further divides into anterior and posterior fascicles. These fascicles divide into the Purkinje fibers, and these are distributed to the ventricular myocardium. The entire ventricular muscle mass contracts synchronously. Finally when depolarization is complete, repolarization of the cells of the ventricular myocardium starts. Its earliest stages are normally not detectable at the body surface, but later stages are represented as T wave.

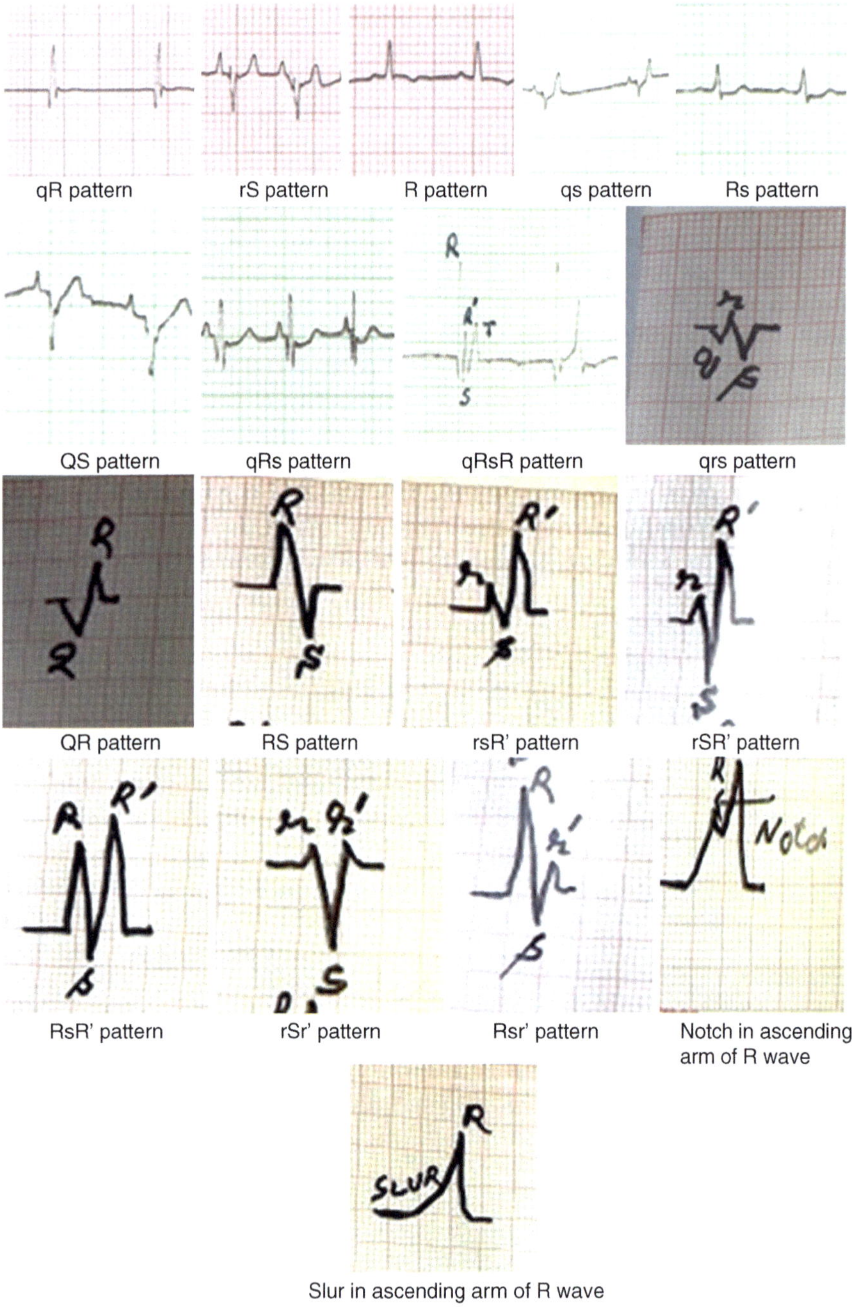

Fig. 3.5 Showing different patterns of “QRS” complex

R and small s wave present), "RS" (both R and S prominent), "qs" (both q and s waves small), "rS" (r wave small and S large), rsR′ (small r and s followed by large R′ wave), rSR′ (r wave small followed by large S and R′ wave), RsR′ (R normal small s followed by large R′), rSr′ (small r and r′ and large S), Rsr′ (normal R small s followed by small r′), notch in ascending arm of the R, or slur in ascending arm of the R wave.

3.6 Clinical Conditions Requiring an Electrocardiographic Examination

An electrocardiographic examination should invariably be taken in dogs with the following clinical conditions to ascertain whether cardiac functioning is normal or compromised:

- Accidents (falls from height, automobile accidents)
- Arrhythmias (detected during auscultation)
- Sudden Shock/syncope/seizures/fainting
- Sudden onset of severe dyspnea/panting
- Chronic persistent coughing/nocturnal coughing
- Cardiac murmurs (detected during auscultation)
- Increased/decreased auscultation area of the heart
- Renal disease (acute or chronic renal failure)
- Endocrinopathies (Addison's disease, Cushing syndrome, thyroid dysfunctions)
- Prolonged urinary obstruction
- Systemic diseases (pyometra, pancreatitis, uremia, neoplasms) affecting the heart
- Acid-base and or ionic imbalances
- Hyperthermia or heat stroke
- Hypothermia
- Electrocution
- Poisoning (including chocolate poisoning)
- Drug toxicities

3.7 Electrocardiographic Physiology

The "P" wave is formed due to depolarization of the atrium. The "PR" segment is created due to short delay occurring after contraction of the atrium and beginning of ventricular contraction. This delay facilitates complete filling of the ventricle. Ventricular depolarization creates "QRS" complex. With ventricular contraction blood is ejected intro the aorta and pulmonary artery. The "ST" segment and "T" wave represents repolarization of ventricles.

3.8 Measurement Details of Different Waves and Intervals in ECG

Details of measurement of the different complexes and intervals in the electrocardiogram are given below.

- P wave: It is measured from beginning to end of "P" wave and from top to bottom. It is positive in leads I, II, III, and aVF.
- P-R segment: It is measured from end of "P" wave to beginning of QRS.
- P-R interval: It is measured from beginning of P wave to beginning of QRS complex.
- QRS complex: It is measured from the beginning to end of QRS complex and from baseline to the top of "R" wave.
- S-T segment: It is measured between end of "S" wave to beginning of "T" wave. It runs along with the baseline before going into the "T" wave.
- T wave: It is measured from the end of S-T segment to the end of "T" wave.
- Q-T interval: It is measured from the beginning of "Q" wave to end of "T" wave.
- R-R interval: It is measured between two R.

Measurement details of the amplitude and duration of different complexes and intervals in the electrocardiogram are illustrated in Fig. 3.6.

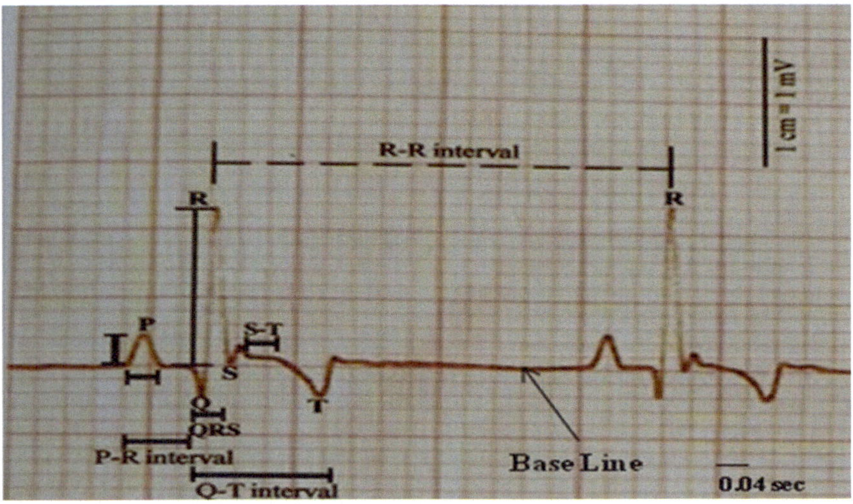

Fig. 3.6 Electrocardiogram (lead II, sensitivity 1, paper speed 25 mm/s) of a healthy German Shepherd showing different complexes and intervals for measurement purpose. The measurement of the complexes and intervals are as "P" 0.2 mV and 0.04 s duration, "P-R" interval 0.08 s, "Q" 0.02 mV, "R" 1.2 mV, "QRS" 0.06 s, "S-T" segment 0.06 s, "T" negative 0.02 mV and of 0.06 s, "R-R" interval 0.64 s, and "Q-T" interval 0.18 s in this figure. (This figure is with the courtesy of Dr. Bendangla Changkija, Assistant Professor, Department of Medicine, Veterinary College, Assam)

3.9 Interpretation of Normal Cardiac Waveforms

Wave forms	Interpretation
"P"	Atrial depolarization. Normally positive in leads II, III and avF.
"P-R" interval	Time from onset of atrial muscle depolarization through conduction over the AV node, bundle of His, and Purkinje fibers.
"QRS" complex	Depolarization of ventricular muscle. Q is first negative deflection, "R" the first positive deflection, and "S" the second negative deflection.
"J" Point	End of QRS complex.
"S-T" segment	Period of phase 2 of action potential.
"T"	Ventricular muscle repolarization.
"Q-T" interval	Total time of ventricular depolarization and repolarization.

3.10 Heart Rate Variability (HR Variability)

It has long been recognized that instantaneous heart rate fluctuates on beat-to-beat basis. With disease and/or aging, the ability of the body is reduced to maintain a constant heart rate at rest. Beat-to-beat variability in heart rate of the dogs is under the influence of many external inputs exercising their influence through parasympathetic (PNS) and sympathetic nervous system (SNS). Parasympathetic nervous system plays a vital role in heart rate variability. Factors such as blood pressure regulation, thermoregulation, respiration, action of the renin-angiotensin system, and circadian rhythms influence R-R intervals. In fact heart rate variability is a statistical analysis of small variations in time intervals between heartbeats. Vasovagal tonus index (VVTI) is a natural logarithm of the variance in R-R interval and therefore is a reliable indicator of heart rate variability. It is calculated according to a formula given by Haggstrom et al. (1996) and Doxey and Boswood (2004).

It has been observed that vasovagal tonus index (VVTI) differed significantly ($P < 0.05$) among different breeds of dog. A significantly ($P < 0.05$) higher VVTI in Boxers (brachycephalic dog) and a lower VVTI in German Shepherd and Labradors (non-brachycephalic dogs) have been observed (Changkija 2007).

The VVTI, a time domain measure of heart rate variability, measures high-frequency variations exclusively owing to variations in vagal tone and the periodicity of respiration in normal dogs. The age and sex of the dogs seem to have no significant influence on the VVTI. It has been shown that the changes in heart rate variability are associated with a wide range of diseases and other factors like obesity, pain, and stress. Some studies have indicated that reduction in heart rate variability is associated with heart diseases, sepsis, diabetes, obesity, chronic pain, and stress. Heart rate variability is also known to be an indicator of emotional states in animals. A link between canine behavioral problems and heart rate variability is also speculated.

3.11 Q-T Interval Variability

Q-T interval is dependent on preceding R-R interval and varies inversely with heart rate. If it is taken into consideration without heart rate, it may be misinterpreted. Since alterations in QT interval length reflect abnormalities of the ventricular repolarization predisposing to occurrence of arrhythmias, Q-T interval must be properly evaluated and corrected in order to improve its diagnostic utility.

A number of correction formulae, derived for analyzing human data, are currently used, some for analyzing anaesthetized animals or animal models for toxicology assessments. Little information, however, is available on how to correct the QT interval for clinical practice, considering different HR ranges (due to different breeds) in conscious healthy dogs. It may be calculated using Bazett's formula (Bazett 1920).

References

Bazett HC (1920) An analysis of time relations of electrocardiograms. Heart 7:353–370

Changkija B (2007) Electrocardiographic studies in dogs with reference to management of cardiac tachyarrhythmia by alternative drugs. Ph.D. thesis submitted to Deemed University, Indian Veterinary Research Institute, Izatnagar

Doxey S, Boswood A (2004) Difference between breeds of dog in a measure of heart rate variability. Vet Rec 154:713–717

Haggstrom J, Hamlin RL, Hansson K, Kvart C (1996) Heart rate variability in relation to severity of mitral regurgitation in Cavalier King Charles Spaniels. J Small Anim Pract 37:69–75

Further Reading

Binkley PF, Nunziata E, Hass GJ, Nelson SD, Cody RJ (1991) Parasympathatic withdrawal is an integral component of autonomic imbalance in congestive heart failure: demonstration in human subjects and verification in a paced canine model of ventricular failure. J Am Coll Cardiol 18:464–472

Bolton GR (1975) Handbook of canine electrocardiography. W.B. Saunders Company, Philadelphia, PA

Calvert CA (1998) Heart rate variability. Vet Clin North Am Small Anim Pract 28:1409–1427

Ettinger SJ, Suter PF (1970) The recognition of cardiac disease and congestive heart failure. In: Canine cardiology. W.B. Saunders, Philadelphia, PA, p 215

Friedberg CK (1966) Diseases of heart, 3rd edn. W.B. Saunders Company, Philadelphia, PA

Guyton AC (1971) Text book of medical physiology, 4th edn. W.B. Saunders Company, Philadelphia, PA

Harada T, Fumiko I, Hamada M, Horie N, Nitta Y, Nitta K, Katsuoka H, Nakamura S (2010) Circadian rhythm of heart-rate variability and autonomic cardiovascular regulation in Parkinson's disease. Auton Neurosci 158:133–140

Katayama M, Kubo T, Mogi K, Ikeda K, Nagasawa M, Takfumi K (2016) Heart rate variability predicts the emotional state in dogs. Behav Process 128:108–112

Rasmussen CE, Vesterholm S, Ludvigsen TP, Haggstrom J, Pederson HD, Moesgarrd SG, Olsen LH (2011) Holter monitoring in clinically healthy Cavalier King Charles Spaniels, Wirehaired Dachshund and Cairn Terriers. J Vet Intern Med 25:460–468

Simonson E, Minn M, Cady LD Jr (1962) The normal Q-T interval. Am Heart J 63:747–753
Stein PK, Bosner MS, Kleiger RE, Conger BM (1994) Heart rate variability: a measure of cardiac autonomic tone. Am Heart J 127:1376–1381
Tilley LP (1985) Essentials of canine and feline electrocardiography, 2nd edn. Lea and Febiger, Philadelphia, PA
Wormald D, Lawrence AJ, Carter G, Fisher AD (2017) Reduced heart rate variability in pet dogs affected by anxiety related behavioural problems. Physiol Behav 168:122–127

A Systematic Reading of an Electrocardiogram

4

After completion of the tracing, the electrocardiogram (ECG) should be filled in properly and stored. Basic information such as name, breed, age, and sex of the dog, owner's name and address with mobile number, date of tracing, clinical complaints, and observations should be filled in. Leads I, II, III, aVR, aVL, aVF, CV5RL, CV6LL, CV6LU, and V10 should be marked on the tracings itself if ECG is taken using a manual machine. Speed (25 or 50 mm/s) and sensitivity (0.5, 1 or 2) should be specified for calculating heart rate and measurements of different wave forms in case of manual machines. Modern automatic machines automatically record these details as per their setting. Taking an ECG tracing is an easy task, but its interpretation is an important and difficult task requiring fundamental knowledge of cardiac anatomy, physiology, and cardiology. Casual approach in interpretation may be unrewarding. An electrocardiogram should be interpreted keeping in view the history and clinical manifestations of the patient. Definite interpretations from ECG regarding anatomic defects or change in physiological states may not be always predictable. Therefore clinician is the best person to interpret an ECG findings in the background of his patient's history, clinical status, and other investigations. A veterinarian should also be conversant with wide variations in the normal and overlapping ECG patterns between various conditions.

4.1 Systematic Approach to ECG

Four important information such as heart rate; its rhythm; amplitude and duration of waves, segments, and intervals; and mean electrical axis can be obtained by systematic stepwise reading of an electrocardiogram. An ECG should also be examined for presence of any miscellaneous criteria.

J. P. Varshney, *Electrocardiography in Veterinary Medicine*,
https://doi.org/10.1007/978-981-15-3699-1_4

1. Calculation of heart rate: Average heart rate is generally calculated in lead II. The numbers of predominant complex (QRS) are counted in 3 s (15 large squares, i.e., 75 small squares if machine is running at the speed of 25 mm/s or 30 large squares, i.e., 150 small squares if machine is running at the speed of 50 mm/s) and is multiplied by 20 to get heart rate per minute. This is the simplest way of calculating heart rate.
2. Determining cardiac rhythm: In normal ECG, each complex occurs with a "P" wave followed by a short "P-R" segment followed by "QRS" followed by, S-T segment and finally T wave. If there is "P" wave for every "QRS" complex and "R-R" or "P-R" intervals are of normal duration, then rhythm is said, to be sinus rhythm. For evaluating heart rhythm, lead II strip is examined. In case of frank arrhythmia, gross irregularities are obvious and can be detected even at first sight, while small irregularities need proper attention warranting measurement of "P-R" and "R-R" intervals. In normal sinus rhythms, these intervals (P-R interval and R-R interval) do not vary much from complex to complex.
3. Measuring the amplitude and duration: Amplitude and duration of waves ("P," "QRS," "T"), segments (P-R segment, S-T segment) and intervals ("P-R" interval, "R-R" interval, "Q-T" interval) are to be measured in each electrocardiogram. Generally these measurements are recorded in lead II. But other leads (I, III, aVR, aVL, and aVF) are also analyzed for confirming the changes in lead II in confusing situations and determining mean electrical axis on frontal plane, etc. Some of this information is available on the ECG strip obtained through automatic ECG machines. Measurement procedure of amplitude, duration of waves, segments, and intervals is given in Chap. 3.
   (a) *"P" wave*—Amplitude and duration of "P" is measured. In dogs "P" wave may be positive, notched, biphasic, or negative depending on the lead (Fig. 4.1). In lead II, normal "P" wave is small, rounded, and positive, or it may be M-shaped. It is negative (−ve) in lead aVR and avL. Sometimes biphasic "P" is seen possibly owing to shift of pace maker site.
   (b) *P-R interval*—It is measured from the beginning of the "P" to the beginning of the "Q" (Fig. 4.2). Generally P-R interval is approximately same from complex to complex. Variation in P-R interval from beat to beat indicates arrhythmia.
   (c) *"Q" wave*—It is a first negative deflection after "P" wave, and it reflects right ventricle depolarization (Fig. 4.3).
   (d) *"R" wave*—In QRS complex it is the first positive deflection indicating depolarization of left ventricle (Fig. 4.4).
   (e) *"S" wave*—It is the negative deflection that follows the R wave and represents depolarization of right ventricle (Fig. 4.5).
   (f) *"QRS" complex*—In canines and felines, lead II is given importance. Generally "R" wave is much more prominent. There are various forms of "QRS" complex in limb leads as shown in Chap. 3.

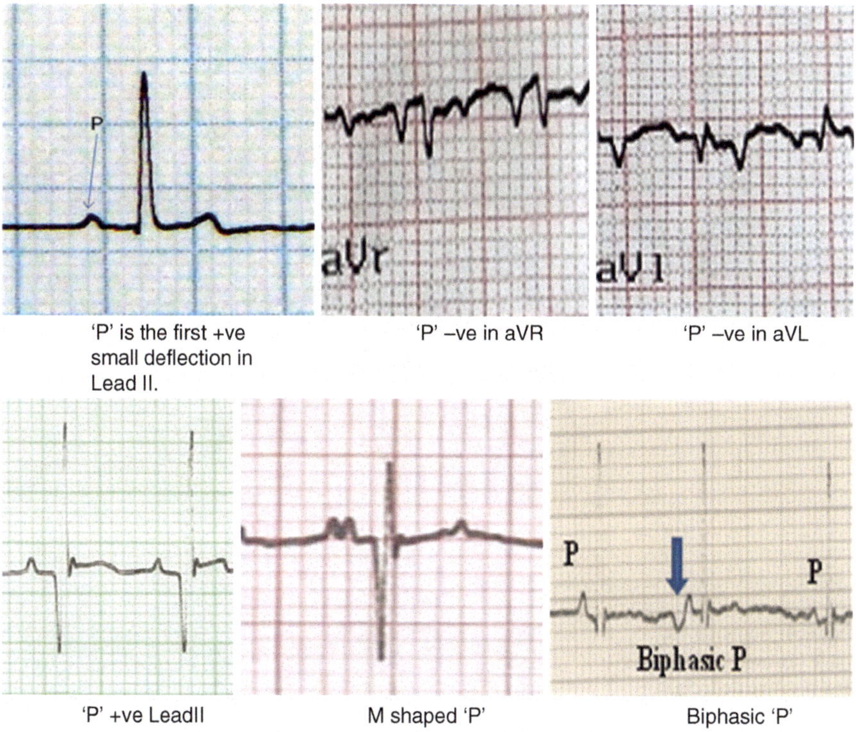

Fig. 4.1 Showing different patterns of "P" waves

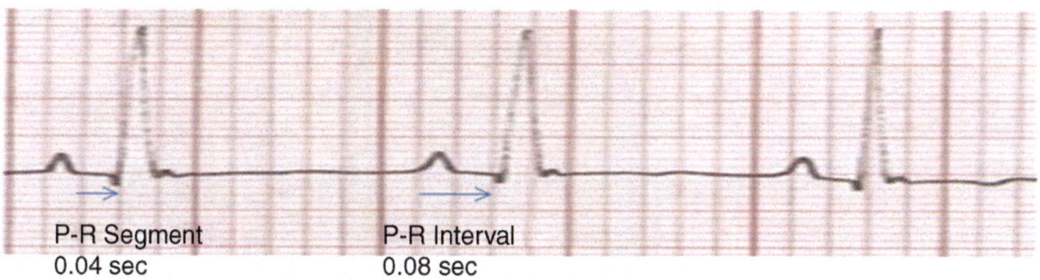

Fig. 4.2 Showing P-R segment and P-R interval. P-R segment is measured from end of "P" to beginning of "QRS," and P-R interval is measured from beginning of "P" to beginning of "QRS" as shown in the figure

(g) *"S-T" segment*—It is the time interval between the end of QRS and the onset of "T" wave (Fig. 4.6). It is generally at the level of baseline. It may be elevated or depressed from the level of baseline. Significant change in the level of S-T segment is considered abnormal.

(h) *"T" wave*—After "QRS" complex, it is the first major deflection representing repolarization of the ventricles (Fig. 4.7). In dogs it may be positive, negative, notched, or biphasic.

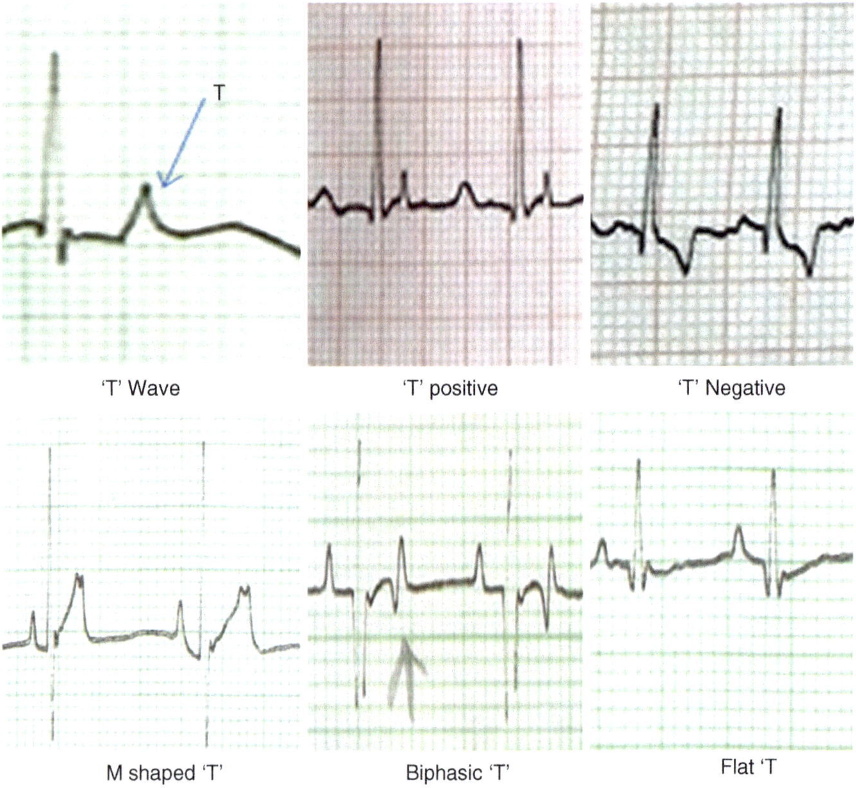

Fig. 4.7 Showing different types of "T" wave. T wave is the first major deflection after "QRS"

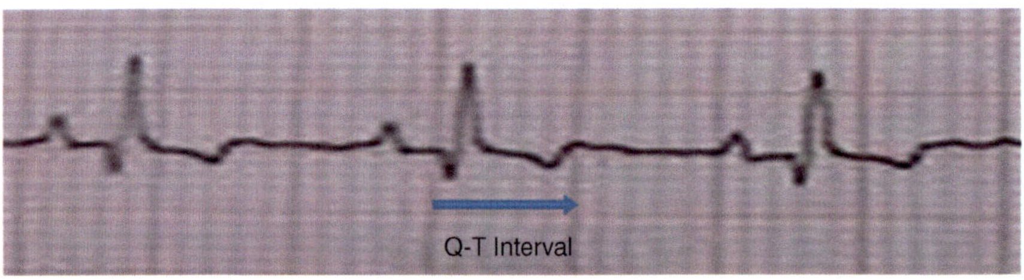

Fig. 4.8 "Q-T" interval is between beginning of "QRS" to the end of "T" wave

(i) *"Q-T" interval*—It is inversely related with the heart rate. Alone it is not of much diagnostic significance in canine medicine. As a thumb rule, it is less than half the preceding "R-R" interval in case of normal sinus heart rates. It represents a time period of ventricular depolarization and repolarization (Fig. 4.8).

4. Determine mean electrical axis: Mean electrical axis (MEA) is an average direction of ventricular vectors during cardiac cycle. It is a determination of the sum

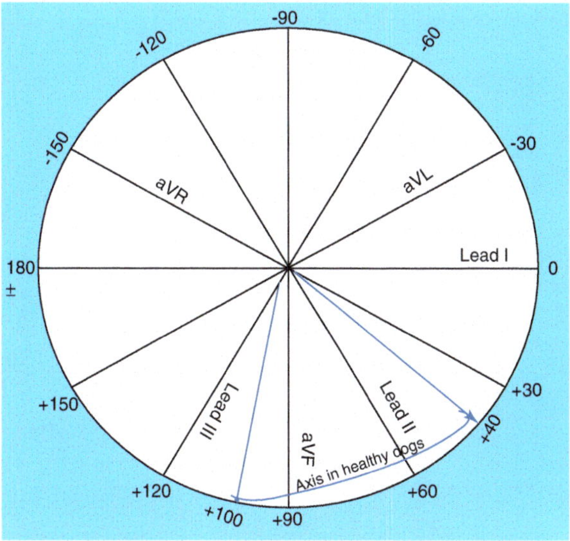

Fig. 4.9 Diagrammatic representation of average direction of ventricular vectors during cardiac cycle

MEA Normal Dogs	+40° to +100° (marked with blue lines)
MEA right axis deviation	+100° to −90°
MEA Left axis deviation	+40° to −90°

of voltage and direction of the heart's ventricular electrical activity in a 360° circle assigned to the frontal plane of the body. An abnormal MEA indicates defects in intraventricular conduction, ventricular size, and/or cardiac axis deviation. It is usually determined in the frontal plane. Mean electrical axis of the dog's heart lies between +40° and +100° (Fig. 4.9). Axis less than +40° to −90° indicates left axis deviation, and more than +100° to −90° indicates right axis deviation. It can be estimated by either of the following ways:

(a) Find the lead (I, II, III, aVR, aVL, and aVF) with largest "R" wave deflection. The positive electrode of this lead points to approximate MEA.
(b) Find the lead (I, II, III, aVR, aVL, or aVF) with isoelectric "QRS"; then identify the lead perpendicular to this lead. If "QRS" in this perpendicular lead is positive, MEA is toward the positive pole. If "QRS" in the perpendicular lead is negative, MEA is in the direction of the negative pole. If all leads appear isoelectric, the frontal axis cannot be determined.
(c) The simplest and practical method of determining mean electrical axis is to measure the net amplitudes in lead I and lead III, and value can be found out from the table given by Tilley (1985) in his book.

Identification of miscellaneous criteria: Examination of ECG strip for miscellaneous criteria includes pattern and amplitude of S wave in leads I, II, III and CV6LU; amplitude of "Q" wave in leads II, III and aVF; and a positive "T" wave in lead V10 (Fig. 4.10). If any of these criteria (deep "S," deep "Q" or positive "T" in V10 lead) is present, right ventricular hypertrophy/enlargement is indicated.

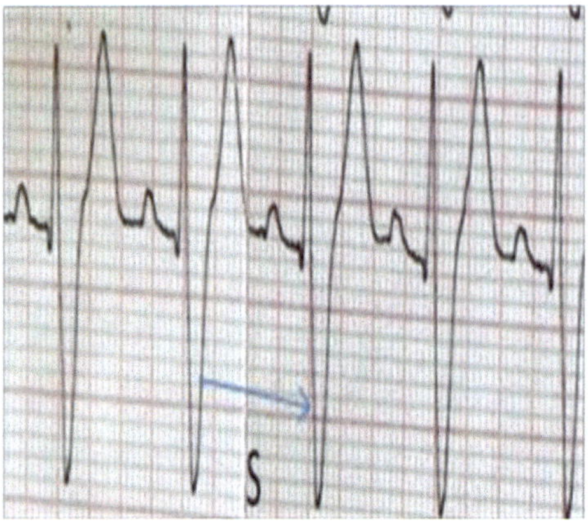

Deep 'S' wave (lead II)

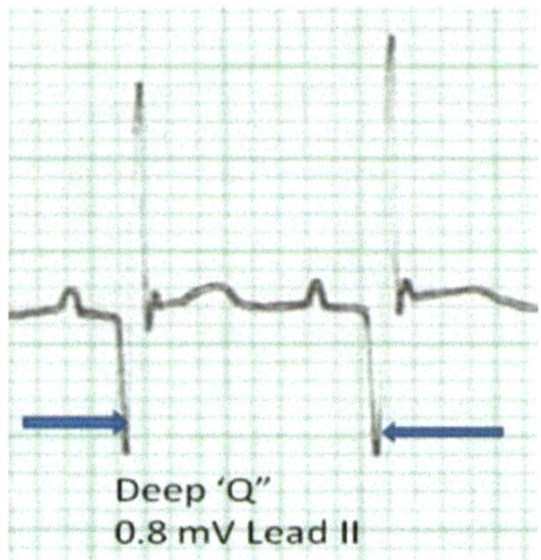

Deep 'Q' wave(Lead II)

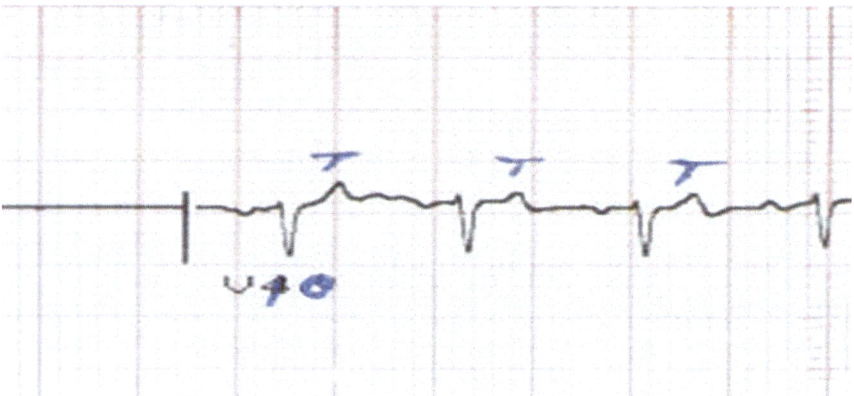

'T' wave + ve in Lead V10

Fig. 4.10 Showing miscellaneous criteria of right ventricular enlargement

4.2 Record Keeping

Electrocardiograms and relevant information of each case should be kept properly for further reference as it may be needed at a later date to review the progress of the case. Electrocardiographic findings should be recorded in writing on the ECG strip as soon as tracing is over. The findings may also be recorded on an electrocardiogram report card or sheet. A format of the report card is given below. It can be modified as per need of the hospital.

Name and Address of the Hospital.

OPD No. Date.

Dogs name.

Breed . Age . Sex. Body weight

Owner's name and address.

Mobile Number.

History of illness.

Clinical diagnosis . Referred by Dr.

Electrocardiogram (Lead II measurements, sensitivity 1, speed 25 mm/second)

Heart Rate : bpm , Rhythm :

'P' wave : mV , seconds , - 'PR' interval : seconds

' QRS' complex : mV, seconds., 'RR' interval : seconds.

' T' wave : mV, positive or negative

' S-T' segment : seconds , elevated mV, depressed mV, or normal

'Q-T' interval : seconds

Presence or absence of any miscellaneous criteria :

Mean electrical axis on frontal plane (MEA):

Other Observations, if any:

ECG interpretation:

Signature

Name and qualification of the Veterinarian

Note: Please correlate the findings clinically

4.3 Clinical Information Obtainable from an Electrocardiogram

Electrocardiogram provides information concerning electrical events occurring in the heart during depolarization and repolarization, their direction, and spread. Information regarding cardiac muscle contraction force, wall thickness, septal defects, and status of heart valves cannot be obtained from an electrocardiogram. Departure from the normal electrocardiogram may lead to speculations about certain changes in the conducting tissues of the heart or in the myocardium. But the conclusions drawn may be illusive unless the electrocardiographic findings are viewed in conjunction with clinical profile of the dog. Information that can be derived from electrocardiogram about cardiac abnormalities can be grouped under four heads, viz., (a) abnormalities of conduction; (b) abnormalities of heart rate, rhythm, or both; (c) abnormalities related to cardiac muscle mass; and (d) disturbances of myocardial metabolism/electrolyte imbalance. Disorders of (c) and (d) groups are reflected in electrocardiogram only when there are severe changes. The normal limits of variation of the electrocardiogram are so wide that the borderline between normal and abnormal is often uncertain and becomes a matter of personal opinion based on experience.

(a) Abnormal conduction: Important conduction abnormalities are sinoatrial block (SA block)/sinoatrial arrest (SA arrest), atrioventricular block (AV block first degree, second degree, or third degree), and bundle branch blocks.
(b) Abnormalities of heart rate, rhythm or both: These abnormalities are generally termed as arrhythmias. They are attributed to stimuli arising in sites/outside the normal "pace maker." Common arrhythmias are grouped in three categories, viz., premature beats (atrial, A-V nodal or ventricular), atrial fibrillation or flutters, and ventricular fibrillations or flutters.
(c) Abnormalities related to cardiac muscle mass: These abnormalities are generally related to the size of the heart and include enlargement of atrium and/or enlargement of ventricles.
(d) Disturbances of myocardial metabolism/electrolyte imbalance: Not detected commonly. Nevertheless severe abnormalities of potassium and calcium are reflected in the electrocardiogram.

Reference

Tilley LP (1985) Essentials of canine and feline electrocardiography: interpretation and treatment, 2nd edn. Lea and Febiger, Philadelphia, PA

Benchmarks for Normal Electrocardiogram

5

To differentiate an abnormal electrocardiogram from the normal one is the most important step in interpreting the electrocardiographic findings. The criteria for normal electrocardiogram provide basis for comparing electrocardiograms of ailing dogs. Much information is available on electrocardiograms of the diseased dogs rather than that of healthy dogs. Large studies are also not available on ECG patterns of healthy dogs in India. ECG of a healthy male Labrador with detailed analysis of measurements of the complexes and intervals is illustrated in Fig. 5.1. Like hemogram and biochemical values, electrocardiogram is not an absolute indicator of health and diseases. Departure from criteria for normal electrocardiogram suggests abnormality but fails to identify structural involvement of the heart or valvular defects. Electrocardiogram is undoubtedly a valuable diagnostic modality in the diagnosis of arrhythmias in dogs and cats and chamber enlargement in dogs.

5.1 Benchmarks for Normal Canine Electrocardiogram

Heart rate varies in dogs. Dogs of small breeds have higher heart rate as compared to dogs of medium and large breeds. In adults heart rate varies from 70 to 160 bpm. In toy breeds heart rate may vary from 80 to 180 bpm. Puppies have the highest heart rate, and it may go up to 220 bpm. Normal heart rhythm is called sinus rhythm where R-R interval is almost constant. When R-R interval is varying and heart rate is within normal range, it is designated as sinus arrhythmia. Varying amplitude of P wave is called wandering pace maker and is generally considered as innocuous. Electrocardiographic measurements are generally taken in lead II. Values of electrocardiographic indices observed in healthy dogs at the hospital are given in Table 5.1.

Electrocardiograms of a healthy Labrador (medium breed) and Pomeranian (small breed) are given in to compare their waves, segments, and intervals Fig. 5.1. These electrocardiograms of healthy adult male dogs show that the heart rate is more and amplitude of R wave is less in Pomeranian as compared to Labrador.

J. P. Varshney, *Electrocardiography in Veterinary Medicine*,
https://doi.org/10.1007/978-981-15-3699-1_5

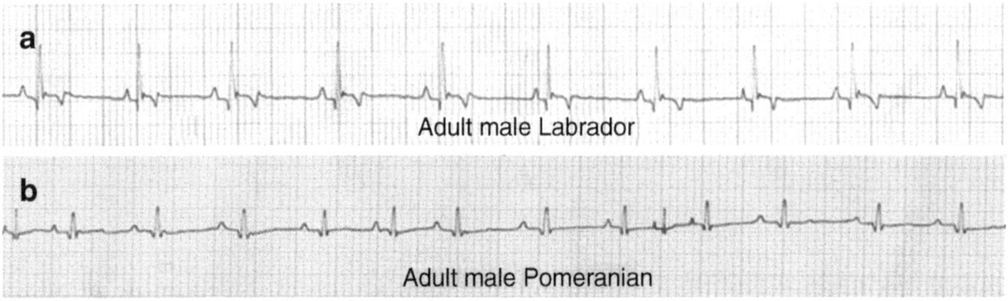

Fig. 5.1 Electrocardiogram of healthy dogs (sensitivity 1, speed 25 mm/s) of medium and small breeds. (**a**) Adult male Labrador showing heart rate, complexes, and intervals (HR 100 bpm, P 0.2 mV, 0.04 s, P-R interval 0.08 s, q 0.2 mV, R 1.2 mV, QRS 0.04 s, S-T 0.08 s, T −ve 0.2 mV, 0.06 s). (**b**) Adult male Pomeranian showing heart rate, complexes, and intervals (HR 160 bpm, P 0.15 mV, 0.03 s, P-R interval 0.09 s, q 0.1 mV, R 0.6 mV, QRS 0.03–0.04 s, S-T 0.14 s, T +ve, very small)

Table 5.1 Values of electrocardiographic parameters in healthy dogs

Electrocardiographic parameters	Range values
Heart rate Small/toy breeds Large breeds Puppies	70–170 bpm 80–170 bpm 70–160 bpm 200–220 bpm
"P" amplitude	0.15–0.40 mV
"P" duration	0.02–0.04 s
"R" amplitude Small/toy breeds Large breeds	0.90–2.8 mV 0.9–2.4 mV 1.20–2.8 mV
"QRS" duration Small breeds Large breeds	0.03–0.05 s 0.03–0.04 s 0.04–0.05 s
"T" amplitude	0.15–0.50 mV
"T" duration	0.04–0.08 s
"P-R" interval	0.08–0.12 s
"S-T" segment	Baseline Depression 0.05 mV Elevation 0.05 mV
S-T segment duration	0.04–0.1 s
"Q-T" interval	0.11–0.23 s depending on heart rate
"R-R" interval	0.35–0.8 s depending on heart rate
MEA	+40° to +100°

5.2 Effect of Breeds, Age, and Sex on ECG Indices

Heart rate in small and toy breeds is higher than that of large breeds. However, heart rate varies from 70 to 170 bpm in healthy dogs. Heart rate is also influenced by age as puppies and younger dogs have higher heart rate as compared to adults. Sex does not seem to significantly influence the heart rate in dogs. P duration in small breeds is slightly short than that of large breeds. Breeds of dog seem to influence the amplitude of R wave. It is comparatively tall in large breeds. Dalmatian and Pomeranians seem to have slightly low amplitude of "T" wave as compared to Boxers and Cocker Spaniels. P-R interval increases with advancing age. S-T segment duration also varies among different breeds. ST segment is longest in Dalmatians and shortest in Cocker Spaniels. S-T segment also increases with advancing age (Changkija 2007) (Table 5.2).

There are variations in the normal values of electrocardiographic parameters in different breeds of dog. Therefore, for research studies, it is always better to generate data of control healthy animals of the breed being investigated so that it can be compared well.

5.3 Effect of Exercise on the Electrocardiogram

Physical activity and exercise influences the activity of cardiovascular system. The extent of change depends on the nature and duration of physical activities or exercise. Heart rate is an important factor that is first affected with exercise. Secondly cardiac arrhythmias and cardiac dysfunction (physiological or pathological) are precipitated with exercise. Post exercise electrocardiography is done in many cases to detect arrhythmias which are otherwise not detected in resting electrocardiography. Increase in QRS duration and change in polarity of T wave from monophasic (negative or positive) to biphasic have been reported in dogs after exercise. Exercise led to an increase in heart rate (from 80 to 100 bpm) and a decrease in Q-T (from 0.24 to 0.20 s) and R-R (from 0.76 to 0.60 s) intervals (Figs. 5.2 and 5.3).

Table 5.2 Electrocardiographic parameters of healthy dogs of different breeds (Changkija 2007)

ECG parameters	Pom./Spitz	German Shepherd	Labrador	Great Dane	Doberman	Cocker Spaniel	Boxer	Dalmatian	Nondescript
Heart rate	166.37 ± 59.27	162.07 ± 52.72	170.49 ± 51.98	134.80 ± 32.5	174.11 ± 70.19	166.14 ± 26.22	142.8 ± 47.54	161.67 ± 50.96	176.42 ± 60.71
P amplitude	0.16 ± 0.07	0.17 ± 0.08	0.18 ± 0.07	0.16 ± 0.11	0.17 ± 0.08	0.19 ± 0.07	0.16 ± 0.04	0.15 ± 0.08	0.18 ± 0.08
P duration	0.032 ± 0.01	0.036 ± 0.01	0.036 ± 0.01	0.034 ± 0.01	0.035 ± 0.01	0.036 ± 0.01	0.035 ± 0.01	0.036 ± 0.01	0.034 ± 0.01
R amplitude	0.90 ± 0.47	1.08 ± 0.51	1.01 ± 0.54	1.29 ± 0.38	0.83 ± 0.38	1.27 ± 0.31	1.52 ± 0.63	0.90 ± 0.50	0.95 ± 0.51
QRS	0.038 ± 0.01	0.039 ± 0.01	0.04 ± 0.01	0.036 ± 0.01	0.04 ± 0.01	0.039 ± 0.01	0.042 ± 0.01	0.04 ± 0.01	0.038 ± 0.01
T amplitude	0.19 ± 0.12	0.21 ± 0.15	0.24 ± 0.14	0.22 ± 0.14	0.24 ± 0.15	0.28 ± 0.12	0.27 ± 0.13	0.12 ± 0.11	0.22 ± 0.14
T duration	0.053 ± 0.02	0.054 ± 0.02	0.061 ± 0.02	0.046 ± 0.02	0.054 ± 0.02	0.07 ± 0.02	0.063 ± 0.02	0.042 ± 0.02	0.055 ± 0.02
P-R interval	0.077 ± 0.02	0.086 ± 0.02	0.085 ± 0.02	0.094 ± 0.03	0.085 ± 0.02	0.073 ± 0.01	0.093 ± 0.03	0.089 ± 0.02	0.081 ± 0.02
S-T segment	0.058 ± 0.03	0.056 ± 0.03	0.057 ± 0.03	0.062 ± 0.03	0.06 ± 0.02	0.046 ± 0.02	0.049 ± 0.02	0.076 ± 0.03	0.053 ± 0.03
Q-T interval	0.142 ± 0.03	0.148 ± 0.03	0.158 ± 0.03	0.158 ± 0.03	0.144 ± 0.04	0.147 ± 0.02	0.153 ± 0.03	0.139 ± 0.04	0.141 ± 0.03
R-R interval	0.40 ± 0.15	0.402 ± 0.14	0.396 ± 0.13	0.471 ± 0.15	0.388 ± 0.15	0.369 ± 0.08	0.447 ± 0.11	0.402 ± 0.16	0.376 ± 0.14
MEA	63.71 ± 44.93	73.9 ± 38.93	57.83 ± 46.12	86.23 ± 18.19	50.05 ± 40.22	78.92 ± 16.61	79.50 ± 12.86	58.13 ± 40.44	62.73 ± 44.78

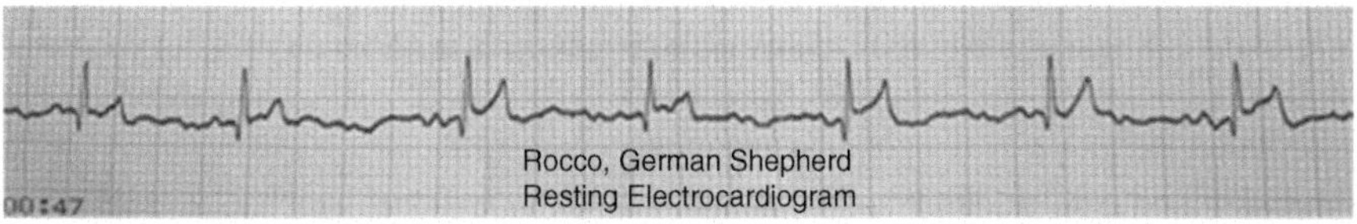

Fig. 5.2 Electrocardiogram (sensitivity 1, speed 25 mm/s) of a 15-month-old male German Shepherd at rest showing heart rate as 80 bpm, QT interval 0.24 s, and R-R interval 0.76 s

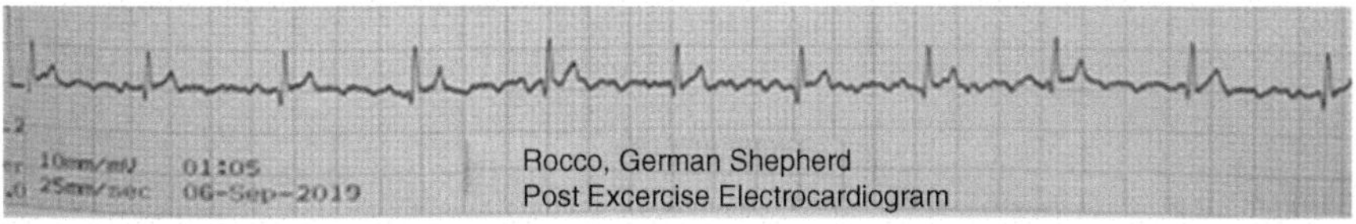

Fig. 5.3 Electrocardiogram (sensitivity 1, speed 25 mm/s) of a 15-month-old male German Shepherd (same dog whose electrocardiogram is at Fig. 5.2) immediately after 20 min exercise showing heart rate as 100 bpm, Q-T interval 0.20 s, and R-R interval 0.60 s

Reference

Changkija B (2007) Electrocardiographic studies in dogs with reference to management of cardiac tachyarrhythmia by alternative drugs. Ph.D. thesis submitted to Deemed University, Indian Veterinary Research Institute, Izatnagar

Further Reading

Bolton GR (1975) Handbook of canine electrocardiography. W.B. Saunders Company, Philadelphia, PA

Constable PD, Hinchcliff KW, Olson JL, Hamlin RL (1994) Athletic heart syndrome in dogs competing in a long-distance sled race. J Appl Physiol 76:433–438

Santos POPR, Santos EA, Reis AC, Santos AMMR, Kuster MCC, Trivilin LO, Aptekmann KP (2018) Effect of exercise on cardiovascular parameters in search and rescue trained dogs. Arq Bras Med Vet Zootec 70:1036–1044

Tilley LP (1985) Essentials of canine and feline electrocardiography: interpretation and treatment, 2nd edn. Lea and Febiger, Philadelphia, PA

6 Abnormal Wave Forms, Segments, and Intervals in Electrocardiogram

When the measurements of different wave forms, their intervals, and/or duration of segments are not in line with the values detailed for normal electrocardiogram, it is said to be an abnormal electrocardiogram. Changes in the measurements of the waves, intervals, or segments are indicative of specific cardiac abnormalities. The details are illustrated below:

Abnormal wave forms	Indications
"P" wave (Fig. 6.1)	
Broad (duration >0.04 s and/or notched)	Left atrium enlargement (P-mitrale)
Tall (amplitude >0.4 mV)	Right atrium enlargement (P-pulmonale)
Tall and broad (>0.4 mV and >0.04 s)	Biatrial enlargement
Variable amplitude of P wave	Wandering pace maker
Absence of P wave	Atrial standstill or silent atrium
Ta wave (increased height of descending arm of "P")	Right atrium enlargement
"QRS" complex (Fig. 6.2)	
Tall "R" (amplitude >2.5 mV in small breeds >3.0 mV in large breeds)	Left ventricular enlargement (LVE) or left bundle branch block (LBBB)
Wide "QRS" (duration >0.05 s small breeds, >0.06 s large breeds)	LVE or LBBB
Deep "S" (amplitude >0.35 mV in leads II, III, aVF; >0.8 mV in lead CV5RL)	Right ventricular enlargement (RVE) or right bundle branch block (RBBB)
Deep "Q" wave (amplitude >0.5 mV in leads II, aVF)	RVE

J. P. Varshney, *Electrocardiography in Veterinary Medicine*,
https://doi.org/10.1007/978-981-15-3699-1_6

Abnormal wave forms	Indications
Low-voltage "QRS" complexes (amplitude <0.5 mV in leads I, II, III, aVF)	Pericardial effusions
	Pleural effusions
	Pneumothorax
	Obesity
	Cardiomyopathy
	Hypothyroidism
	Loose lead contact with skin
Notched "QRS" with normal duration	Normal or minor atrioventricular defect
"R"-alternans (varying amplitude)	Pericardial effusion
	Alternating bundle branch block
	Supraventricular tachycardia
"J" wave (Fig. 6.3)	
Deflection at "R" ST junction	Hypothermic dogs
	Normal dogs
S-T segment/J point (Fig. 6.4)	
Elevation (>0.15 mV), i.e., above the baseline	Hypoxia
	Pericarditis
	Infarction
	Secondary change (ventricular hypertrophy, VCPs, conduction disturbance)
	Left ventricular epicardial injury.
	Transmural myocardial infarction
	Digoxin toxicity
Depression (>0.2 mV), i.e., below the baseline	Infarction, ischemia
	Conduction disturbance
	Cardiac trauma
	Hyper- or hypokalemia
	Secondary change (VCPs, ventricular hypertrophy, conduction disturbance)
	Digitalization
	Myocardial infarction/injury
	False depression
Slurring or coving	Left ventricular hypertrophy
P-R interval (Fig. 6.5)	P-R interval is inversely proportional to heart rate
Increase (>0.13 s)	First-degree heart block
Q-T interval (Fig. 6.6)	
Increased (>0.25 s)	Hypocalcemia
	Hypothermia
	Hypokalemia
	CNS disorders
	Ventricular hypertrophy
	Conduction disorders
	Strenuous exercise
	Quinidine toxicity
	Ethylene glycol poisoning
	Secondary to prolonged QRS duration

Abnormal wave forms	Indications
Short (<0.15 s)	Hypercalcemia
	Hyperkalemia
	Digitalis toxicity
"T" wave (Fig. 6.7)	"T" wave changes are mostly nonspecific
Sudden change in polarity	Hypoxia
	Abnormal conduction
	Ventricular enlargement
	Metabolic disorders
Height >25% of "R"	Hyperkalemia
Small biphasic	Hypokalaemia
Large T wave	Myocardial hypoxia
	Ventricular enlargement
	Intraventricular conduction abnormalities
	Hyperkalemia
	Metabolic diseases
	Respiratory diseases
	Normal variation
Tented T wave	Hyperkalemia
T-alternans (amplitude varying)	Pericardial effusion
	Alternating bundle branch block
	Supraventricular tachycardia
"U" wave (Fig. 6.8)	Small rounded deflection after T wave.
	It was first described by Einthoven (1906).
	Represents last phase of ventricular repolarization.
	Usually monophasic, positive, or negative.
	Characteristic of hypokalemia in dogs (Tai Fu et al. 1984).
	U wave with same polarity of T has also been seen in normal human beings (Goesing et al. 2009).
"R-R" interval (Figs. 6.9, 6.10, and 6.11)	
Variable	Arrhythmia
Pause between R-R <twice of normal R-R interval	Sinus block
Pause between R-R >twice of normal R-R interval	Sinus arrest

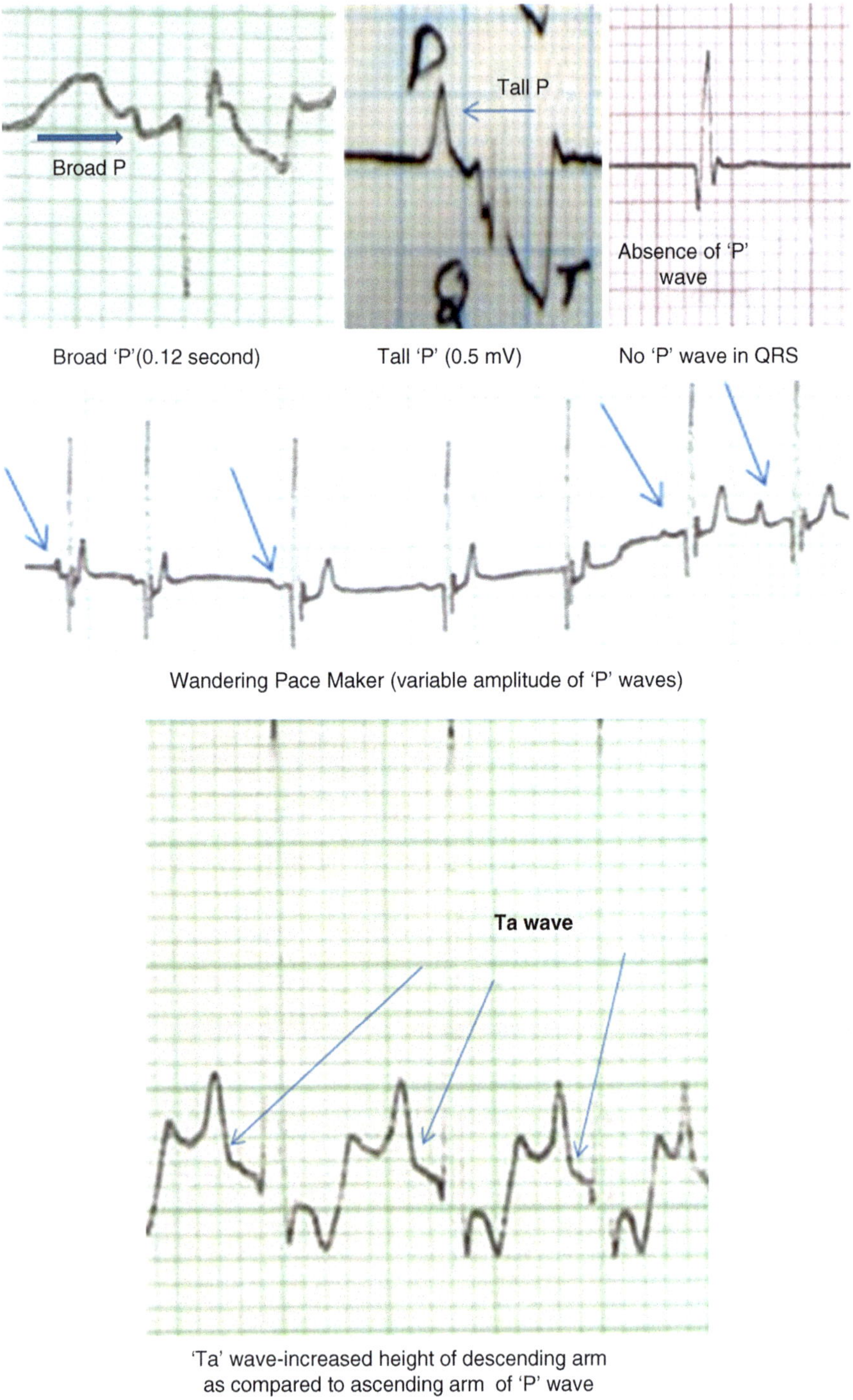

Fig. 6.1 Showing various abnormal forms of "P" wave in dogs

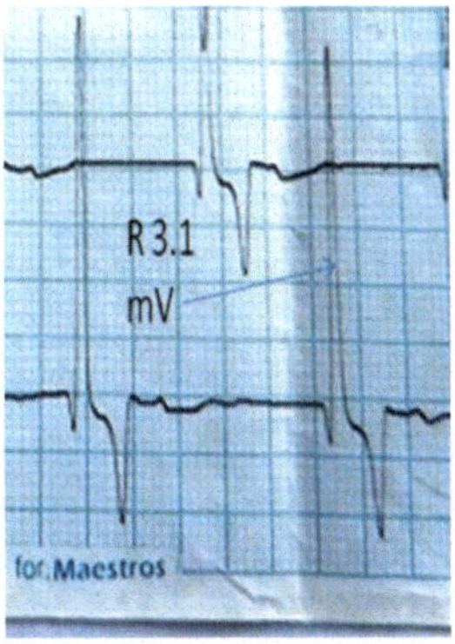

Tall 'QRS' (3.1 mV)

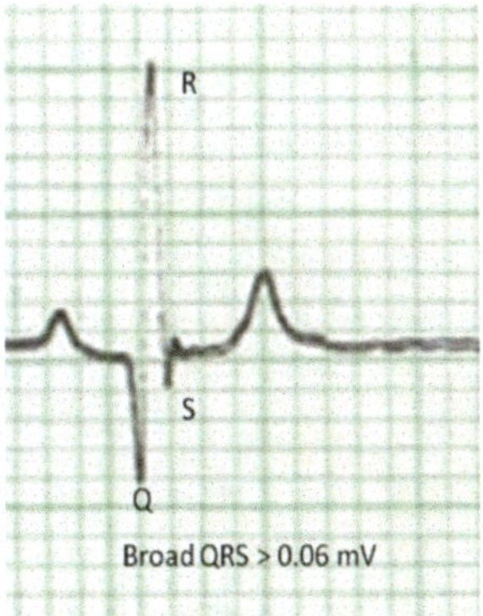

Broad 'QRS' > 0.06 mV

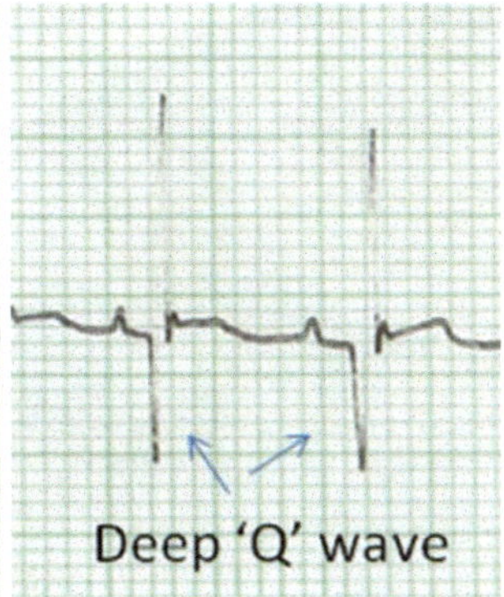

Deep 'Q' wave (0.85 mV)

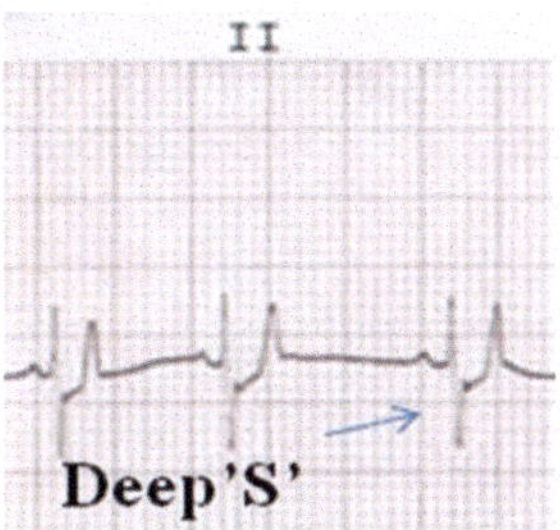

Deep'S' (0.7 mV)
Lead II

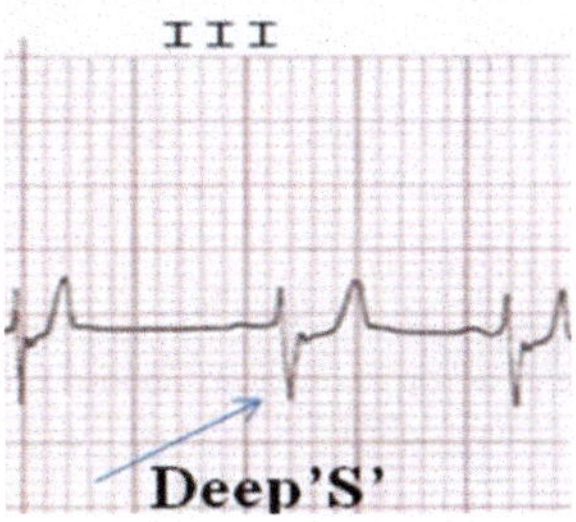

Deep'S' (0.6 mV)
Lead II I

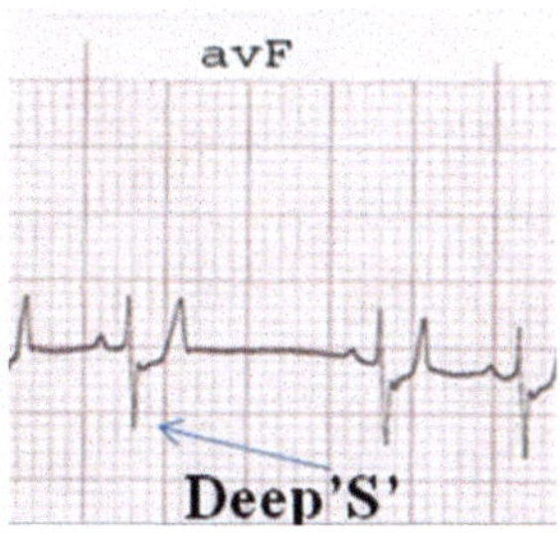

Deep 'S' (0.6 mV)
lead aVF

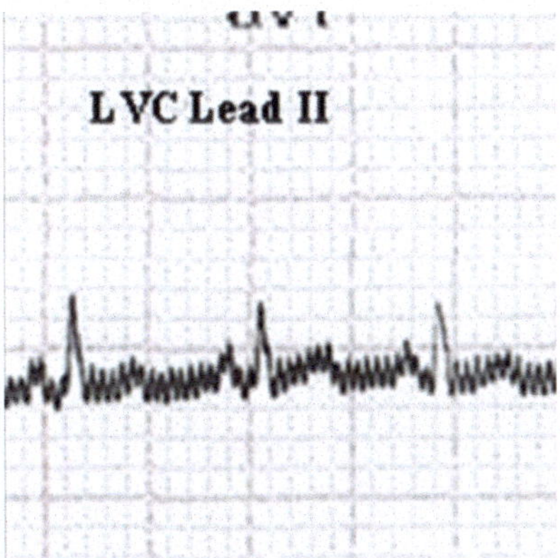

Low voltage 'QRS (0.2 mV)
Lead II

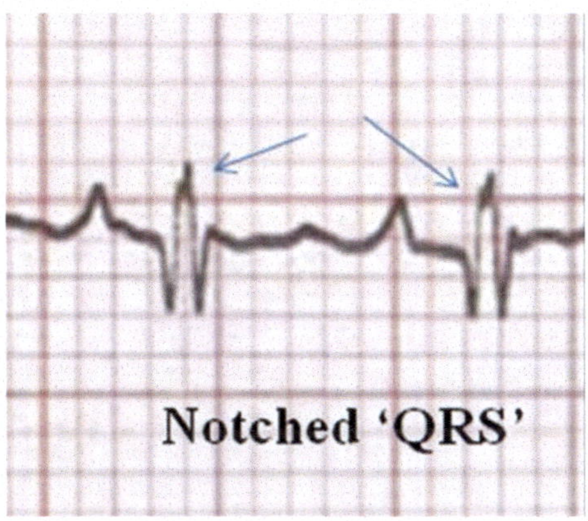

Notched 'QRS' Lead II

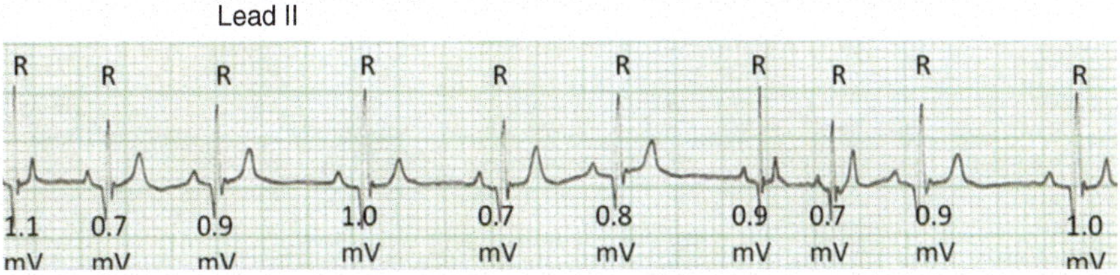

Amplitude of 'R' wave is varying from complex to complex (0.7 mV to 1.1 mV). 'P'

Fig. 6.2 Showing various abnormal forms of "QRS" in dogs

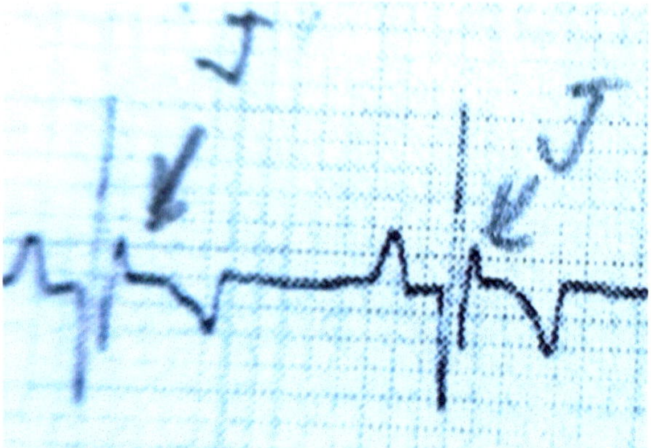

Fig. 6.3 Showing "J" wave in a dog

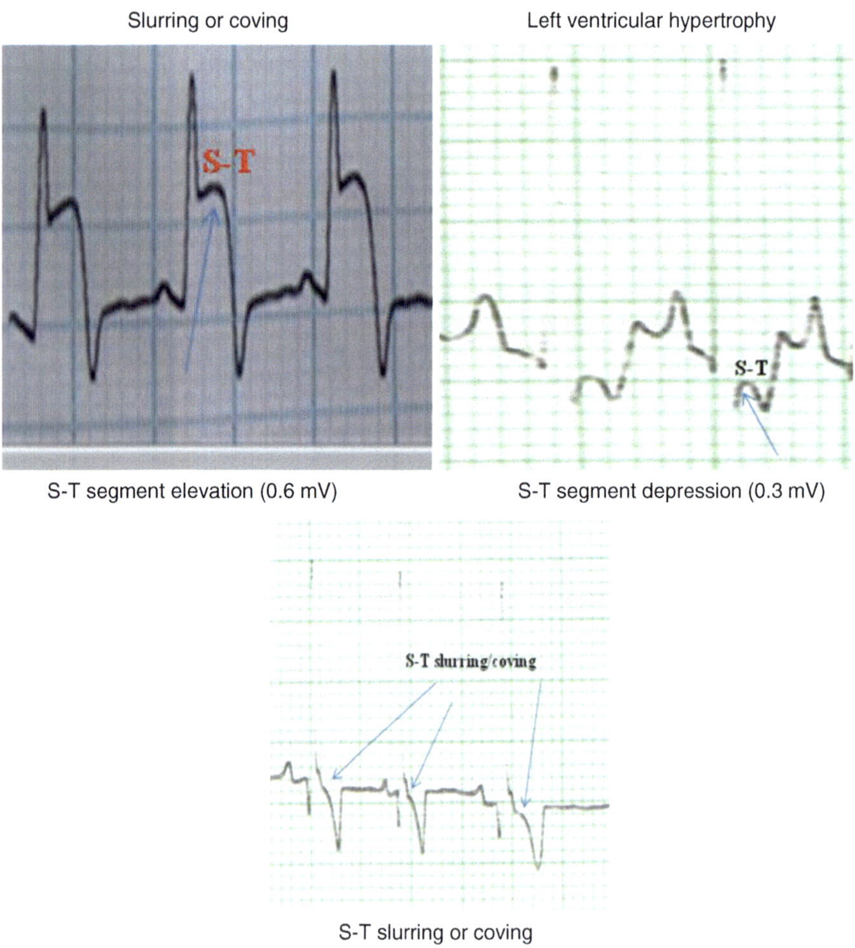

Fig. 6.4 Showing various abnormalities of S-T segment in dogs

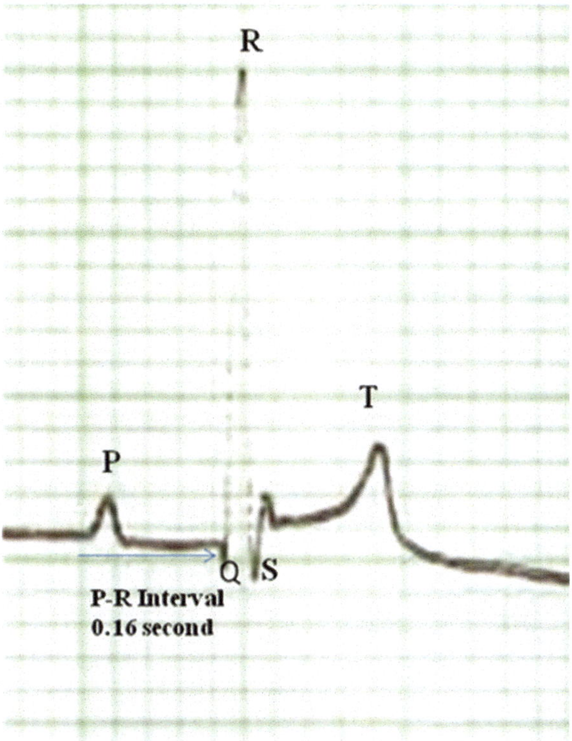

Fig. 6.5 P-R interval is from beginning of "P" to beginning of "Q." It is 0.16 s (four squares at the speed of 25 mm/s) indicating first-degree heart block

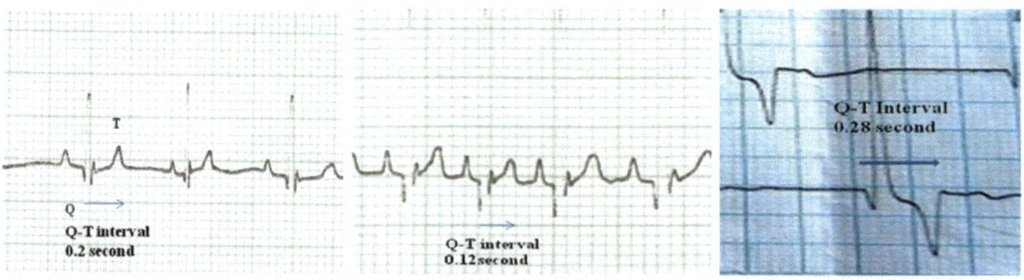

'Q-T interval (beginning of 'Q' to end of 'T' duration (0.2 second)

Short 'Q-T interval (0.12 secnnd)

Prolongrd Q-T interval (0.28 second)

Fig. 6.6 Showing variations (normal, short, and long) in "Q-T" interval in dogs

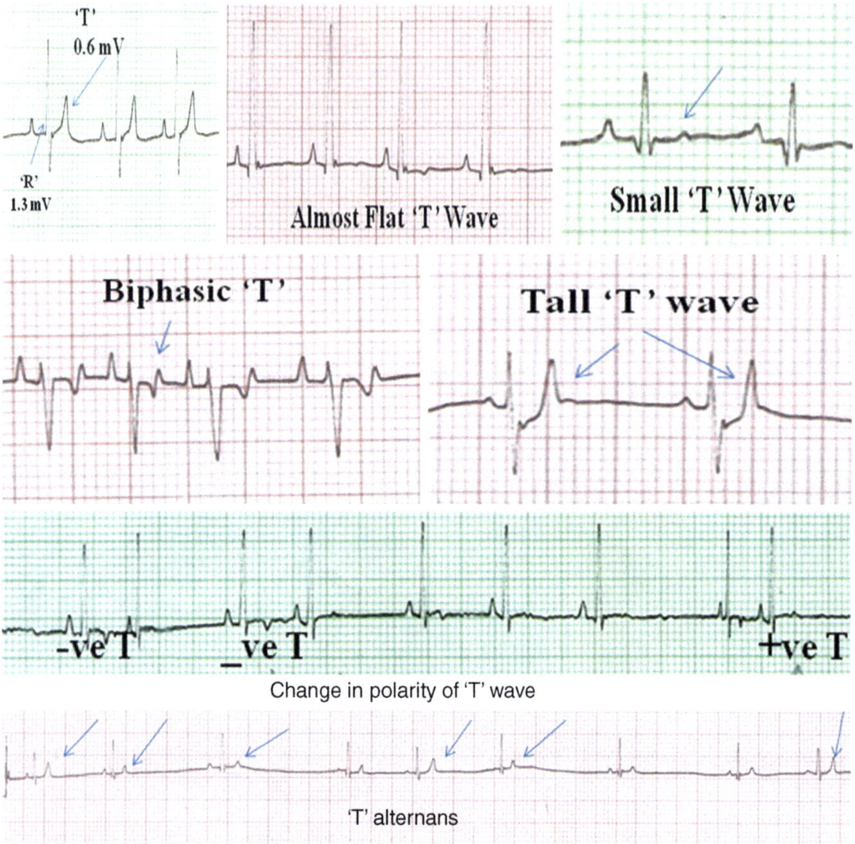

Fig. 6.7 Showing various abnormalities of "T" wave in dogs

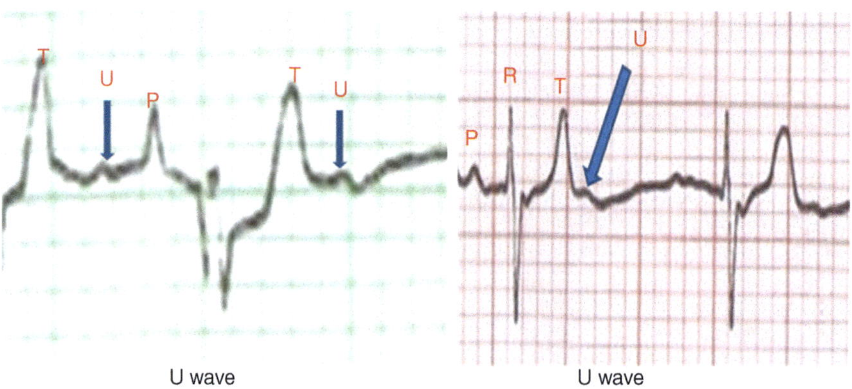

Fig. 6.8 Showing "U" wave in dogs. U wave is a small round deflection after T wave

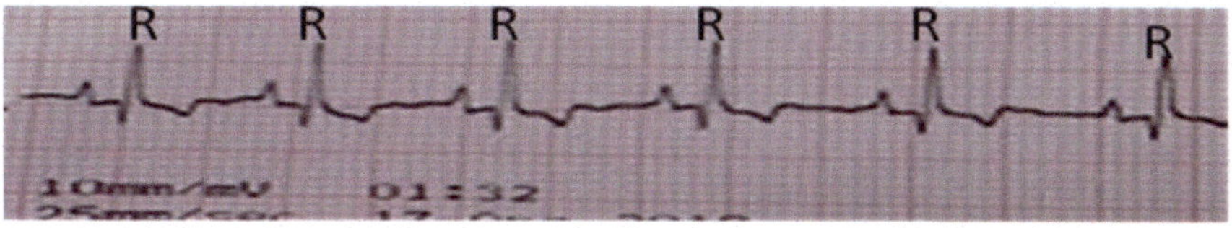

Fig. 6.9 Showing almost equal R-R interval (0.44–0.48 s) between different QRS complexes

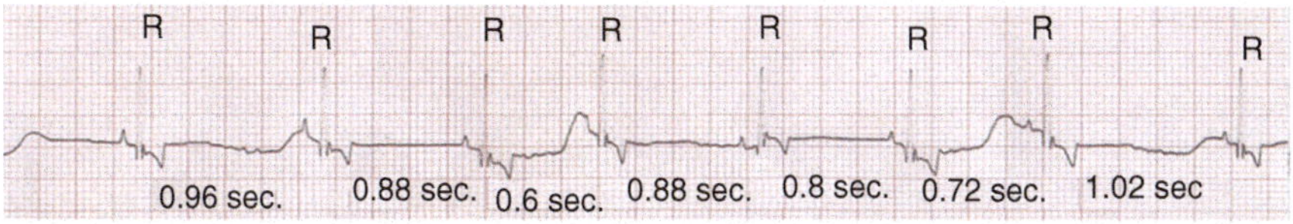

Fig. 6.10 R-R interval is varying from complex to complex but is less than twice of the minimum R-R interval indicating arrhythmia. The seventh R-R interval (1.02 s) is greatest but is less than twice of the minimum R-R interval (0.6 s) of the third complex indicating sinus block

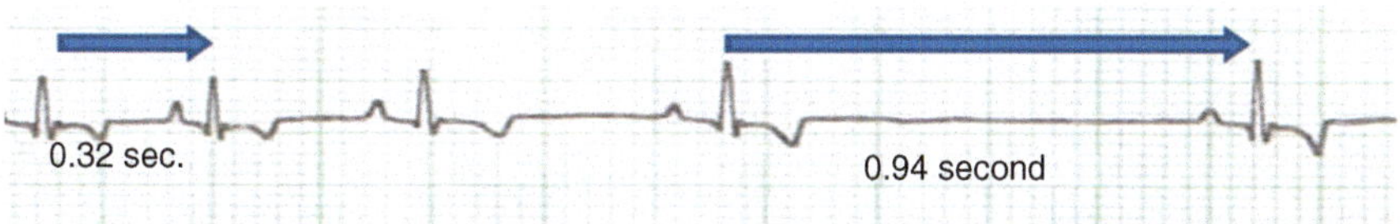

Fig. 6.11 R-R interval of 0.94 s is >double of normal R-R intervalof 0.32 s. It is indicating sinus arrest

References

Einthoven W (1906) Le telecardiogramme. Arch Int Physiol 4:132–164

Goesing M, Haueisen J, Liehr M, Sehlosser M, Figulla HR, Leder U (2009) Detection of U wave activity in healthy volunteers by high resolution magnetocardiography. J Electrcardiol 43:43–47

Tai Fu L, Kato N, Takahashi N (1984) Hypopotassemia induced U wave in electrocardiogram (an experimental study for possible mechanism). Basic Res Cardiol 79:494–502

Further Reading

Bolton GR (1975) Handbook of canine electrocardiography. W.B. Saunders Company, Philadelphia, PA

Tilley LP (1985) Essentials of canine and feline electrocardiography, 2nd edn. Lea and Febiger, Philadelphia, PA

7 Atrial and Ventricular Enlargement Patterns and Clinical Associations

Cardiac enlargement is seen in many cardiac diseases. It can either be due to enlargement of atrium or ventricle or both. Electrocardiogram can provide confirmatory evidence of the side or chamber of the heart that is affected. Electrocardiographic features of the enlargement of different heart chambers are given below:

7.1 Normal Atrial Pattern

- "P" wave amplitude <0.4 mV
- "P" wave duration <0.04 s

7.1.1 Right Atrial Enlargement Pattern

"P" wave indicates atrium depolarization and its initial half stroke signifies right atrium. Hence increase in amplitude of "P" wave is suggestive of right atrium enlargement.

- "P" wave amplitude >0.4 mV (Fig. 7.1).
- "P" wave duration 0.04 s or less.
- "T_a" wave (Fig. 7.2) slight depression of base line following "P" may sometimes indicate right atrial enlargement.

7.1.2 Left Atrial Enlargement Pattern

Later half of the "P" wave is produced by depolarization of left atrium. Hence prolonged duration of "P" wave is suggestive of left atrium enlargement.

- "P" wave duration >0.04 s (Fig. 7.3)
- Notched and wide "P" wave

J. P. Varshney, *Electrocardiography in Veterinary Medicine*,
https://doi.org/10.1007/978-981-15-3699-1_7

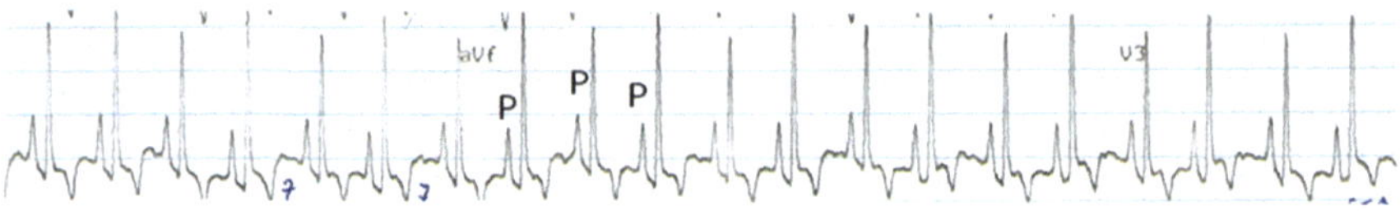

Fig. 7.1 Electrocardiogram of a dog (lead II, sensitivity 1, speed 25 mm/s) showing increased amplitude of "P" wave (0.5 mV) suggesting right atrial enlargement

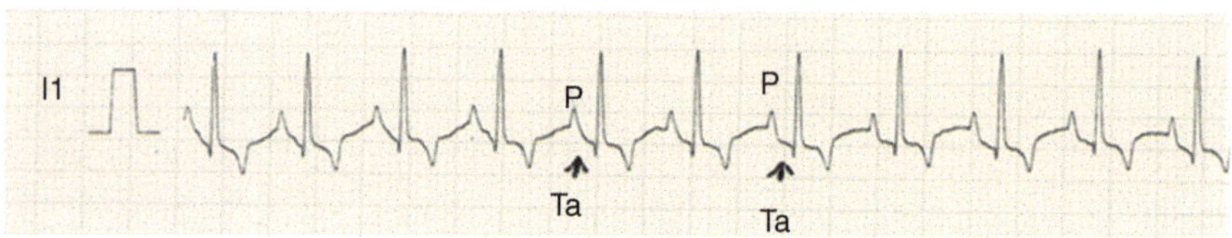

Fig. 7.2 Electrocardiogram of a Pomeranian dog (lead II, sensitivity 1, speed 25 mm/s) showing depression of baseline following P, i.e., descending arm of "P" longer than ascending arm of P wave. It is called T_a wave. Sometimes it indicates right atrium enlargement

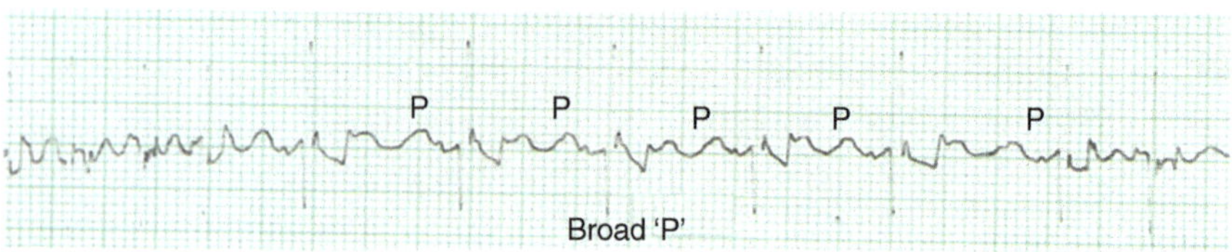

Fig. 7.3 Electrocardiogram of a dog (lead II, sensitivity 1, speed 25 mm/s) showing broad "P" (0.08 s) suggesting left atrium enlargement

7.1.3 Biatrial Enlargement Pattern

Total "P" wave is the contribution of depolarization of both right and left atrium, therefore an increase in "P" wave's height and duration is suggestive of biatrial enlargement.

- "P" wave amplitude >0.4 mV and duration >0.04 s (Fig. 7.4)
- Notching or slurring of "P" may be present

7.2 Normal Ventricular Pattern (Fig. 7.5)

- Mean electrical axis within +40° to +100°
- No "S" wave in lead I
- "R" wave in lead II is taller than that of lead I
- "R" wave in lead CV_6LL is larger than "S"

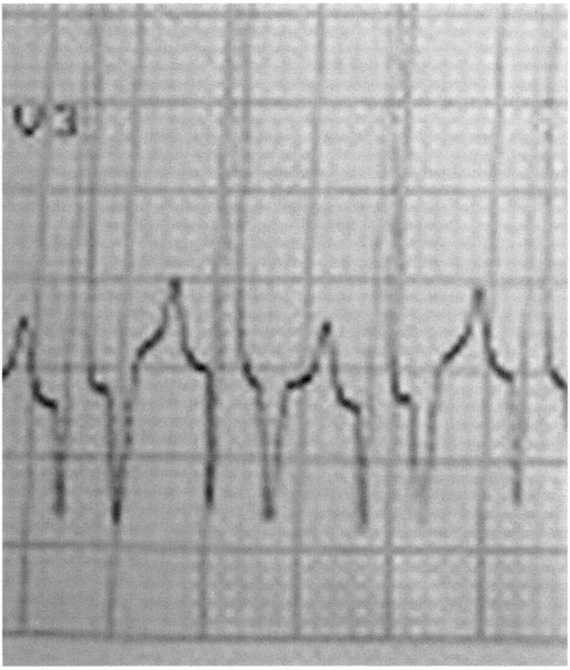

Fig. 7.4 Electrocardiogram of a dog (lead II, sensitivity 1, speed 25 mm/s) showing increased amplitude (>0.4 mV) and prolonged duration of "P" (0.08 s) suggesting biatrial enlargement

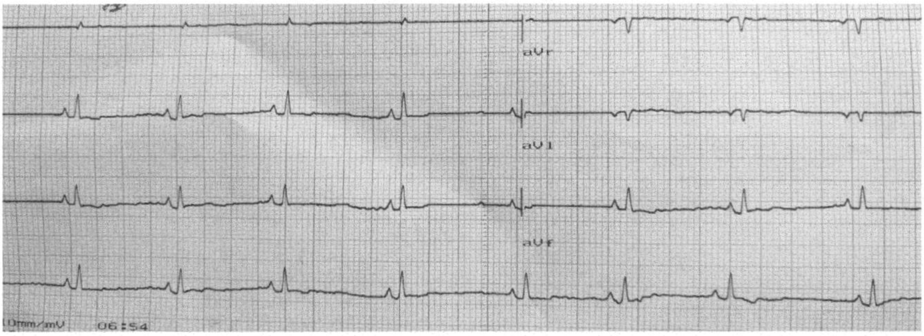

Fig. 7.5 Electrocardiogram of a dog (lead, I, II, III, aVR, aVL, aVF, sensitivity 1, speed 25 mm/s) showing no "S" wave in lead I and "R" wave taller in lead II and III than lead I suggesting normal ventricular pattern

7.2.1 Right Ventricular Enlargement Pattern

"Q" and "S" waves represent right ventricular depolarization. Hence changes related to these waves are suggestive of right ventricle enlargement. Only one criterion is not sufficient to say right ventricle enlargement. For confirming right ventricular enlargement, at least three criteria should be present in the ECG. The following are the accepted criterion that indicates change in the size of right ventricle:

- Deviation of mean electrical axis (>+100°) on frontal plane
- Presence of "S" wave in lead I (more than 0.05 mV) (Fig. 7.6)
- Presence of "S" wave in lead II (more than 0.35 mV) (Figs. 7.7 and 7.8)
- Presence of "S" wave in lead III (Fig. 7.6)
- Presence of "S" wave in lead aVF (Fig. 7.6)
- Presence of "S" wave in lead CV_6LL (more than 0.8 mV) (Fig. 7.6)
- Presence of "S" wave in lead CV_6LU (more than 0.7 mV)
- Large "S" wave with normal "R" in lead CV_6LU alters the R:S ratio. Ratio of <0.8 is suggestive of right ventricular enlargement
- W-shaped "Q-S" in lead V_{10}
- Positive "T" wave in lead V_{10} in dogs (Fig. 7.9) other than Chihuahua breed is also suggestive of right ventricular enlargement
- Large "Q" wave (amplitude more than 0.5 mV) in lead II, III, and aVF (Fig. 7.10)

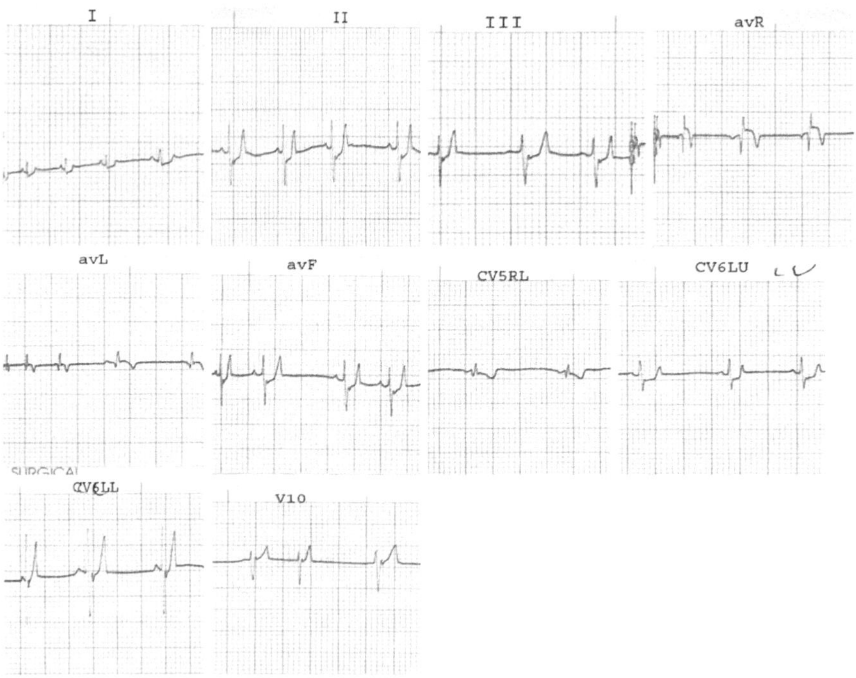

Fig. 7.6 Electrocardiogram of an adult male Yorkshire Terrier showing "S" wave in all leads (lead I 0.05 mV, lead II 0.65 mV, lead III 0.6 mV, aVR 0.3 mV, aVL 0.15 mV, aVF 0.6 mV, CV_6LU 0.35 mV, CV_6LL 0.9 mV), T +ve in lead V_{10} and R:S ratio 0.75 (<0.8) suggesting right ventricle enlargement

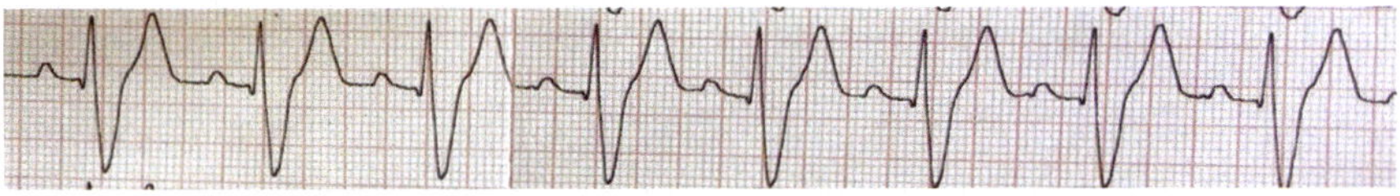

Fig. 7.7 Electrocardiogram (lead II, sensitivity 1, speed 25 mm/s) of an 18-year-old male Pomeranian dog showing heart rate as 80 bpm, sinus rhythm, deep "S" wave (lead II 1.2 mV), suggesting right ventricular enlargement

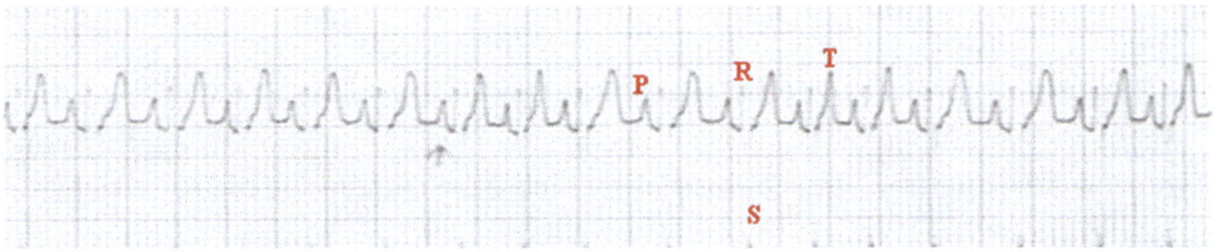

Fig. 7.8 Electrocardiogram (lead II, sensitivity 1, speed 25 mm/s) of a male adult dog showing deep S wave (2.0 mV) suggesting right ventricle enlargement. The base line for "S" is poorly defined

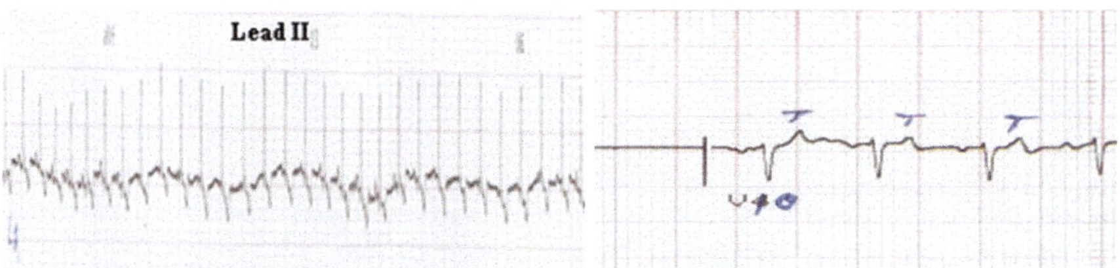

Fig. 7.9 Electrocardiogram (lead II, sensitivity 1, speed 25 mm/s) of an adult male dog with dirofilariasis (*Dirofilaria immitis*) showing no abnormality except tachycardia in lead II but +ve "T" wave in lead V_{10} suggesting right ventricular enlargement

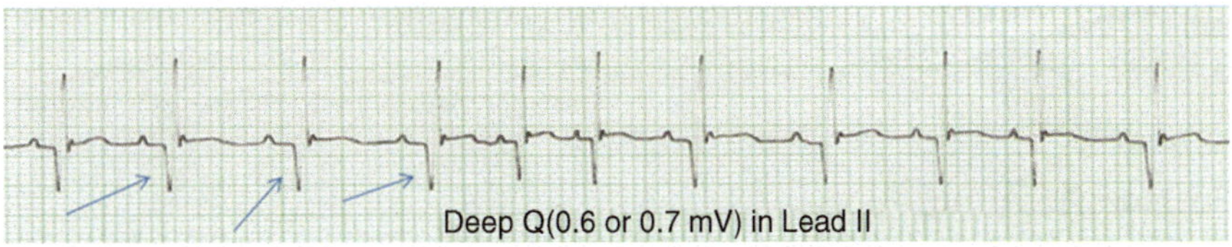

Fig. 7.10 Electrocardiogram (lead II, sensitivity 1, speed 25 mm/s) of an adult German Shepherd dog with ascites showing deep "Q" wave (0.0.6–0.7 mV) in lead II suggesting right ventricular enlargement. Ascites in this dog was of cardiac origin

7.2.2 Left Ventricular Enlargement Pattern

"R" wave represents left ventricular depolarization. Hence changes related to "R" wave is suggestive of left ventricle enlargement. The following are the accepted criteria that indicates change in the size of left ventricle. For confirming LVE only one criterion may not be sufficient.

- Deviation of mean electrical axis (less than +40°) on frontal plane. Axis deviation alone is not diagnostic. When axis deviation is associated with increased voltage, it is suggestive of LVE.
- Tall "R" wave (>2.5 mV in lead II and aVF in small breeds and aged dogs, >3.0 mV in lead II and aVF in large breed dogs) is suggestive of LVE (Fig. 7.11). Increased amplitude of "R" is also seen in cases with volume overload. In young emaciated or narrow chested dogs simply increased amplitude of "R" is not sufficient criteria of LVE.
- Tall "R" wave in chest lead CV_6LU (amplitude more than 3.0 mV).
- Tall "R" wave in chest lead CV_6LL (amplitude more than 2.5 mV).
- Broad "QRS" (duration of QRS more than 0.05–0.06 s depending on the breed of the dog) (Fig. 7.11). Severe left ventricular enlargement is generally associated with broad and prolonged "QRS."

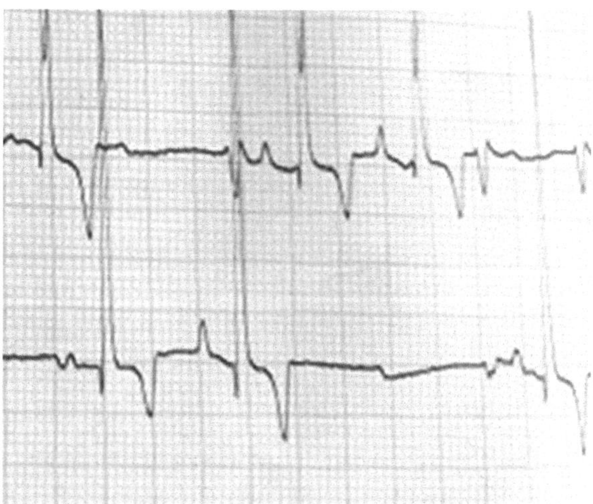

Fig. 7.11 Electrocardiogram (lead II, sensitivity 1, speed 25 mm/s) of a male Labrador dog suffering from heart failure showing increased amplitude of "R" wave (3.6 mV), broad "QRS" (0.08 s) suggestive of left ventricular enlargement

- Enlarged “T” wave >25% larger than the “R” wave is also seen in dogs with LVE (Fig. 7.12).
- Slurring or coving of S-T segment is also indicative of LVE (Fig. 7.13).
- Elevation (Fig. 7.14) or depression (Fig. 7.15) of S-T segment is also seen in dogs with LVE.
- Many cases of LVE may show more than one abnormality (Fig. 7.16) in the electrocardiogram.

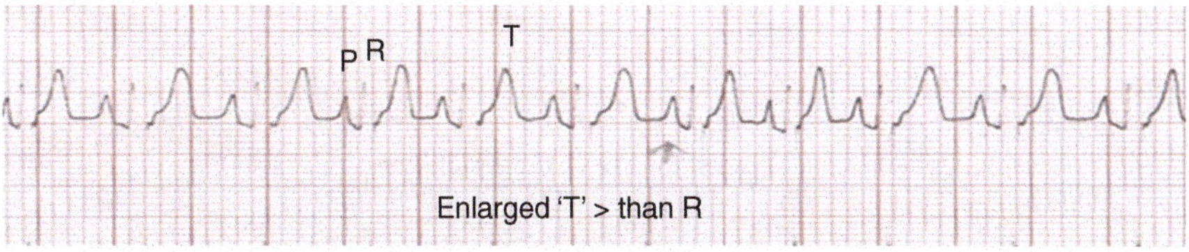

Fig. 7.12 Electrocardiogram (lead II, sensitivity 1.0, speed 25 mm/s) of a dog showing enlarged T wave (>25% of R)—another criteria for left ventricular enlargement

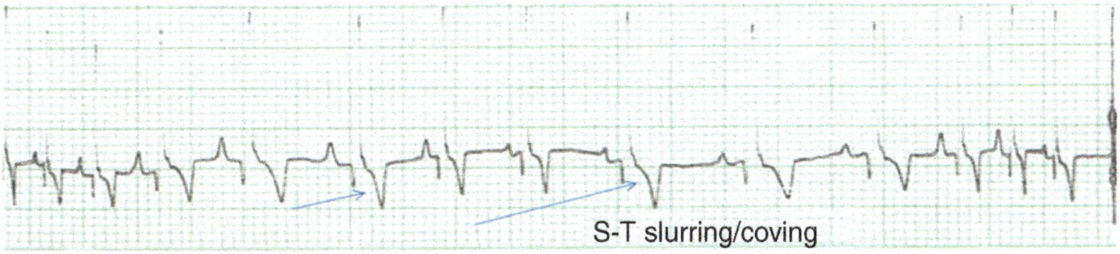

Fig. 7.13 Electrocardiogram (lead II, sensitivity 1.0, speed 25 mm/s) of a dog showing S-T slurring/coving—another criteria for left ventricular enlargement

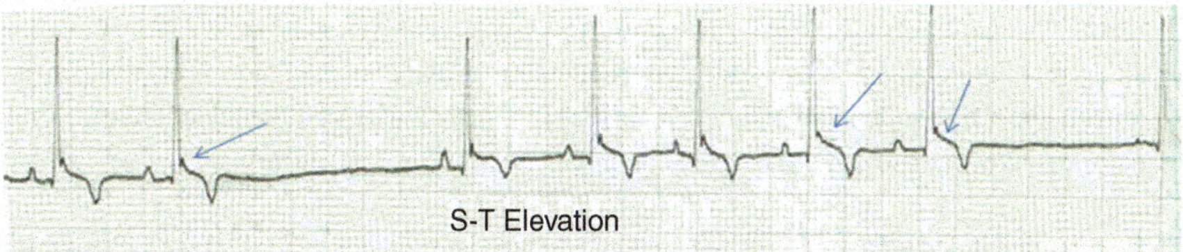

Fig. 7.14 Electrocardiogram (lead II, sensitivity 1.0, speed 25 mm/s) of a dog showing S-T elevation—another criteria for left ventricular enlargement

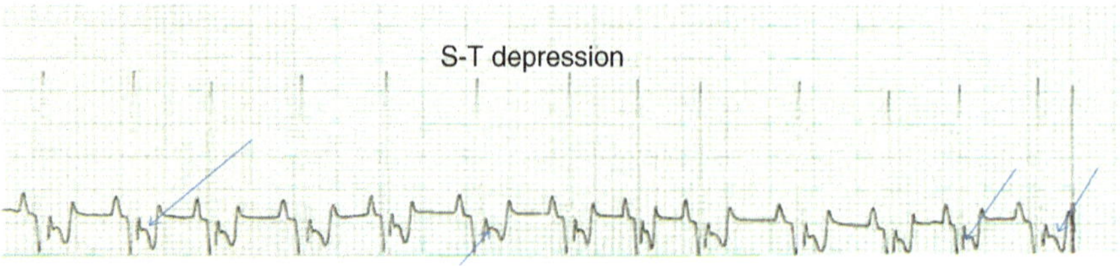

Fig. 7.15 Electrocardiogram (lead II, sensitivity 1.0, speed 25 mm/s) of a dog showing S-T depression—another criteria for left ventricular enlargement

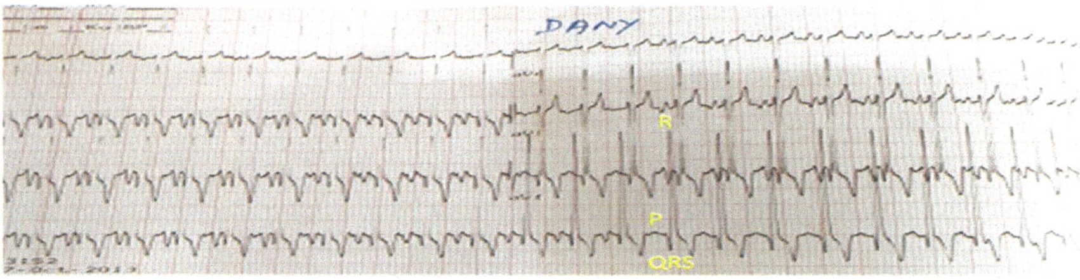

Fig. 7.16 Electrocardiogram (lead II, sensitivity 1.0, speed 25 mm/s) of a male Rottweiler dog suffering from congestive heart failure showing broad "P" (0.06 s), increased amplitude of "R" (3.5 mV), broad "QRS" (0.06–0.07 s) and S-T coving at few places suggesting left heart enlargement

7.2.3 Biventricular Enlargement Pattern

- It is difficult to diagnose simultaneous enlargement of both ventricles by electrocardiography.
- Diagnosis of left ventricular enlargement is more accurate.
- Right axis deviation on frontal plane and left ventricular enlargement in precordial chest leads in humans is suggestive of biventricular enlargement.
- Deep "S" wave and tall "R" wave (in lead CV_6LL and CV_6LU), LVE with right axis deviation, broad QRS with increased amplitude of "R" (in lead CV_6LU), increased amplitude of "Q" wave (in lead I, II, III, and aVF) along with left ventricular enlargement are suggestive of biventricular enlargement in dogs (Fig. 7.17).
- Severe heart enlargement in radiographs with almost normal electrocardiogram may also indicate biventricular enlargement.

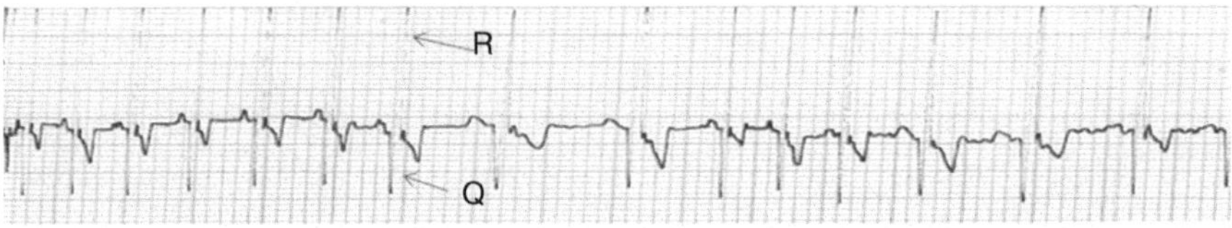

Fig. 7.17 Electrocardiogram (lead II, sensitivity 0.5, speed 25 mm/s) of a male German Shepherd dog with congestive heart failure showing tachyarrhythmia (HR 260 bpm with variable R-R interval), deep "Q" wave (2.0–2.4 mV), increased amplitude of "R" wave (4.2–4.4 mV), and broad "QRS" (0.06 s) suggestive of biventricular enlargement

7.3 ECG Wave Enlargement Patterns and Clinical Associations

Alterations in amplitude or duration of wave forms in electrocardiogram can suggest enlargement of a particular chamber of the heart or conduction disturbance. Enlargement of a particular heart chamber is seen in many diseases. Common clinical associations of these chamber enlargement pattern are briefly given below:

ECG findings	Clinical associations
Left atrial enlargement (Fig. 7.18)	Mitral valve insufficiency
	Cardiomyopathy
	Patent ductus arteriosus (PDA)
	Subaortic stenosis (SAS)
	Ventricular septal defect (VSD)
Right atrial enlargement (Fig. 7.19)	Tricuspid valve insufficiency
	Chronic respiratory disease (CRD)
	Inter-atrial septal defect
	Pulmonary stenosis (PS)
Left ventricular enlargement (Fig. 7.20)	Dilated cardiomyopathy
	Mitral valve insufficiency
	Aortic insufficiency (AI)
	Patent ductus arteriosus (PDA)
	Ventral septal defects (VSD)
	Sub aortic stenosis
Left ventricular hypertrophy	Hypertrophic cardiomyopathy
	Subaortic stenosis
Right ventricular enlargement (Fig. 7.21)	Pulmonary stenosis (PS)
	Tetralogy of Fallot
	Tricuspid valve insufficiency
	Severe dirofilariasis (heartworm disease)
	Heartworm disease
	Pulmonary hypertension

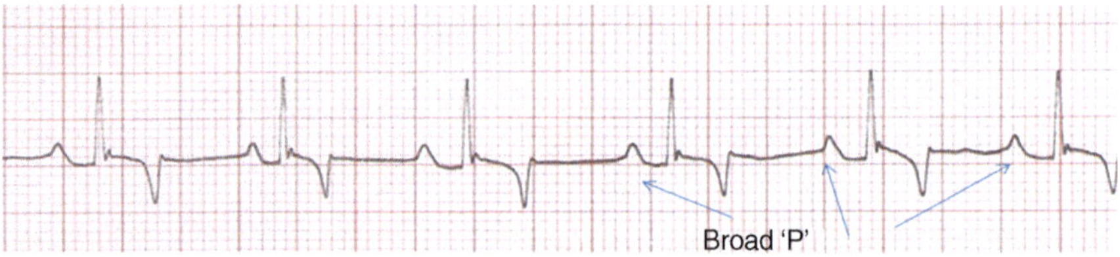

Fig. 7.18 Electrocardiogram (lead II, sensitivity 1.0, speed 25 mm/s) of a 5-year-old male bulldog showing broad "P" (0.06–0.08 mV) suggesting left atrial enlargement

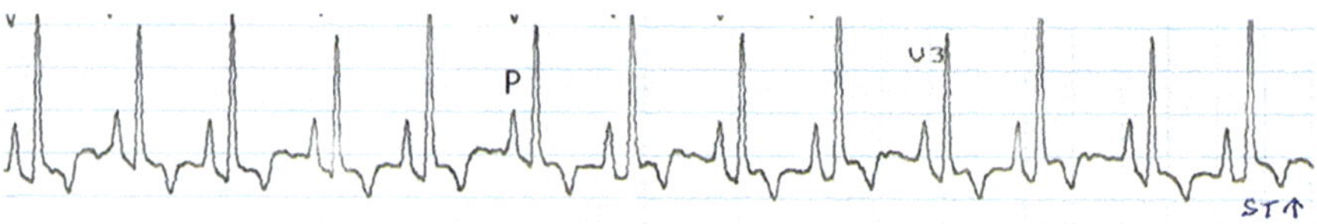

Fig. 7.19 Electrocardiogram (lead II, sensitivity 1.0, speed 25 mm/s) of a dog showing tall "P" wave suggestive of right atrium enlargement

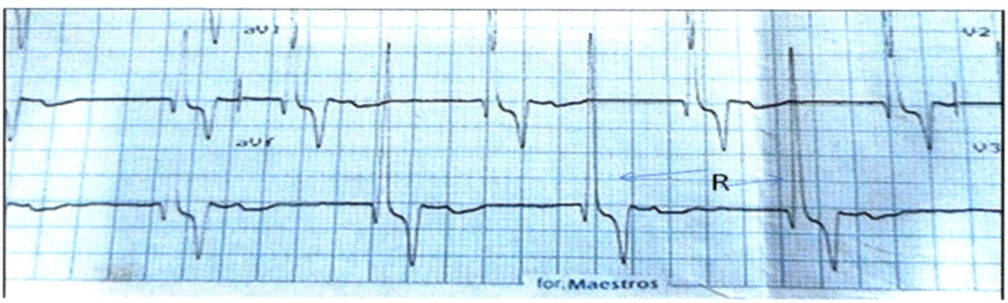

Fig. 7.20 Electrocardiogram (lead II, sensitivity 1, speed 25 mm/s) of a Labrador showing tall "R" (3.4 mV) suggesting left ventricle enlargement/hypertrophy

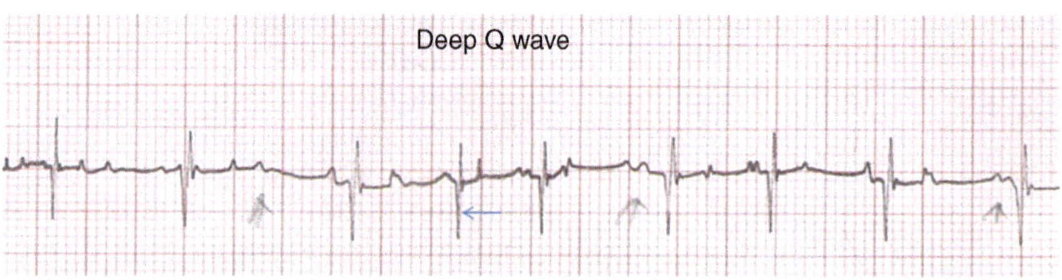

Fig. 7.21 Electrocardiogram (lead II, sensitivity 1, speed 25 mm/s) of a dog showing deep Q (0.7 mV) suggesting right ventricular enlargement

Further Reading

Bolton GR (1975) Handbook of canine electrocardiography. W.B. Saunders Company, Philadelphia, PA

Tilley LP (1985) Essentials of canine and feline electrocardiography, 2nd edn. Lea and Febiger, Philadelphia, PA

Intraventricular Conduction Abnormality and Bundle Branch Blocks

8

Current normally flows from sinoatrial node to ventricular myocardium through atrioventricular node (AV node), common bundle of His, bundle branches (left and right), fascicles, and the Purkinje fibers at a certain speed. Any delay or block in the conduction pathways below the bundle of His leads to an intraventricular conduction abnormality popularly known as bundle branch blocks (BBB). Intraventricular conduction system consists of the right bundle branch, left bundle branch, and anterior and posterior fascicle of the left bundle branch. Any of the pathways may be affected. A block or delay may occur in any one, two, or three pathways at a time leading to delayed depolarization causing changes in "QRS" configuration. Well-accepted electrocardiographic features of bundle branch blocks are as follows.

8.1 Left Bundle Branch Block (LBBB)

When block in conduction pathway occurs at the level of left bundle branch, the following changes may be reflected in an electrocardiogram.

- Mean electrical axis (MEA) on frontal plane may remain within normal limits without any significant change.
- "QRS" complex in different leads (lead I, II, III, aVF, CV_6LL, and CV_6LU) is broad and positive (Fig. 8.1).
- Sometimes a small "Q" wave is seen in different leads (lead II, III, and aVF). It is suggestive of incomplete LBBB (Fig. 8.1).
- Because of slow conduction due to block, QRS complex may become broad and sloppy (Fig. 8.2).
- Left bundle branch block is also suggested by the presence of a small "Q" wave in lead I, CV_6LL, and CV_6LU.

J. P. Varshney, *Electrocardiography in Veterinary Medicine*,
https://doi.org/10.1007/978-981-15-3699-1_8

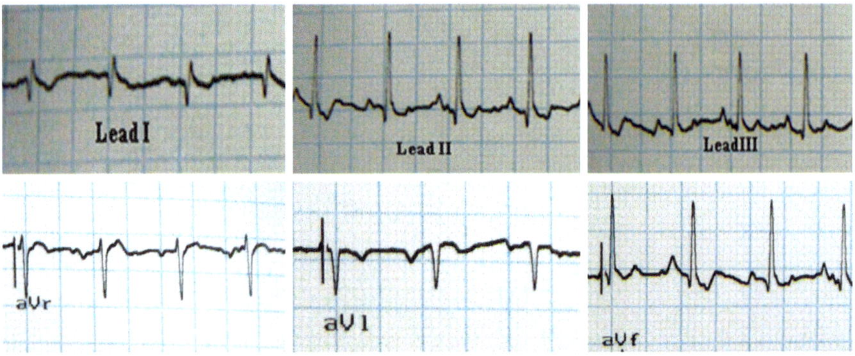

Fig. 8.1 Electrocardiogram (lead I, II, III, aVR, aVL, and aVF, sensitivity 1.0, speed 25 mm/s) of a Rottweiler showing wide (0.08 s) and positive "QRS" in lead I, II, III, and aVF; and a small q in lead I, II, III, and aVF suggesting incomplete left bundle branch block

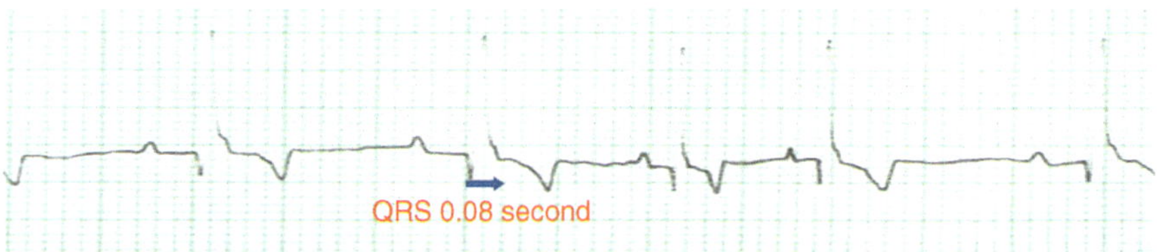

Fig. 8.2 Electrocardiogram (lead II, sensitivity 1.0, speed 25 mm/s) of a 12-year-old female dog showing wide and positive (0.08 s) "QRS" with normal amplitude of R (1.6 mV) and small q (0.15 mV) suggesting incomplete left bundle branch block

- Inverted "QRS" in lead aVR, aVL, and CV5RL is another indication of left bundle branch block.
- Left bundle branch block is indicated when above electrocardiographic features are present in the electrocardiogram with the absence of left ventricular enlargement in radiographs (lateral, ventrodorsal, and dorsoventral views).

8.2 Left Anterior Fascicular Block

When block in conduction pathway occurs at the level of left anterior fascicle, the following changes may be reflected in an electrocardiogram.

- Duration of "QRS" complex does not prolong. It remains within normal range.
- MEA may show marked left axis deviation on frontal plane.
- Small "Q" and tall "R" waves may be seen in leads I and aVL (Fig. 8.3).
- "S" wave may become enlarged (increased amplitude) in leads II, III, and aVF

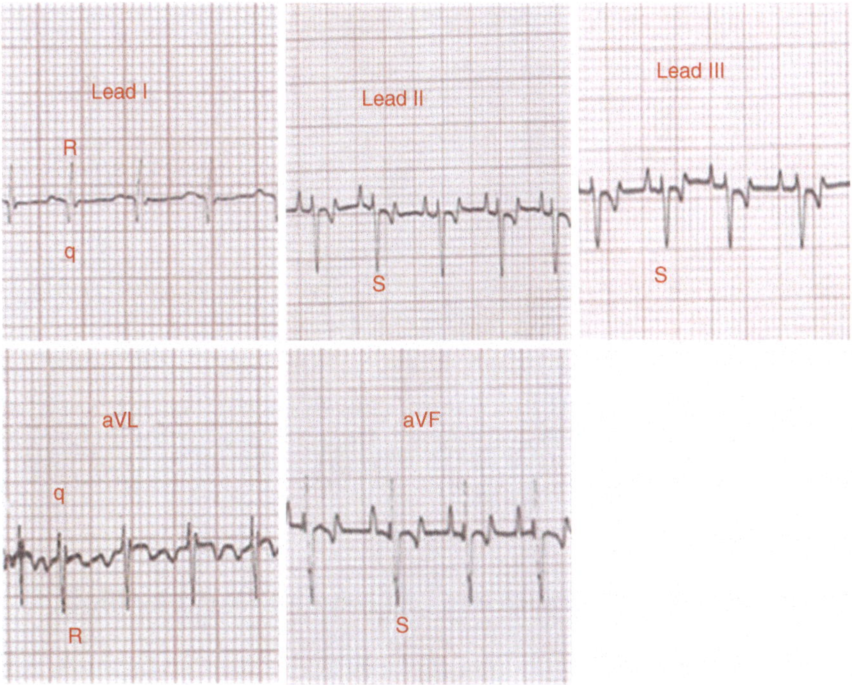

Fig. 8.3 Electrocardiogram (lead I, II, III, aVL, and aVF, sensitivity 1.0, speed 25 mm/s) of a 10-year-old male Pomeranian dog showing small q and R wave in lead I and aVL; normal QRS; and deep S in lead II, III, and aVF suggesting left anterior fascicular block

8.3 Right Bundle Branch Block (RBBB)

When block in conduction pathway occurs at the level of right bundle branch, the following changes may be reflected in an electrocardiogram. The presence of a single feature in electrocardiogram may be illusive. For definite diagnosis of RBBB at least two or more electrocardiographic features should be there in the electrocardiogram (Fig. 8.4).

- "QRS" becomes wide due to delayed conduction (0.07 s or more)
- MEA may show marked right axis deviation on frontal plane (more than +104°).
- "QRS" complex which is normally negative in leads aVR, aVL, and CV5RL is altered and becomes positive.
- "QRS" complex may show a wide RSR′ or rsR′ pattern.
- A wide "S" wave may be observed in leads I, II, III, aVF, CV6LL, and CV6LU.
- Lead V10 may show a "S" wave or W pattern.

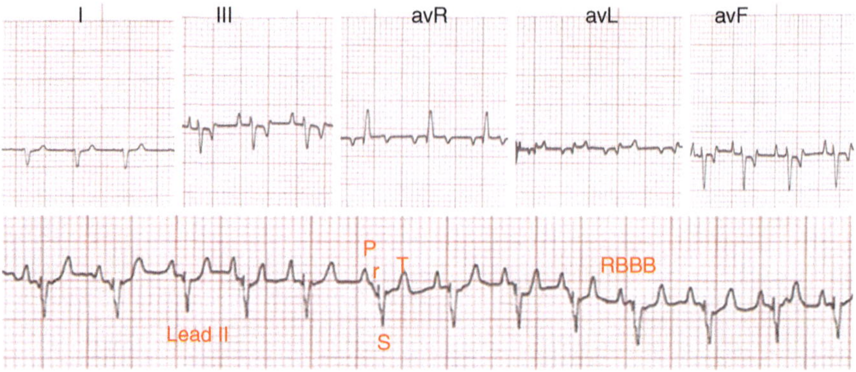

Fig. 8.4 Electrocardiogram (lead I, II, III, aVR, aVL, and aVF, sensitivity 1.0, speed 25 mm/s) of a 11-year-old female Pomeranian dog suffering from babesiosis showing wide (0.07 s) "QRS" in lead I, II, III, and aVF; QRS +ve in aVR and aVL; QRS having small r and large S wave in lead I, II, III, and aVF suggesting right bundle branch block

8.4 Right Bundle Branch and Left Anterior Fascicular Block

When block in conduction pathway occurs at the level of right bundle branch and left anterior fascicle, the following changes may be reflected in an electrocardiogram.

- ECG may show a wide "QRS."
- MEA reflects marked left axis deviation (<+40°) on frontal plane.
- "S" wave may be broad and large in leads I, II, III, aVF, and CV6LU.
- Lead I and aVL may show a tall "R" wave and small "Q" wave.
- Wide rsR′, RSR′ pattern (M shaped) of "QRS" complex may be seen in CV5RL lead.

Further Reading

Bolton GR (1975) Handbook of canine electrocardiography. W.B. Saunders Company, Philadelphia, PA

Tilley LP (1985) Essentials of canine and feline electrocardiography, 2nd edn. Lea and Febiger, Philadelphia, PA

ECG Patterns Associated with Electrolyte Imbalances, Drug Toxicities and Physical and Chemical Agents

9

Intracellular concentration of K^+ is approximately 30 times higher than the extracellular concentration in cardiac cells, while concentration of Na^+ and Ca^{2+} is higher in extracellular fluid. Sodium and potassium pumps are active transport mechanisms responsible for resting membrane potential, depolarization, and repolarization of cardiac muscle cell. Hence appropriate concentrations of Na^+, K^+, and Ca^{2+} are important for normal depolarization and repolarization of cardiac cells. Any imbalance in the concentration of these electrolytes may lead to change in the pattern of electrocardiogram. Electrolyte imbalances are commonly seen in dehydration, severe vomiting, diarrheas, or renal failure. Electrocardiographic changes in response to electrolyte imbalances are detailed below.

9.1 ECG Changes Associated with Electrolyte Imbalance

Electrolyte imbalances	Alterations in electrocardiogram
Hyperkalemia	– Large spiked "T" wave
	– Short Q-T interval
	– Flat or absence of "P" wave
	– Wide "QRS"
	– Depression of S-T segment
	– Prolonged P-R interval
	– Sinoventricular rhythm
	– Slowing of heart rate
	– Decrease in R amplitude
	– Complete heart block in severe hyperkalemia

J. P. Varshney, *Electrocardiography in Veterinary Medicine*,
https://doi.org/10.1007/978-981-15-3699-1_9

Electrolyte imbalances	Alterations in electrocardiogram
Hypokalemia	– Progressive S-T segment depression
	– Small biphasic "T" wave
	– Prolonged Q-T interval
	– Appearance of U wave
	– Tachyarrhythmia
	– Delayed and abnormal repolarization
	– Supraventricular and ventricular arrhythmias
	– Increase in R amplitude
	– Increased "P" wave amplitudes and durations (in severe hypokalemia)
Hypercalcemia	– Short Q-T interval. It may be fused with upstroke of "T" wave
Hypocalcemia	– Prolonged Q-T interval
	– Tachyarrhythmia
	– Q-T interval is correlated with plasma calcium levels
Hypermagnesemia	– Prolonged P-R interval
	– Widening of QRS complexes
	– Heart block
Hypomagnesemia	– Decreased resting membrane potential of myocardial cells
	– Increased excitability of Purkinje fibers
	– Wide QRS complexes
	– S-T segment depression
	– Peaked "T" wave
	– Atrial fibrillation
	– Supraventricular tachycardia
	– Ventricular tachycardia
	– Ventricular fibrillation

9.2 ECG Changes Associated with Drug Toxicities

A wide variety of drugs are used in the management of diseases in routine clinical practice in canine medicine. Many drugs are known to alter electrocardiographic pattern in dogs and cats, while many others may influence cardiovascular system owing to their potential side effects and may lead to changes in the electrocardiogram. The drugs, commonly employed in canine medicine, influencing cardiac functioning as side effects, are given below with their possible effects on electrocardiogram. Knowledge of such electrocardiographic alterations is important in routine management of clinical cases in dogs.

Drugs	ECG changes
Amitriptyline	– Tachycardia
	– Arrhythmias
	– Conduction disturbances
Acetylpromazine	– Sinus bradycardia

Drugs	ECG changes
Atropine	– Bradycardia followed by tachycardia is seen when atropine is given intravenously in doses higher than 0.015 mg/kg
	– Atrial premature complexes (APCs), ventricular premature complexes (VPCs), second-degree AV block, or sinus tachycardia may be seen when atropine is given in doses less than 0.015 mg/kg intravenously
Amitraz	– Bradycardia
Amiodarone	– Arrhythmias
	– Conduction disturbances
	– Bradycardia or a systole
Atenolol	– Bradycardia
	– Impaired atrioventricular conduction
Adrenaline	– Tachycardia
	– Arrhythmia
Aminophylline	– Tachycardia
	– Arrhythmia
Amphetamine	– Tachycardia
	– Arrhythmia
Aminoglycosides	– Tachycardia
	– Arrhythmias
Barbiturates	– Ventricular bigeminy
	– Ventricular arrhythmias
Bretylium	– Ventricular tachycardia
Clindamycin and lincomycin (rapid IV in sensitive dogs)	– Cardiovascular collapse
Clomipramine	– Cardiac arrhythmias
	– Tachycardia
Cimetidine	– Tachycardia (occasionally)
Calcium gluconate (rapid IV infusion)	– Bradycardia
	– Shortening of Q-T interval
	– S-T elevation
Cyclophosphamide	– Cardiac enlargement pattern
	– Low-output heart failure
Digoxin	– Prolongation of P-R interval
	– AV block (second or third degree)
	– Sinus bradycardia or sinus arrest
	– Ventricular premature complexes
	– Accelerated junctional rhythm
	– Ventricular tachycardia
	– Paroxysmal atrial tachycardia with blocks
	– Atrial fibrillation with slow ventricular rate
Doxorubicin	– Atrial premature complexes
	– Ventricular tachycardia
	– AV block
	– Supraventricular arrhythmias
	– Heart block
Diltiazem	– Prolong AV node conduction time

Drugs	ECG changes
Diazepam	– Tachycardia
	– Normal QRS
Ephedrine	– Tachycardia
Glycopyrrolate	– Increased heart rate
	– Ventricular premature complexes (less frequent)
Halothane	– Sinus bradycardia
	– Ventricular arrhythmias
Hydralazine	– Reflex tachycardia
	– Hypotension
Isoproterenol	– Ectopic complexes
	– Tachycardia
Isopropamide	– Tachycardia
Imipramine	– Sinus tachycardia
	– AV bundle branch block
Lidocaine	– AV block
	– Sinus arrest
	– Pace maker suppression
	– Asystole
Morphine	– Bradycardia
Nicotine	– Tachycardia
Norepinephrine	– Sinus tachycardia or atrial tachycardia
Neostigmine (ophthalmic)	– Bradycardia
Neostigmine methylsulfate	– Cardiac arrest
Organic phosphates	– Bradycardia or tachycardia
Oxyphenonium	– Bradycardia followed by tachycardia and arrhythmias
Prochlorperazine	– Tachycardia
Propantheline bromide	– Tachycardia
PGF2-alpha	– Tachycardia
	– Ventricular tachycardia
Prazosin	– Hypotension
	– Tachycardia (less frequent than with hydralazine)
Propranolol	– Pace maker suppression
	– Atrioventricular (AV) block
Physostigmine (ophthalamic)	– Bradycardia
Quinidine	– Increased heart rate
	– Q-T prolongation
	– AV blocks
	– Ventricular tachyarrhythmia
	– Sinus arrest
	– Wide QRS
	– Pace maker suppression
	– Asystole
	– Ventricular fibrillation
	– Heart block (first, second, and third degree)
Salbutamol	– Tachycardia
	– Palpitation

Drugs	ECG changes
Thiopental	– Cardiac arrest
Terbutaline	– Direct stimulation of SA node may cause sinus tachycardia
	– Increased coronary circulation
	– Enhanced myocardial perfusion
	– Ventricular tachycardia
	– Multifocal ventricular tachycardia
	– Sinus tachycardia
	– Atrioventricular (AV) block
Verapamil	– Pace maker suppression
	– Atrioventricular block
Xylazine	– Sinus bradycardia
	– Sinus arrest
	– Sinoatrial block
	– AV block
	– Ventricular tachyarrhythmia

9.3 ECG Changes Associated with Physical and Chemical Agents

Temperature variations in the body and ingestion of many chemicals (environmental pollutants or other toxicants) may also affect cardiovascular system. Effect of some of these physical and chemical agents on ECG profile is given below:

Physical/chemical agents	ECG changes
Hyperpyrexia (heat stroke)	– S-T segment and T wave abnormalities
	– Tachycardia
	– Ventricular premature complexes
Hypothermia	– Reduced heart rate (bradycardia)
	– "J" wave
Arsenic toxicity	– Tachycardia
	– S-T segment depression
	– Q-T interval prolongation
	– "T" wave inversion
Lead poisoning	– Sinus bradycardia
	– Occasional arrhythmias
	– S-T segment changes
	– "T" wave changes
Carbon monoxide poisoning	– S-T segment changes
	– Conduction abnormalities
	– Arrhythmias
Bufo poisoning (toad poisoning)	– Arrhythmias
	– Tachycardia
	– Ventricular fibrillation

Physical/chemical agents	ECG changes
Viper envenomation (viper bites)	– Atrial tachycardia
	– Ventricular fibrillation/flutter
	– Ventricular premature complex
	– Ventricular tachycardia
	– Sinus tachycardia
	– Sinus bradycardia
	– Normal heart rate with sinus rhythm
Cardiac glycoside poisoning (oleander toxicity)	– Arrhythmias
Phorate poisoning	– Sinus rhythm
	– Tachycardia
Theobromine + caffeine (chocolate poisoning)	– Arrhythmias
	– Ventricular premature complexes
	– Tachycardia

Further Reading

Bolton GR (1975) Handbook of canine electrocardiography. W.B. Saunders Company, Philadelphia, PA

Tilley LP (1985) Essentials of canine and feline electrocardiography, 2nd edn. Lea and Febiger, Philadelphia, PA

Cardiac Arrhythmias 10

Cardiac arrhythmias, disturbances in rhythm, and rate of the heart are frequently encountered in canine practice. There is hardly any dog that has not experienced arrhythmia in his or her lifetime. Any impulse that generates outside the sinoatrial node (usual pace maker) causes arrhythmias. The abnormal impulses generating outside the usual pace maker (in the SA node) are termed as ectopic impulses. The ectopic impulses may originate from the atrium, junction, or ventricle and are referred to as atrial, junctional, supraventricular, or ventricular ectopic beats on the basis of their seat of origin. When the ectopic beat is occurring earlier than the next expected normal sinus impulse, it is termed as premature beat or premature complex. The ectopic beat occurring late or after the normal sinus impulse is called escape rhythm. The ectopic beat or abnormal impulse may occur singly or in multiples. Occurrence of premature impulses in three or more is generally referred to as episode of tachycardia. Brief bouts of tachycardia are called paroxysmal tachycardia. When bouts of tachycardia are prolonged, it is called sustained tachycardia. Asymptomatic arrhythmias are benign having no clinical significance and require no particular therapeutic attention, but arrhythmias associated with clinical manifestations (symptomatic arrhythmias) are serious enough to threaten the life and need proper differential diagnosis and immediate therapeutic intervention. As has been emphasized earlier, electrocardiogram is an effective tool in differentiating the type of arrhythmias. Surveys done in other countries revealed that the incidence of arrhythmias in dogs varied from 3.17% to 42% in different circumstances. In India large-scale studies on the prevalence of cardiac arrhythmias are lacking. A study at this hospital has revealed the prevalence of cardiac arrhythmias as 3.04% in a population of 20,000 canine cases (Varshney et al. 2013). The arrhythmias were grouped into three categories as sinus arrhythmia, abnormalities of impulse formation, and abnormalities of impulse conduction with a prevalence rate as 1.295%, 1.415%, and 0.33%, respectively (Table 10.1).

J. P. Varshney, *Electrocardiography in Veterinary Medicine*,
https://doi.org/10.1007/978-981-15-3699-1_10

Table 10.1 Prevalence of arrhythmias in dogs (Varshney et al. 2013)

Types of arrhythmias	No. of cases of arrhythmias	Prevalence among arrhythmias % ($n = 608$)	Prevalence of arrhythmias among population of 20,000 new cases
1. **Sinus arrhythmias**	**259**	**42.59**	**1.295**
Sinus bradycardia	19	3.12	
Sinus tachycardia	165	27.13	
Sinus arrhythmias	69	11.34	
Wandering pace maker	6	0.98	
2. Abnormalities of impulse formation	**283**	**46.54**	**1.415**
(a) Atrial arrhythmias	212	34.86	
Atrial flutter	8	1.31	
Atrial fibrillation	129	21.21	
Atrial premature complex	13	2.13	
Atrial tachycardia	51	8.38	
AV junctional premature complex	6	0.98	
AV junctional tachycardia	5	0.82	
(b) Ventricular arrhythmias	71	11.67	
Ventricular flutter	4	0.65	
Ventricular fibrillation	6	0.98	
Ventricular premature complex (VPC)	39	6.41	
Ventricular tachycardia	18	2.96	
Ventricular asystole	4	0.65	
3. Abnormalities of impulse conduction	**66**	**10.85**	**0.330**
(a) A-V Block	45	7.40	
First degree	27	4.46	
Second degree	14	2.30	
Third degree	4	0.65	
(b) SA block	12	1.97	
(c) Atrial standstill	3	0.49	
(d) Sick sinus syndrome	6	0.98	

Researches in canine cardiology have indicated that an arrhythmia may arise from either cardiac or non-cardiac origin. Causes of arrhythmias of cardiac origin are congenital or hereditary diseases of the heart, cardiomyopathy, congestive heart failure, ischemia, trauma, neoplasm, pericardial effusions, or valvular heart diseases, while non-cardiac causes of arrhythmias include respiratory problems (respiratory arrest, pulmonary edema, pneumonia, brachycephalic airway disease, pulmonary shunting, etc.), hypoxic state (anemia, hypervolemia, shock), neurologic defects (cerebral lesions, spinal cord lesions, increased intracranial pressure, vago-sympathetic disorders), gastrointestinal tract diseases (acute gastric dilatation, pancreatitis, peritonitis, volvulus), endocrine abnormalities (hypothyroidism, hyperthyroidism, diabetes mellitus, hypoadrenocorticism, hyperadrenocorticism, hypocalcemia, hypercalcemia), urogenital diseases (renal failure, pyometra, acute prostatitis), and blood and lymphatic abnormalities (anemia, lymphosarcoma, hemangiosarcoma). Arrhythmias are also observed during sedation, anesthesia; administration of sympathomimetic, sympatholytic, parasympathomimetic, or parasympatholytic drugs; electrolyte therapy; cancer therapy; pain; hypothermia; hyperthermia; exercise; or excitement/fear.

Diagnosis of arrhythmia is easy by systematic evaluation of the ailing dog and its electrocardiogram. Use of multiple leads and a long strip is of great assistance in evaluating the wave forms of an arrhythmic dog. A precise diagnosis of arrhythmia is essential to determine its cause and severity, adopting rational therapeutic approach.

10.1 Classification of Cardiac Arrhythmias

As per definition arrhythmias are due to disturbances in cardiac rate/rhythm or both or conduction disturbances. They may be of different origin (atrial, AV junctional, ventricular) or may be due to conduction defect. Keeping these variation in view, arrhythmias can be grouped under various categories. Though arrhythmias can be classified in various ways, no additional information is derived except similarity of a character.

10.1.1 Arrhythmias due to Variation in Heart Rate

Arrhythmias can be categorized on the basis of heart rate (increase or decrease in heart rate without any disturbance of heart rhythm) as sinus bradycardia (heart rate is lower than the normal), or sinus tachycardia (heart rate is higher than the normal). In this category the similar character is the heart rate. Under the category of sinus bradycardia, there are various types of arrhythmias such as sinoatrial standstill (no “P” wave), ventricular escape rhythm, complete heart block, sinus arrest, advanced second-degree heart block, and junctional escape rhythm. Sinus tachycardia is characterized by heart rate higher than the normal limit. Various types of arrhythmias such as atrial tachycardia, atrial fibrillation/flutters, junctional tachycardia, and ventricular tachycardia are in the category of sinus tachycardia.

10.1.2 Arrhythmias due to Rhythm Irregularities

Another way of looking at arrhythmias may be whether the rhythm is normal or not. Disturbance in the heart rhythm without change in heart rate is termed as sinus arrhythmia. Wandering pace maker, atrial premature beat, sinus arrest, junctional premature beat, junctional tachycardia, ventricular premature beat, heart blocks (first and second degree), bundle branch blocks, and Wolff-Parkinson-White syndrome are the arrhythmias under this category.

10.1.3 Arrhythmias due to Variation in Heart Rate as well as Rhythm Irregularities

Arrhythmias such as bradyarrhythmia, sinus arrest, sick sinus syndrome, and second-degree heart block (high grade) are characterized by slow heart rate and irregular rhythm and therefore can be grouped under arrhythmias with slow heart rate with irregular rhythm. On the other hand, atrial or supraventricular premature contractions, paroxysmal atrial or supraventricular tachycardia, atrial flutter/fibrillation, ventricular premature contraction, and paroxysmal ventricular tachycardia are characterized by fast heart rate with irregular rhythm and therefore can be grouped under arrhythmias with fast heart rate with irregular rhythm. These types of arrhythmias have disturbances of both heart rate (increased or decreased heart rate) and rhythm (irregular rhythm).

10.1.4 Arrhythmias due to Abnormal Impulse Generation in the Seat of Origin

Arrhythmias can also be grouped on the basis of abnormalities in impulse generation in the sinoatrial node (SA node), atrioventricular junction (AV junction), or ventricle and are termed as supraventricular, AV junctional, or ventricular arrhythmias, respectively. Sinus arrest, atrial premature complexes, atrial tachycardia, atrial fibrillation, and atrial flutter are under the category of supraventricular arrhythmia. The category of AV junctional arrhythmias includes AV junctional premature complexes, AV junctional tachycardia, and AV junctional escape rhythm. Ventricular arrhythmias arising due to abnormality of impulse formation in ventricle consist of ventricular premature complexes, ventricular tachycardia, ventricular fibrillation, ventricular flutter, ventricular asystole, and ventricular escape rhythm.

10.1.5 Arrhythmias due to Abnormal Impulse Conduction

Abnormal impulse conduction also leads to arrhythmias. In this category impulse is formed in a normal way, but its conduction is affected. Sinoatrial block (SA block), atrial standstill, and atrioventricular block (AV block) fall under this category.

10.1.6 Arrhythmias due to Abnormal Impulse Generation and Conduction

Arrhythmias may also occur when there is disturbance of impulse formation and its conduction. Parasystole and Wolff-Parkinson-White syndrome are the arrhythmias which are associated with the disturbance of impulse formation and its conduction.

10.1.7 Arrhythmias due to Differing Pace Maker Site

Arrhythmias can also be grouped on the basis of their origin, whether they have originated from pace maker in the SA node or other than the SA node. Sinus arrhythmia, sinus arrest, sinoatrial block, wandering pace maker, sinus tachycardia, and sinus bradycardia are due to pace maker disturbance in the SA node and therefore can be grouped as sinoatrial arrhythmias. When the pace maker lies at a site/focus other than SA node, arrhythmias originated from such focus are called ectopic arrhythmias. The focus of ectopic beat may be supraventricular or ventricular and are accordingly termed as supraventricular or ventricular arrhythmias. In supraventricular arrhythmias the foci of ectopic beat lie above the ventricle; therefore arrhythmias such as atrial premature beat, junctional premature beat, atrial fibrillation/flutter, and junctional tachycardia are called supraventricular arrhythmias. When foci of ectopic beat lie in ventricular mass (pace maker is found in the bundle of His, bundle branches, and Purkinje fibers), these arrhythmias are termed as ventricular arrhythmias. Ventricular premature beats, ventricular tachycardia, and ventricular fibrillations/flutters are thus called ventricular arrhythmias.

10.1.8 Conduction Disturbances

Atrioventricular (AV) conduction time is the interval between onset of atrial activation and the onset of ventricular activation. In ECG, it is represented by P-R interval. P-R interval denotes the total travel time taken by the depolarization wave from the SA node to the AV node. P-R intervals in clinically healthy dogs vary from 0.08 to 0.12 s. P-R interval is of small duration in small breed dogs as compared to that of giant breed dogs. There are four types of conduction disturbances, namely, heart blocks or AV blocks, sinoatrial standstill (SA standstill), bundle branch blocks (BBB), and Wolff-Parkinson-White syndrome (WPW syndrome). The heart blocks are further grouped as first-degree, second-degree, and third-degree heart blocks depending on the disturbance or delay in the conduction of supraventricular impulse through the AV node and bundle of His. The bundle branch blocks (BBB) are also further divided into two categories, viz., right bundle branch block (RBBB) and left bundle branch block (LBBB) depending on the site of the bundle branch affected.

10.2 Factors Precipitating Arrhythmias

Arrhythmias are precipitated by various cardiac or non-cardiac factors as detailed below.

10.2.1 Cardiac Factors

Concurrent cardiac abnormalities may precipitate both atrial and ventricular arrhythmias.

(a) Atrial Arrhythmias—Atrial arrhythmias are precipitated by many cardiac diseases such as valvular insufficiencies (mitral valve insufficiency, tricuspid valve insufficiency), cardiomyopathies (dilated cardiomyopathy, myocardial fibrosis), congenital malformation, ischemia, intra-atrial catheterization, and cardiac tumors; drugs (anesthetics, digitalis, bronchodilators); and increased sympathetic tone.

(b) Ventricular Arrhythmias—Ventricular arrhythmias are precipitated by many cardiac diseases such as cardiomyopathies (Boxer cardiomyopathy, Doberman Pinscher cardiomyopathy), myocarditis, pericardial disease (pericarditis), cardiac tumors, ischema, cardiac trauma/chest trauma, congenital heart diseases, dirofilariasis, degenerative valvular diseases with myocardial fibrosis, ventricular dilation, pace maker wire, and cardiac catheterization and drugs (anesthetics, sympathomimetic, tranquilizers, anticholinergic, and digitalis).

10.2.2 Non-cardiac Factors

There are many factors other than cardiac factors that may precipitate arrhythmias. These factors are listed below for both atrial and ventricular arrhythmias.

(a) Atrial Arrhythmias—Atrial arrhythmias are precipitated by many non-cardiac factors such as catecholamine, electrolyte imbalance, acidosis, alkalosis, electric shock, thoracic surgery, severe anemia, thyrotoxicosis, and hypoxia.

(b) Ventricular Arrhythmias—Ventricular arrhythmias are also precipitated by many non-cardiac factors such as abnormal potassium level, hypoxia, fever, heat stroke/sun stroke, hypothermia, acidosis, alkalosis, electric shock, pancreatitis, uremia, septicemia, thyrotoxicosis, toxemia, abdominal or chest tumors, lower respiratory tract diseases (lung diseases), gastric dilation and volvulus, anxiety, pain, vagal stimulation, sympathetic stimulation, and viper envenomation.

10.3 Diagnostic Criteria for Arrhythmias

Diagnosis of arrhythmias is based on changes in electrocardiographic features with respect to presence or absence of "P" wave, its morphology, uniformity, and regularity; morphology and pattern of "QRS" complex; and variations in the measurement of P-R and R-R intervals.

Site of the arrhythmia can be differentiated by the configuration of QRS complex and P wave. Normal QRS complex indicates that ventricular depolarization is normal; hence rhythm disturbances are supraventricular (originating above the bundle branches in the the atria, AV node, or bundle of His). Wide QRS complexes preceded by related P waves suggest abnormality of the ventricular conduction pathway (bundle branch block or ventricular enlargement). Wide QRS complex unassociated with P wave indicates that the beat is ventricular in origin. Further differentiation of arrhythmias is based on the presence of P wave and its relation to QRS complex. Sometimes identification of P wave in tachyarrhythmia and low-voltage complexes may become difficult. Differentiation of supraventricular tachycardia becomes impossible as it is difficult to evaluate "P" wave because of increased heart rate. It becomes possible only when heart rate slows down sufficiently to evaluate the presence of P waves. P wave unrelated to QRS complexes suggests junctional or ventricular arrhythmia.

10.4 Electrocardiographic Features of Arrhythmias

Arrhythmias are associated with variations in the electrocardiogram. The electrocardiographic alterations with respect to different types of arrhythmias are illustrated below.

Type of arrhythmias	Electrocardiographic features
Normal sinus rhythm (Fig. 10.1)	Heart rate remains within range.
	R-R intervals remain almost the same.
	There is a P wave for every QRS complex.
	P-R interval is almost constant.
	The variations in P-P interval are <0.12 s.
Sinus arrhythmia (Fig. 10.2)	Heart rate remains within range.
	Pauses are shorter than twice the normal R-R interval.
	There is a P wave for every QRS complex.
	P-R interval is almost constant.
	There are variations in R-R interval.
	Sinus rhythm is more irregular with more P-P variations.
Sinus arrest (Fig. 10.3)	Pauses are equal to or greater than twice the normal R-R interval.
Sinoatrial block (Fig. 10.4)	Pauses are exactly twice or <twice of normal R-R interval.

Type of arrhythmias	Electrocardiographic features
Wandering pace maker (Fig. 10.5)	There are wide variations in the amplitude of "P" waves.
	P-R interval may vary from normal to short in duration.
Sinus tachycardia (Fig. 10.6)	There is marked increase in heart rate as compared to normal heart rate.
	Heart rhythm is regular.
	R-R intervals are normal or may vary slightly.
	There is "P" wave for every QRS complex.
	P-R interval is almost constant.
	Severe tachycardia may be associated with P wave hidden in preceding T wave.
	Configuration of both "P" and "QRS" is normal.
Sinus bradycardia (Fig. 10.7)	There is marked decrease in heart rate as compared to normal heart rate.
	Heart rhythm is almost regular.
	There is a "P" wave for every QRS.
	P-R interval is prolonged.
	"P" wave amplitude may vary (wandering pace maker).
	All "QRS" complexes are alike.
Supraventricular premature complex	Heart rate may be normal.
	Impulses originate above the AV node (either in the atrium or the AV junction).
	Rhythm is irregular and is broken by a premature beat followed by pause.
	P-R interval is either slightly shorter or longer.
	There is a P wave for every QRS complex.
	P wave of premature beat may be hidden in preceding "T" wave.
	Premature "P" may be altered.
	In case of junctional (nodal) premature beat, "P" is inverted, i.e., negative.
	In case of atrial premature beat, "P" wave is positive but different from normal P.
Atrial premature contractions (Figs. 10.8 and 10.9)	Configuration of premature (P′ wave) P′ is different.
	QRS-T complex is normal for normal P wave.
	QRS-T complex is absent for premature (P′) wave.
	P-R interval of APC may be prolonged with altered morphology of QRS.

Type of arrhythmias	Electrocardiographic features
Atrial tachycardia or paroxysmal atrial tachycardia (Figs. 10.10 and 10.11)	Tachycardia may be intermittent or continuous.
	Tachycardia may be regular (R-R interval same).
	Premature (P′) wave is positive in lead II.
	There is a P wave for every QRS in a normal complex.
	Configuration of P′ wave is different.
	P′ wave may be hidden in preceding T wave.
	P′ wave during tachycardia is positive but different from P wave when heartbeat is normal.
Atrial fibrillation (Fig. 10.12)	Heartbeat is exceeding the normal limit.
	Heart rhythm is irregular but may not be apparently visible when heart rate is rapid.
	There are no P waves or P-R intervals.
	P wave is replaced by fine baseline undulations ("f" wave).
	QRS is normal without P wave.
Atrial flutter (Fig. 10.13)	Heart rate is normal or rapid.
	Rhythm is regular or irregular.
	There are no P-R intervals.
	Ventricles respond in a periodic manner (2:1, 3:1, or 4:1).
	Normal P wave is replaced by coarse sawtooth oscillations.
Sino atrial standstill (Fig. 10.14)	Sometimes heart rate is low.
	Rhythm is regular or irregular.
	P wave is absent.
	QRS is without P wave.
	There are no P-R intervals.
	QRS is normal if it arises near the bundle of His or bizarre if it arises in ventricular muscle mass.
Junctional (nodal) escape rhythm (Fig. 10.15)	Hear rate is normal or slow.
	Rhythm is irregular with long pauses in isolated escape beat.
	Rhythm is regular when junctional rhythm persists.
	There is a P wave for every QRS complex.
	P wave occurs before, during, or just after QRS complex.
	P wave is negative.
	P-R interval is slightly shorter when present.
	QRS complexes are normal.

Type of arrhythmias	Electrocardiographic features
AV junctional tachycardia (Figs. 10.16 and 10.17)	Hear rate is normal or rapid.
	Rhythm is usually normal.
	Tachycardia may be in bursts.
	There is a P wave for every QRS complex.
	P wave may be hidden in preceding T wave.
	P-R interval may be constant.
	P waves are negative, while QRS complexes are normal.
	AV junctional beats occurring in high numbers (more than 60 bpm) are termed as AV junctional tachycardia.
Ventricular premature beat (Fig. 10.18)	Heart rate is generally normal.
	Normal rhythm is broken by premature beats followed by pause.
	There is no P wave for premature beats.
	QRS is related to P wave in normal beats.
	Abnormal beats do not have consistent P-R interval.
	QRS complex of premature beat is bizarre.
	If there is one VPC, it is called single.
	Two VPCs in succession are called pair.
	Three VPCs in succession are called run.
	If all VPCs are similar, they are from a unifocal origin.
	When VPCs vary in conformation, they are of multifocal origin.
	If VPC is alternating (i.e., one beat sinus and other VPC), it is called ventricular bigeminy.
	When two beats are sinus and third beat is VPC or one beat is sinus and two beats are VPC, it is called trigeminy.
	When major deflection of VPC is negative in lead II, the focus of ectopic beat is considered in the left ventricle.
	When major deflection of VPC is positive in lead II, the focus of ectopic beat is considered in the right ventricle.
Ventricular fibrillations/flutters (Figs. 10.19 and 10.20)	There is no coordinated heartbeat.
	Specific P waves and QRS complexes are absent.
	There is a series of baseline undulations.
	Coarse undulations are called flutters.
	Fine undulations are called fibrillations.
Ventricular escape beat (Fig. 10.21)	Heart rate is within range or slow.
	Heart rhythm is irregular when there is isolated ventricular escape beat.
	Idioventricular rhythm is regular.
	P wave may or may not occur.
	There is no P wave for escape beats.
	P wave may be normal but QRS complex is bizarre.

Type of arrhythmias	Electrocardiographic features
R-on-T phenomenon (Fig. 10.22)	Reported first in humans by Smirk (1949).
	A VPC is occurring on T.
	It has been reported in dogs also (Engel et al. 1978).
	There is a superimposition of an ectopic beat (R) on
	T wave of preceding beat.
Ventricular tachycardia (Fig. 10.23)	Four or more VPCs in a row are called ventricular tachycardia.
	Enhanced ventricular rhythm (60–100 bpm).
	HR is usually slow than sinus tachycardia.
Ventricular asystole (Figs. 10.24 and 10.25)	There is no ventricular rhythm (QRS).
	Configuration of P is normal with severe third-degree AV block.
	No atrial and ventricular rhythm.
Atrioventricular heart block (AV blocks)	
First-degree AV blocks (Fig. 10.26)	Heart rate is usually normal.
	Heart rhythm is sinus.
	There is a P wave for every QRS complex.
	P-R interval is consistent but increased (more than 0.13 s).
	P wave and QRS complex is normal.
Second-degree AV blocks	Heart rate is normal or slow.
	Rhythm is broken by absence of one or several QRS complexes.
	There is a P wave for every QRS complexes, but some P waves are without QRS complex.
	Normal complexes have consistent P-R interval.
	Normal P wave and QRS complexes are of normal configuration.
	Second-degree AV blocks are of two types (Mobitz types I and II).
Second-degree AV blocks (Mobitz type I) (Fig. 10.27)	There is a progressive prolongation of the P-R interval until a non-conducted P′ wave.
Second-degree AV blocks (Mobitz type II) (Fig. 10.28)	Atrial rate is higher than ventricular rate.
	Rhythm is irregular owing to absence of one or more QRS complexes.
	P wave is usually normal.
	P-R interval is constant and may be normal or prolonged.
Third-degree AV blocks (Fig. 10.29)	Ventricular heart rate is slow.
	Heart rhythm is regular.
	There are more P waves than QRS complexes.
	P-R interval is varying.
	There is no consistent relationship between atrial and ventricular beats.
	Configuration of P waves is normal.
	QRS complexes near normal or bizarre.
Sick sinus syndrome (Fig. 10.30)	It is due to abnormalities of SA node.
	It is also called bradycardia-tachycardia syndrome.
	Tachycardia and bradycardia are alternating.

Type of arrhythmias	Electrocardiographic features
T-on-P phenomenon	There is sinus tachycardia.
	QT interval is prolonged (mainly due to ST segment).
	T is closely followed by next P or P is buried in the downstroke of preceding T.
Ventricular preexcitation	Heart rate is usually normal.
	Rhythm may be sinus or arrhythmic.
	There is a P wave for every QRS complex.
	P-R interval is constant but short.
	P wave is normal.
	QRS complexes may be wide and aberrant.
	Upstroke of the R wave (delta wave) is slurred (Fig. 10.31).
Wolff-Parkinson-White syndrome	Heart rate is extremely high.
	P wave is unrecognizable.
	QRS may be normal, wide with delta wave.
Lown-Ganong-Levine preexcitation	P-Q interval is shortened.
	QRS complex is normal.
Atrioventricular accessory pathway arrhythmia	It is an uncommon but fatal arrhythmia.
	Labrador and Golden Retrievers are predisposed.
	Electrical impulse travels in reverse direction.
	Congestive heart failure is due to tachycardia.
	Heart rate is very high exceeding 240.
	When rhythm is sinus, ECG is normal.

Table 10.2 Summarized electrocardiographic features of supraventricular and ventricular premature complexes

Electrocardiographic features	Supraventricular premature complex	Ventricular premature complex
QRS morphology	No change and is almost similar to normal beat	It is different from normal beat
QRS width	QRS is narrow	QRS is wide
Can occur as fusion beat	Not occurring as fusion beat	May occur as fusion beat
Can occur as interpolated beat	No	Yes
Association of atrial (P) and ventricular (QRS) activation	Yes	No
Compensatory pause present	Pauses are not fully compensatory	Pauses are fully compensatory
When ectopic beats occur in series as tachycardia	Tachycardia is regular	Tachycardia is slightly irregular
When premature beat follows a premature "P" wave	It is supraventricular premature complex	It is not a ventricular premature complex

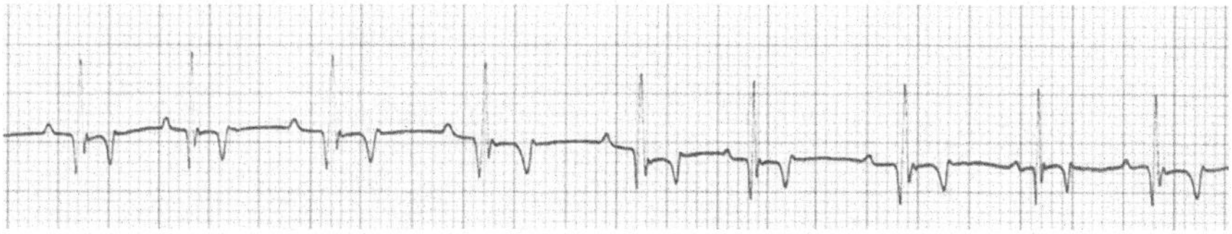

Fig. 10.1 Electrocardiogram (lead II, sensitivity 1, speed 25 mm/s) of a 4-year-old German Shepherd showing heart rate of 120 bpm, constant P-R interval, "P" for every "QRS," and sinus rhythm suggesting normal sinus rhythm

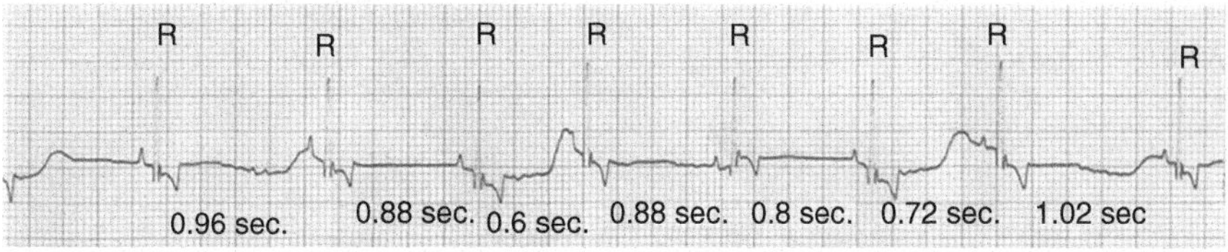

Fig. 10.2 Electrocardiogram (lead II, sensitivity 1, speed 25 mm/s) of a 6-year-old Labrador showing heart rate of 84 bpm, "P" for every "QRS," and variable R-R interval (0.6–1.02 s) suggesting sinus arrhythmia

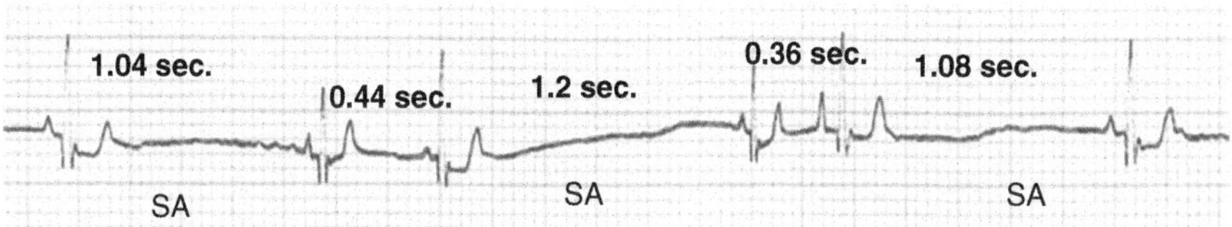

Fig. 10.3 Electrocardiogram (lead II, sensitivity 1, speed 25 mm/s) of an 8-year-old Pomeranian dog with chronic coughing showing heart rate of 75 bpm and variable R-R intervals with pauses greater than twice of the normal R-R interval suggesting sinus arrest

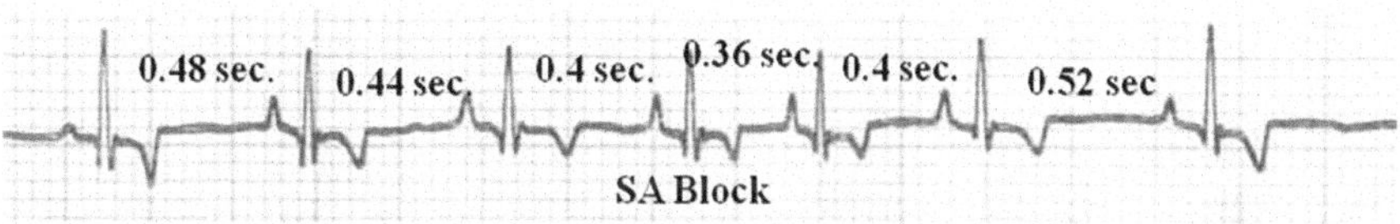

Fig. 10.4 Electrocardiogram (lead II, sensitivity 1, speed 25 mm/s) of a 3-year-old male German Shepherd dog with bilateral epistaxis showing heart rate of 140 bpm and variable R-R intervals with pauses less than twice of the normal R-R interval suggesting sinoatrial block

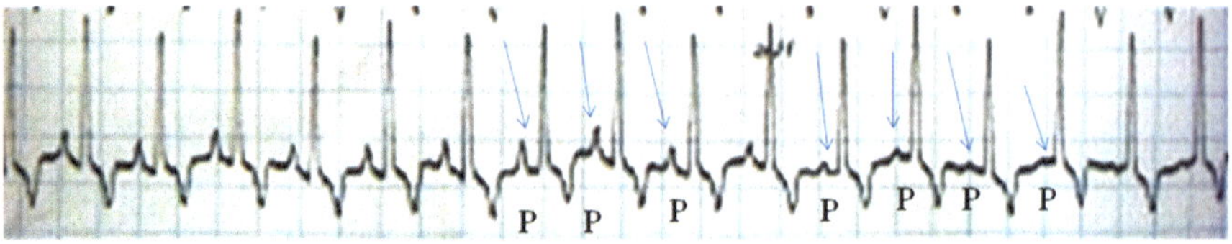

Fig. 10.5 Electrocardiogram (lead II, sensitivity 1, speed 25 mm/s) of an 11-year-old male German Shepherd dog showing heart rate of 160 bpm and varying amplitude of "P" waves (arrows) and little variations in "P-R" interval suggesting wandering pace maker

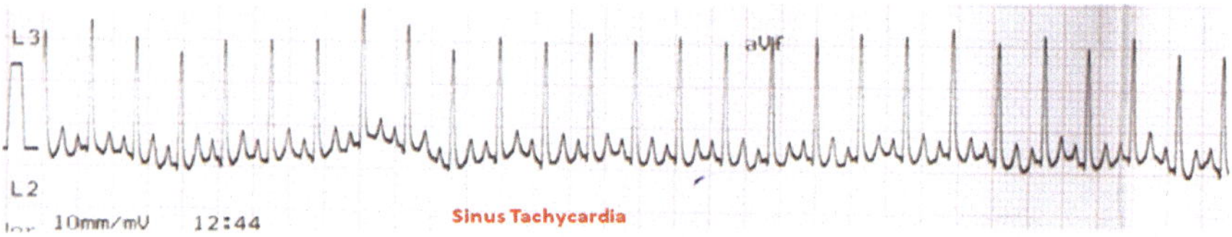

Fig. 10.6 Electrocardiogram (lead II, sensitivity 1, speed 25 mm/s) of a 6-year-old male Great Dane dog with hyperthermia (temperature 107.4 °F) showing heart rate of 240 bpm, sinus rhythm, "P" for every "QRS," almost constant P-R interval (0.05/0.06 s), and normal configuration of P and QRS suggesting sinus tachycardia

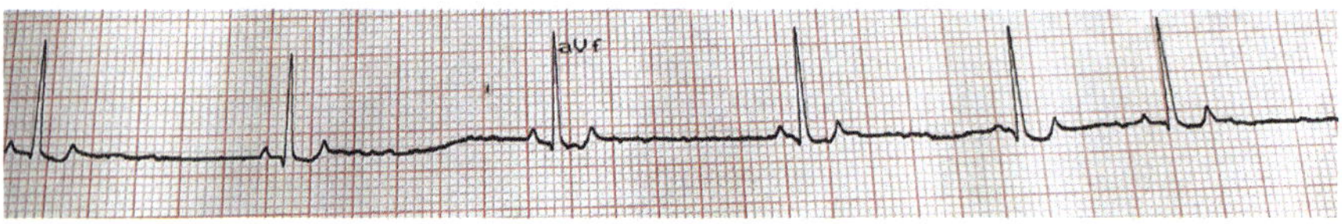

Fig. 10.7 Electrocardiogram (lead II, sensitivity 1, speed 25 mm/s) of a 7-year-old male Labrador dog showing heart rate of 60 bpm, sinus rhythm, "P" for every "QRS," almost constant P-R interval (0.10 s), normal and similar configuration of "QRS" suggesting sinus bradycardia

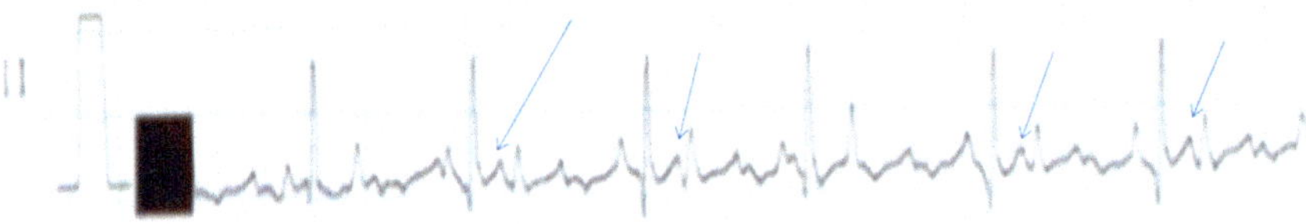

APC

Fig. 10.8 Electrocardiogram (lead II, sensitivity 1, speed 25 mm/s) of a 2.5-year-old male Cocker Spaniel dog showing ventricular heart as 100 bpm and atrial premature complexes (arrows)

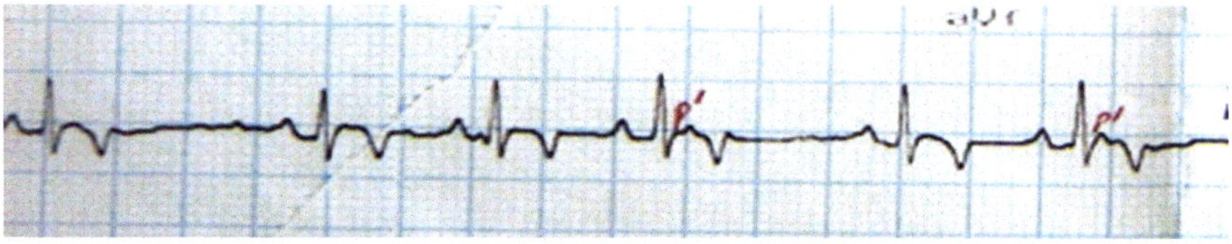

Fig. 10.9 Electrocardiogram (lead II, sensitivity 1, speed 25 mm/s) of a 3-year-old female Labrador dog showing atrial premature complex (APC). The P′ before "T" is the premature complex

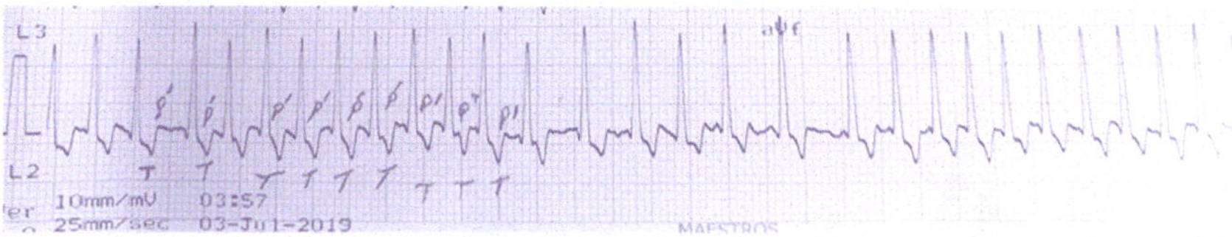

Fig. 10.10 Electrocardiogram (lead II, sensitivity 1, speed 25 mm/s) of an adult 3-year- and 2-month-old French Mastiff showing atrial tachycardia. The atrial rate (P and P′) averages 260 bpm with P′ wave different in configuration from that of sinus P waves

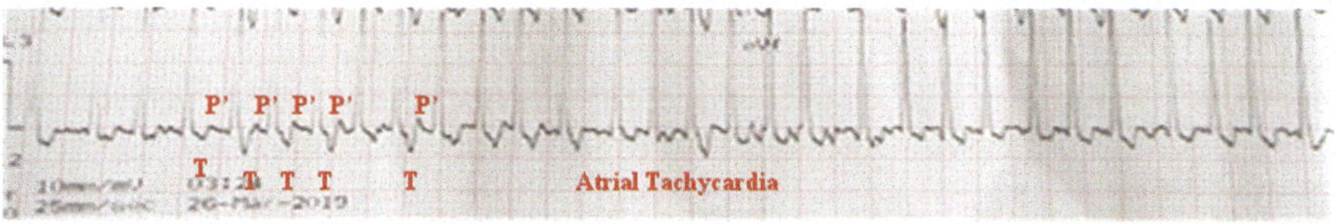

Fig. 10.11 Electrocardiogram (lead II, sensitivity 1, speed 25 mm/s) of a 6-year- and 4-month-old female Golden Retriever female with liver cirrhosis showing atrial tachycardia. Note all P waves are premature (P′)

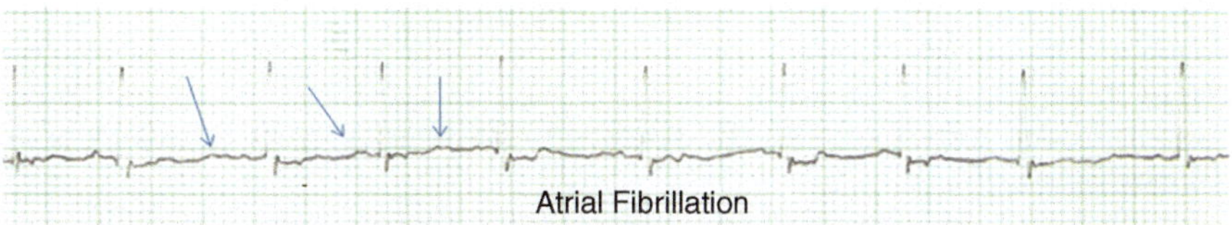

Fig. 10.12 Electrocardiogram (lead II, sensitivity 1, speed 25 mm/s) of a 3-year-old male Great Dane dog showing fine sawtooth oscillations ("f" wave), arrows, replacing well-formed "P" wave suggesting atrial fibrillation

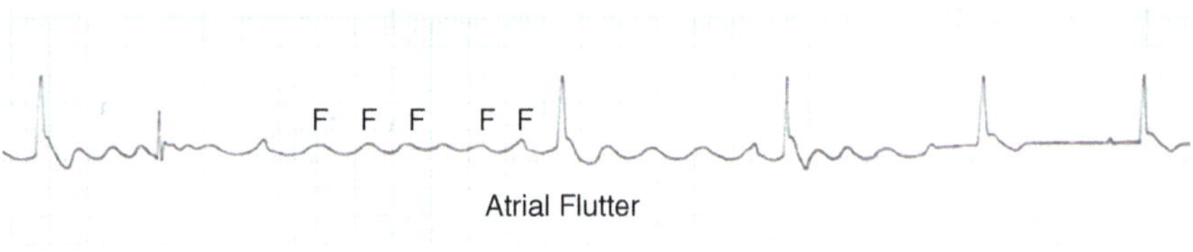

Fig. 10.13 Electrocardiogram (lead II, sensitivity 1, speed 25 mm/s) of a 7-year-old male Great Dane dog showing coarse sawtooth wave ("F" wave), suggesting atrial flutter

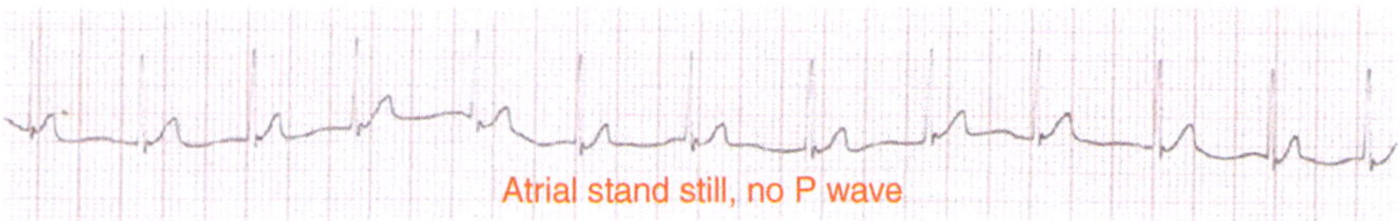

Fig. 10.14 Electrocardiogram (sensitivity 1, speed 25 mm/s) of a 4-year-old Boxer showing absence of "P" wave in all complexes suggesting atrial standstill

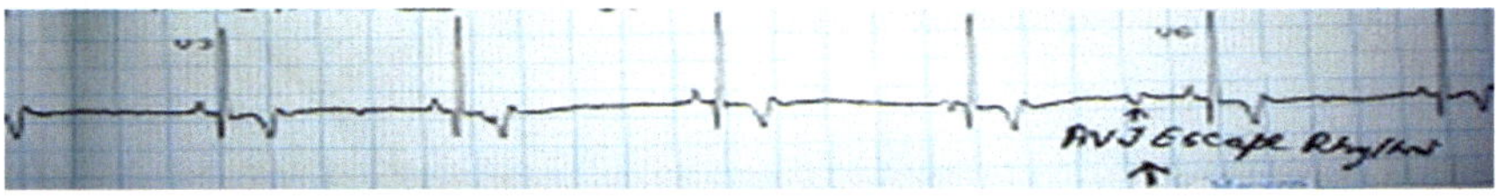

Fig. 10.15 Electrocardiogram (sensitivity 1, speed 25 mm/s) of a 3.5-year-old Pomeranian dog showing negative "P" wave (arrow) suggesting AV junctional premature complex (junctional escape rhythm)

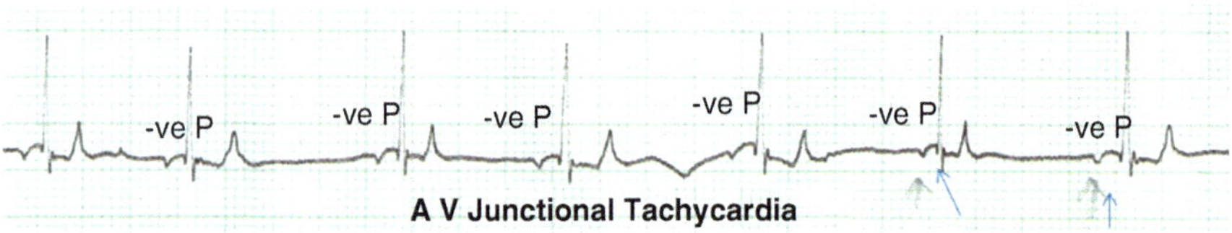

Fig. 10.16 Electrocardiogram (lead II, sensitivity 1, speed 25 mm/s) of a 5-year-old male Lhasa Apso dog showing atrioventicular junctional tachycardia at a rate of 140 bpm (>60 bpm). Note all "P" waves are negative

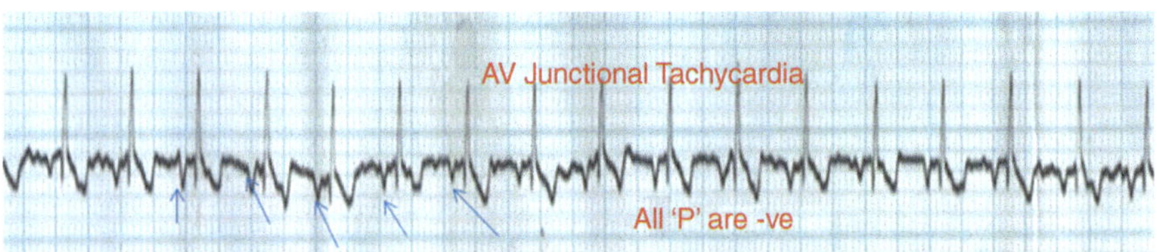

Fig. 10.17 Electrocardiogram (lead II, sensitivity 1, speed 25 mm/s) of an 11-year-old female Cocker Spaniel dog showing atrioventicular junctional tachycardia at a rate of 160 bpm. Note all "P" waves are negative

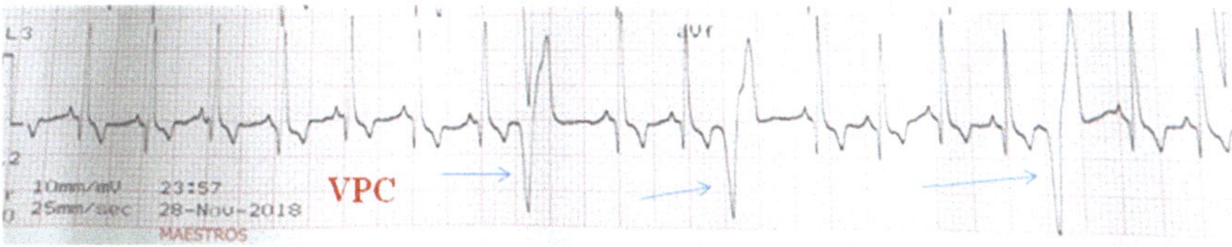

Fig. 10.18 Electrocardiogram (lead II, sensitivity 1, speed 25 mm/s) of a 2.5-year-old male German Shepherd showing Ventricular Premature complexes. Heart rate is 160 bpm. Normal rhythm is broken by VPC (arrow). Abnormal premature beat does not have "P" wave, its QRS is bizarre, and "T" wave is opposite of QRS

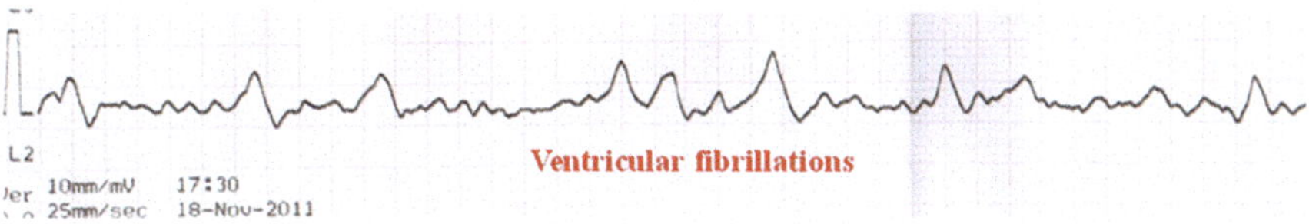

Fig. 10.19 Electrocardiogram (lead II, sensitivity 1, speed 25 mm/s) of a 13-year-old male Doberman Pincher showing series of fine baseline undulations and absence of specific complexes suggesting ventricular fibrillations

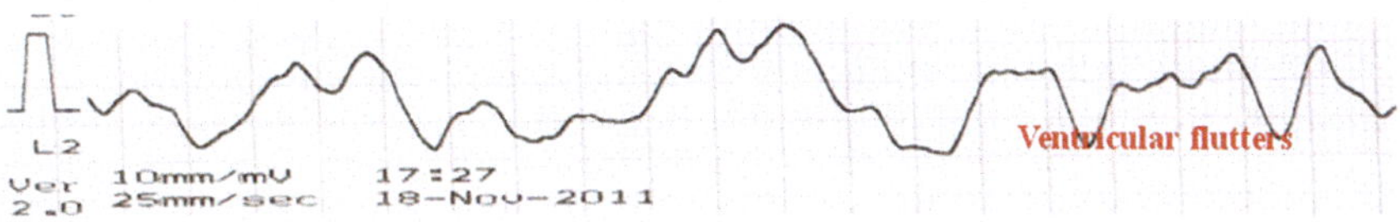

Fig. 10.20 Electrocardiogram (lead II, sensitivity 1, speed 25 mm/s) of a 7.0-year-old male Great Dane showing coarse baseline undulations and absence of specific complexes suggesting ventricular flutters

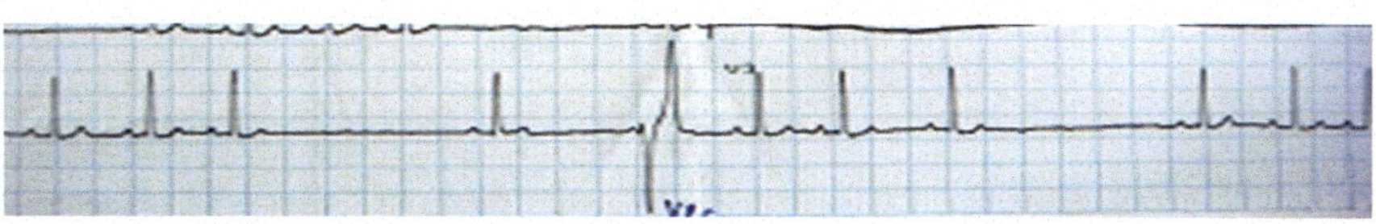

Fig. 10.21 Electrocardiogram (lead II, sensitivity 1, speed 25 mm/s) of a 10-year-old female Dachshund showing ventricular escape beat

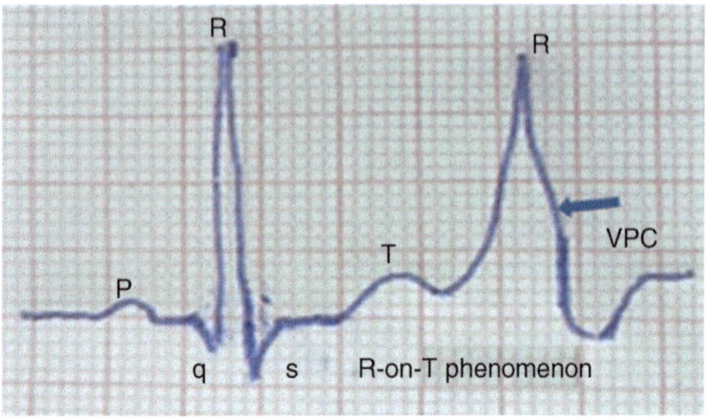

Fig. 10.22 Hand drawing showing "R"-on-"T" phenomenon. It is actually a ventricular premature complex. First complex is sinus complex, and second complex is a ventricular premature complex where ectopic beat "R" is superimposed on "T" of the first beat

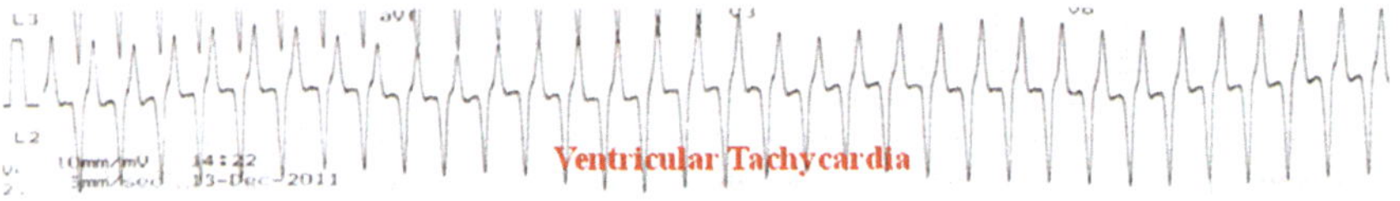

Fig. 10.23 Electrocardiogram (lead II, sensitivity 1, speed 25 mm/s) of an 8-year-old male Doberman showing ventricular tachycardia at a rate of 240 bpm. There is no normal complex in the strip

Fig. 10.24 Electrocardiogram (lead II, sensitivity 1, speed 25 mm/s) of a 5-year-old male Rottweiler showing ventricular asystole with severe AV block. Ventricular activity is seen only in the beginning of the strip. Thereafter only P waves are seen without any ventricular activity

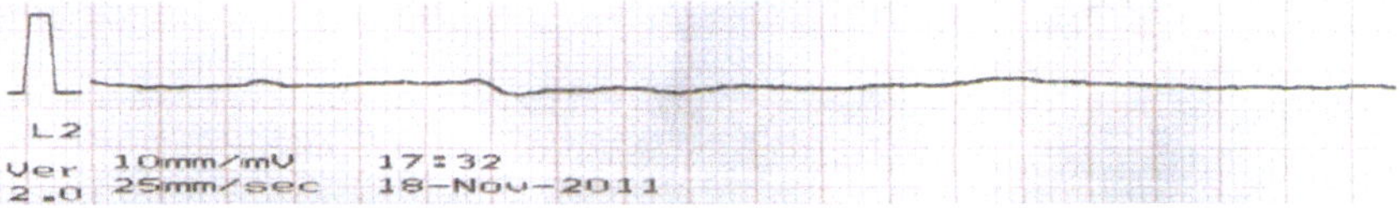

Fig. 10.25 Electrocardiogram (lead II, sensitivity 1, speed 25 mm/s) of a 7-year-old male German Shepherd showing ventricular asystole with severe complete AV block. There is no atrial and ventricular activity

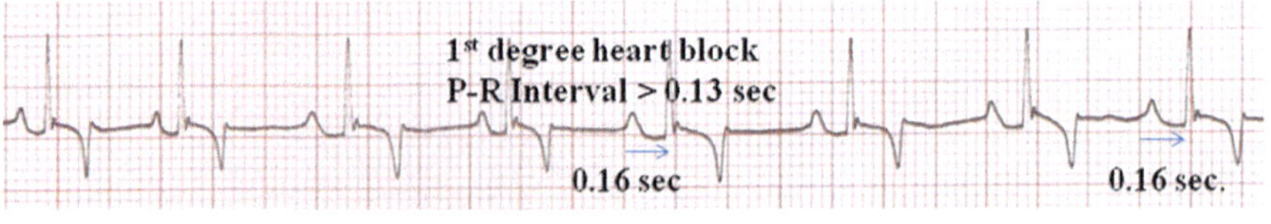

Fig. 10.26 Electrocardiogram (lead II, sensitivity 1, speed 25 mm/s) of a 5-year-old male bulldog showing first-degree AV block. P-R interval is >0.13 s

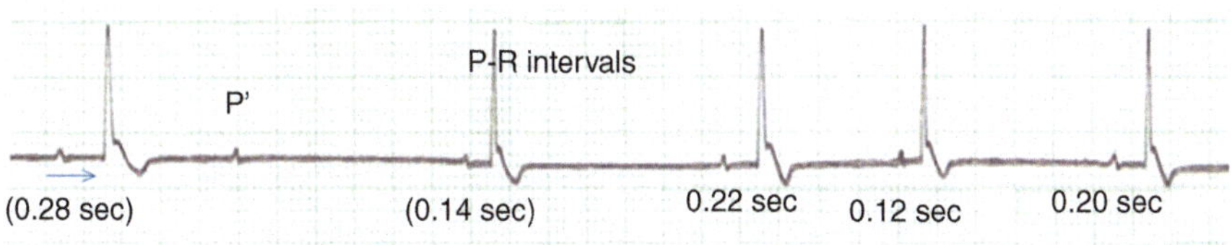

Fig. 10.27 Electrocardiogram (lead II, sensitivity 1, speed 25 mm/s) of a 5-year-old female Spitz dog showing second-degree AV block (Mobitz type I or type A). P′ without QRS complex and P-R interval is varying and is progressively prolonged before the blocked P′

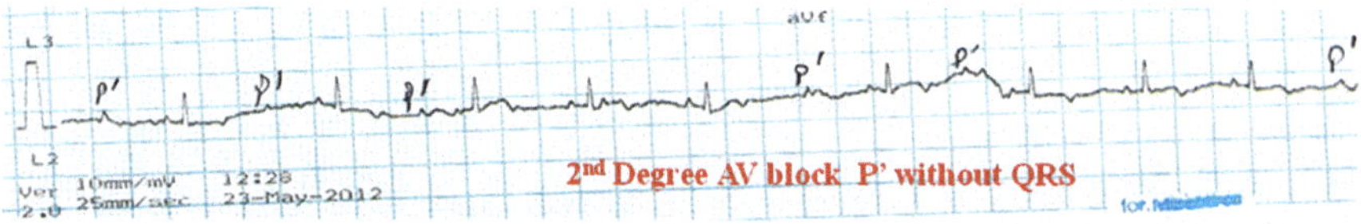

Fig. 10.28 Electrocardiogram (lead II, sensitivity 1, speed 25 mm/s) of a 4-month-old nondescript pup showing second-degree AV block (Mobitz type II or type B). P′ without QRS complex and PR interval preceding the blocked impulse is uniform. There is a fixed relationship between atrium and ventricle of 2:1 (2 P:1 QRS)

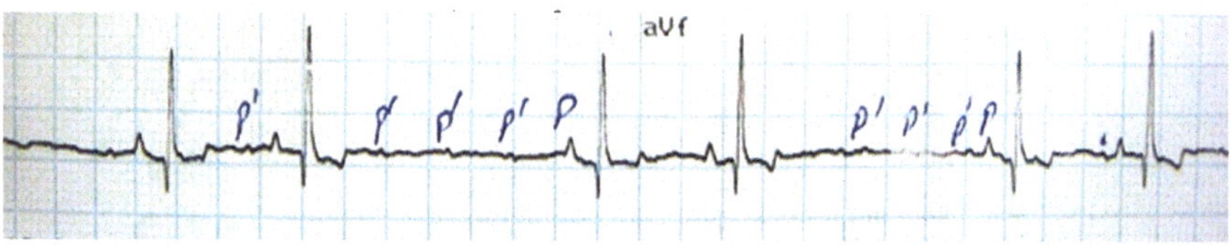

Fig. 10.29 Electrocardiogram (lead II, sensitivity 1, speed 25 mm/s) of a 2-year-old dog showing third-degree AV Block. P′ waves in series are blocked

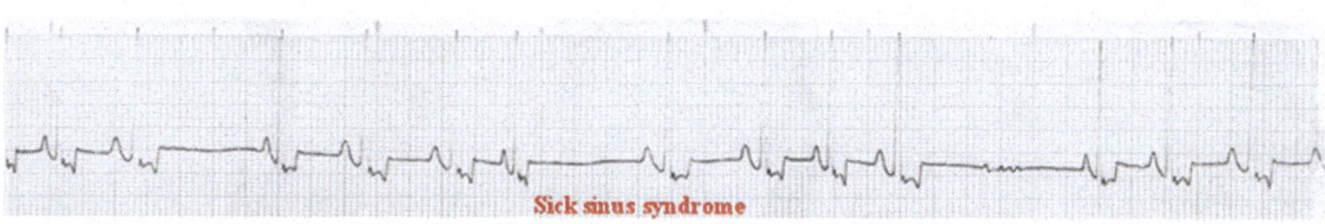

Fig. 10.30 Electrocardiogram (sensitivity 1, speed 25 mm/s) of a 4.5-year-old female Spitz dog with babesiosis (*B. gibsoni*) and ehrlichiosis (*E. canis*) showing sick sinus syndrome (tachycardia-bradycardia pattern)

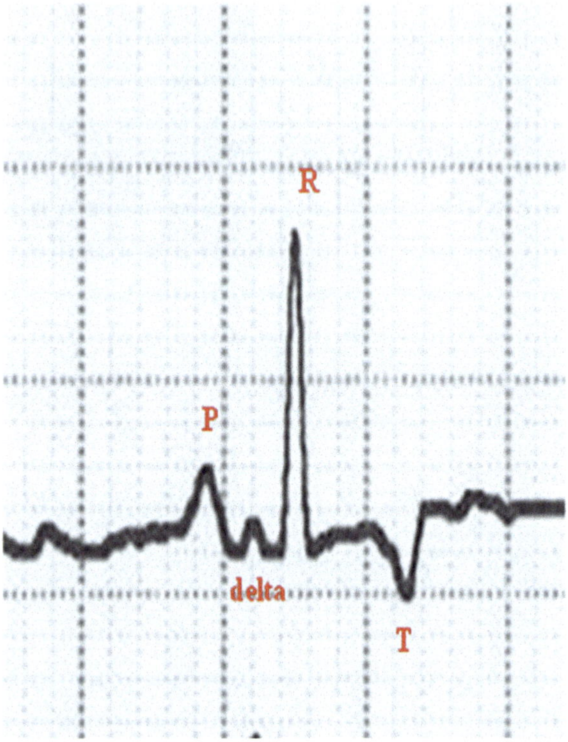

Fig. 10.31 Electrocardiogram (sensitivity 1, speed 25 mm/s) of a 13-year-old male nondescript dog with seizure episode showing delta wave in between P and R, sinus rhythm with normal heart rate (100 bpm), normal P (0.15 mV, 0.04 s), short P-R interval (0.06), normal R (0.8 mV, 0.04 s), normal S-T (0.08 s), and normal T wave (0.15 mV, 0.04 s) suggesting ventricular preexcitation. It is due to premature activation of a portion of ventricle by the SA node impulse without going to the AV node. When ventricular preexcitation is without tachycardia, it does not require any specific treatment

10.5 Clinical Manifestations of Arrhythmias

Symptoms in dogs with arrhythmias are not always classical. Generally whatever symptoms appear, these are mostly nonspecific. Sometimes there is no symptom. Tiring on exercise, weakness, dullness, and disorientation are observed in some arrhythmic dogs. These signs are also seen in so many diseases.

Arrhythmias	Clinical signs
Bradycardia	No appreciable symptom
	Weakness
	Exertion on exercise
	Severe bradycardia may have signs of forward failure (low cardiac output)
Sinus arrest	Marked weakness
	Syncope
	Sudden death

Arrhythmias	Clinical signs
Atrial standstill	Weakness
	Signs related with forward failure (low cardiac output)
	Syncope
	Sudden death
Second- and third-degree AV blocks	Weakness
	Fatigue
	Lethargy
	Third-degree blocks are serious
	Syncope
	Sudden death
Sinus tachycardia	May be asymptomatic
	Weakness
	Breathing difficulty
Atrial premature complex	No sign in isolated APCs
Sustained or paroxysmal atrial tachycardia	Episodes of weakness
	Ataxia
	Seizures
	Syncope
	Sudden death
Atrial flutter/fibrillation	Syncope
	Marked weakness
	Easy exertion
	May lead to congestive heart failure
	Irregular heart rate
	Ascites
	Dyspnea
	Coughing
Ventricular premature complex/ventricular tachycardia	No sign in isolated VPC
	Shock
	Congested mucous membranes
	Breathing difficulty
	Signs of congestive heart failure
	Weakness
	Dullness
	Exercise intolerance
	Seizure
	Collapse
	Coma
	Death
Bundle branch blocks	No clinical signs
	Signs of compromised heart are evident when blocks are associated with myocardial disease or left ventricular hypertrophy

10.6 Clinical Conditions Associated with Arrhythmias and Conduction Disturbances

Arrhythmias and conduction disturbances may be associated with various conditions or may be observed in cardiac or non-cardiac diseases. A brief account of arrhythmias and their associated conditions is given below.

Type of arrhythmia	Associated clinical conditions
Normal sinus rhythm	Normal rhythm of the heart varies with respiration (exercise increases and vagal stimulation decreases heart rate)
Sinus tachycardia	Physical exercise
	Pain
	Fever
	Hyperthyroidism
	Shock
	Anemia
	Infection
	Hypoxia
	Congestive heart failure
	Electric shock
	Drugs such as vasodilators, atropine, or epinephrine
Sinus bradycardia	Increased vagal tone
	Hypothyroidism
	Renal failure
	Tranquilizers (phenothiazine group), digitalis, quinidine, morphine
	Lesions of the central nervous system
	Hypothermia
	Cardiac arrest
Sinus arrhythmia	Respiration associated
Wandering pace maker	Sinus arrhythmia
	Normal in most cases
Atrial premature complexes	Atrial enlargement
	Right atrium hemangiosarcoma
	Tricuspid dysplasia
	Mitral valve insufficiency
	Patent ductus arteriosus
	General anesthesia
	Digitalis toxicity
	Diuretics such as furosemide causing hypokalemia
Atrial tachycardia	Atrial enlargement
	Digitalis toxicity
	Wolff-Parkinson-White syndrome
Atrial flutter	Atrial enlargement
	Ruptured chordae tendineae
	Atrial septal defect
	Tricuspid dysplasia
	Chronic mitral valvular fibrosis
	Wolff-Parkinson-White syndrome
	Common during cardiac catheterization

Type of arrhythmia	Associated clinical conditions
Atrial fibrillation	Atrial enlargement
	Patent ductus arteriosus
	Mitral insufficiency
	Tricuspid dysplasia
	Pulmonic stenosis
	Ventricular septal defects
	Digitalis toxicity
	Anesthesia
	Dirofilariasis
	Cardiac trauma
	Hyper tropic cardiomyopathy
	May occur without any cardiac disease
Atrioventricular junctional premature complexes	Digitalis toxicity
	Atrial enlargement
	Valvular insufficiency
	Right atrium hemangiosarcoma
	Tricuspid dysplasia
	Mitral insufficiency
	Patent ductus arteriosus
	General anesthesia
	Furosemide causing hypokalemia
Atrioventricular junctional tachycardia	Digitalis toxicity
	Cardiac disease
Escape rhythms	Secondary to abnormalities of impulse formation or conduction
	Digitalis toxicity
	Increased vagal tone
	Sick sinus syndrome
	Diseases causing sinus bradycardia
	Sinus arrest or AV block
Ventricular premature complexes	Seen in many healthy dogs
	Congestive heart failure
	Myocardial infarction
	Cardiac tumor
	Pericarditis
	Cardiomyopathy
	Traumatic myocarditis
	Boxer cardiomyopathy
	Doberman cardiomyopathy
	Hypoxia
	Anemia
	Uremia
	Pyometra
	Gastric dilatation-volvulus
	Parvovirus infection
	Pancreatitis
	Toxicities of digitalis, epinephrine, atropine, or anesthetic agents

Type of arrhythmia	Associated clinical conditions
Ventricular tachycardia	Same conditions as VPCs
Ventricular fibrillation	Electric shock
	Anoxia
	Shock
	Aortic stenosis
	Myocardial infarction or trauma
	Hypokalemia
	Hypocalcemia
	Alkalosis
	Heart surgery
	Hypothermia
	Myocarditis
	Increased sympathetic tone
	Reaction to halothane, ultrashort barbiturates, or digitalis
Ventricular asystole	Ventricular fibrillation
	Complete AV block
	Severe acidosis
	Severe hyperkalemia
Sinus arrest	Stimulation of vagus nerve (surgical manipulation, thoracic neoplasm, cervical neoplasm)
	Atrial dilatation
	Atrial fibrosis
	Hemangiosarcoma
	Toxicity of quinidine, propranolol, or digitalis
	Electrolyte imbalances
	Sick sinus syndrome
Atrial standstill	Underlying heart diseases
	Neuromuscular diseases associated with cardiomyopathy (myasthenia gravis, Duchenne muscular dystrophy in man)
	Temporary atrial standstill occurs with digitalis toxicity and hyperkalemia
	Terminal atrial standstill is associated with dying heart or severe hyperkalemia
	Persistent atrial standstill seen in English Springer Spaniels and Shih Tzu breeds in association with facioscapulohumeral-type muscular dystrophy
	Addison's disease
	Renal insufficiency (oliguric renal failure or obstructive uropathy)
	Untreated diabetic ketoacidosis
	Transfusion of stored blood
	Excessive potassium infusion
	Shock
	Excessive use of potassium-sparing diuretics
	Metabolic acidosis

Type of arrhythmia	Associated clinical conditions
First-degree AV block	May occur in healthy dogs
	Moderate to severe digitalis toxicity
	Reaction to propranolol, quinidine, or procainamide
	Conditions leading to hyper-/or hypokalemia
	Drugs like morphine or fentanyl
	Inflammation of the AV node
Second-degree AV block	May occur in normal young dogs
	Supraventricular tachycardia
	Microscopic idiopathic fibrosis
	Hereditary stenosis of the bundle of His in Pugs
	Digitalis toxicity
	Fentanyl toxicity
	Low doses of intravenous atropine
	Intravenous xylazine
	Quinidine toxicity
	Electrolyte imbalances
Third-degree AV block	Isolated congenital AV block
	Aortic stenosis
	Ventricular septal defect
	Severe digitalis toxicity
	Infiltrative cardiomyopathy
	Idiopathic fibrosis
	Hypertrophic cardiomyopathy
	Hyperkalemia or hypokalemia
	Myocardial infarction
	Bacterial endocarditis
	Arteriosclerosis
	Repeated cardiac punctures
	Aortic body tumor
	Severe disease of the AV node
WPW syndrome	Congenital and acquired cardiac defects
Sick sinus syndrome	Diseases affecting the SA node artery
	Replacement of the SA node with fibrous tissue
	Genetic inheritance
	Digitalis toxicity
Electrical alternans	Pericardial effusion
	Heart base tumors
	Metastatic carcinoma
	Right heart failure
	Benign idiopathic pericardial effusion
	Bundle branch blocks
	Supraventricular tachycardia
	Severe hypocalcemia can cause isolated "T" wave alternans
Right bundle branch block	Eccentric or concentric hypertrophy of the right ventricle
Left bundle branch block	Severe myocardial disease

10.7 Clinical Management of Arrhythmias

Persisting severe cardiac arrhythmia decreases cardiac output and changes arterial blood pressure and coronary circulation leading to deterioration of health, exercise intolerance, and eventual death. If the dog is already having compromised cardiac functioning, arrhythmias cause more severe and pronounced hemodynamic changes. Therefore, management of arrhythmias is very vital in dogs with symptomatic arrhythmias to restore normal hemodynamics and save the life of the ailing dog. Nowadays various anti-arrhythmic agents are available for the treatment of arrhythmia in dogs. Arrhythmia management consists of physiological manipulations (ocular pressure, carotid sinus pressure); alteration of receptors (alpha and beta adrenergic, dopaminergic, histamine, purinergic); correction of acid-base, electrolyte, and fluid imbalances; oxygen therapy; drug therapy (vagomimetic, anticholinergic, sympathomimetic, alpha and beta adrenergic blocking agents, anti-arrhythmic drugs, calcium channel blocking agents, inotropic agents); and other therapies (rest, blood transfusion, steroids, surgery) as per individual requirement of the case.

Because of narrow therapeutic margin of anti-arrhythmic drugs, these drugs may cause both beneficial and deleterious effects. After ingestion the drugs are absorbed, distributed, and eliminated depending on their plasma concentration. Prolonged administration of anti-arrhythmic drugs requires dose titration, When anti-arrhythmic drugs are used with antacids, their absorption is decreased. Oral lidocaine therapy is totally ineffective because of pre-systemic elimination.

General considerations for clinical management of arrhythmias

1. A precise diagnosis of arrhythmia is a must.
2. Arrhythmias of non-cardiac etiology owing to acid-base or electrolyte imbalances, hypothermia, hypovolemia, or hypoxia require treatment of primary condition. In such cases anti-arrhythmic drugs are not required, and the aim is to address primary condition first.
3. Development of arrhythmias on administration of digitalis, quinidine, procainamide, xylazine, or acetylpromazine therapy requires discontinuance of these drugs.
4. Anti-arrhythmic drugs need to be selected as per the underlying etiology of the arrhythmia.
5. Anti-arrhythmic drugs should be used in proper dose schedule.
6. Synergistic or antagonistic effects of anti-arrhythmic drugs should be kept in mind while using combination therapy.
7. Management of some arrhythmias may require anti-arrhythmic drugs in addition to other therapeutic approaches. For example, dogs with sick sinus syndrome having bradyarrhythmia-tachyarrhythmia may require a permanent cardiac pace maker for bradyarrhythmia and anti-arrhythmic drugs for tachycardia.

A summarized view of various arrhythmias and their management is given below for the benefit of clinicians.

10.7.1 Sinus Bradycardia

Usually treatment for sinus bradycardia is not attempted until and unless sinus bradycardia is associated with clinical signs such as weakness or collapse. Cases with acute bradycardia may respond to atropine, glycopyrrolate, dopamine, or epinephrine. If atropine fails, isoproterenol can be used to increase heart rate. Primary abnormality needs to be attended first. In long-standing cases, pace maker implant may be the only solution.

10.7.2 Sinus Tachycardia

Usually no specific treatment is required for the management of sinus tachycardia without any symptom. Initially attention is to be paid to underlying primary problem. If congestive heart failure is there, digitalis/digoxin is the drug of choice.

10.7.3 Sinoatrial Arrest

When sinoatrial arrest is associated with no clinical signs, no treatment is recommended. If syncope occurs in brachycephalic dogs and suspected to be due to vagal stimulation, atropine and glycopyrrolate are the drugs of choice.

10.7.4 Sick Sinus Syndrome

Transvenous pacemaker implant along with anticholinergic therapy is recommended.

10.7.5 Atrial and Junctional Premature Complexes

Digitalis/digoxin is the drug of choice for the management of supraventricular premature beats. In most dogs with supraventricular premature beats, there may be severe mitral insufficiency leading to congestive heart failure or making them prone to it. Dogs with mitral insufficiency need to be digitalized irrespective of signs of cardiac decompensation. Quinidine and procainamide should be avoided in the management of supraventricular arrhythmias, though these are effective in controlling supraventricular premature beats but cause depressing action on the myocardium.

If supraventricular premature beats arise after digitalization, digitalis administration should be discontinued.

10.7.6 Atrial, Junctional, Supraventricular Tachycardia

In these arrhythmias, vagal maneuver, digitalis, propranolol, verapamil, and quinidine are used.

10.7.7 Atrial Fibrillation

Quinidine, digoxin, propranolol, and verapamil are the drugs of choice for the management of atrial fibrillation. Digoxin is commonly preferred for all supraventricular arrhythmias owing to its ability to improve congestive heart failure and control heart rate by delaying conduction through AV node. If ascites are present, diuretics are to be considered with low-sodium diet and rest. When heart rate and congestive heart failure are controlled, quinidine is added in the treatment protocol to restore chaotic atrial rhythm to sinus rhythm. Some clinicians prefer diltiazem (calcium channel blocker) for the management of atrial fibrillation. A combined therapy with digoxin and diltiazem may be a better option than either of these drugs used alone.

10.7.8 Atrial Flutter

Atrial flutter comes under the category of supraventricular tachycardia, and digitalis/digoxin is the drug of first choice for its clinical management. After stabilizing heart rate, quinidine is added to restore sinus rhythm.

10.7.9 Ventricular Premature Complexes

Lidocaine or procainamide is used in acute cases. Procainamide, quinidine, propranolol, and phenytoin are used in the management of chronic cases subject to response to drugs used in acute phase. Quinidine or lidocaine therapy is highly effective, but response may vary from dog to dog. When quinidine and procainamide have failed, arrhythmia can be controlled with Dilantin. When VPC occurs >20/min and present as doublets, triplets, multiform, as runs (>3 in sequence) or presence of R-on-T phenomenon, treatment should be started immediately.

10.7.10 Ventricular Tachycardia

Lidocaine, propranolol, quinidine, or combination therapy is used.

10.7.11 Ventricular Fibrillation

It may be a terminal event of a severe illness. Internal or external cardiac massage should be started without loss of time. Dog with ventricular fibrillation should be given oxygen therapy immediately and should be put on a rapid sodium bicarbonate drip. Administration of epinephrine converts fine fibrillation to coarse ones which are suitable for defibrillation. After heart massage, sodium bicarbonate, and epinephrine, defibrillation may again be attempted. Calcium may also be considered for improving myocardial tone.

Bretylium tosylate may be helpful in the management of ventricular fibrillation as its anti-arrhythmic properties are due to its antiadrenergic property.

10.7.12 Atrial Standstill (Hyperkalemia)

In dogs showing atrial standstill due to hyperkalemia, treatment should be initiated with sodium bicarbonate, intravenous saline, calcium gluconate, and treatment of primary disorder. Hyperkalemia is generally treated with intravenous dextrose saline, intravenous or intramuscular hydrocortisone, intramuscular desoxycorticosterone acetate, intravenous sodium bicarbonate, and regular intravenous or subcutaneous insulin.

10.7.13 Persistent Atrial Standstill

Atropine sulfate or glycopyrrolate is used initially followed by pace maker transplant. Terbutaline may also be considered.

10.7.14 Atrioventricular Block

First- or second-degree heart block does not warrant any treatment in asymptomatic dogs. If heart block of any degree appears after digitalis administration, it should be discontinued temporarily till the arrhythmias are corrected. In case of exaggerated vagal tone, these arrhythmias appear without any apparent cause. In this situation atropine may provide the proof of vagal stimulation and may be treated accordingly. Atropine, glycopyrrolate, transvenous pace maker, or infusion of isoproterenol is considered. Incomplete heart blocks may also be treated with diphenylhydantoin.

Complete heart block poorly responds to medical management. Isoproterenol is the drug of choice for complete heart block. It does not relieve the atrioventricular block, but is used to increase ventricular rate. Some medicines such as digitalis, quinidine, and potassium are contraindicated in heart block situations. Any anti-arrhythmic drug (even lidocaine or diphenylhydantoin) should not be used in case of complete heart block.

10.7.15 Ventricular Escape Beats and Rhythms

If these changes are due to hyperkalemia, correction of hyperkalemia will improve the rhythm. In case complete heart block is of idiopathic nature, atropine, isoproterenol, or an artificial pace maker will increase ventricular rate.

10.7.16 Atrioventricular Accessory Pathway Arrhythmia

A new approach called radiofrequency catheter ablation (RFCA) has been developed by Wright et al. (2018) to treat atrioventricular accessory pathway arrhythmia in dogs. The technique has been borrowed from human medicine. RFCA techniques uses radiofrequencies to destroy aberrant circuits and restore normal functioning of the heart.

10.7.17 Anti-arrhythmic Drugs

Various drugs are available for the management of arrhythmias. These drugs have been classed into various categories based on their electrophysiological effects on the cardiac cells as per Vaughan Williams classification introduced in 1970 (Willium 1970). Class I drugs slow down conduction and decrease cardiac muscle automaticity and excitability. Class II drugs inhibit the effect of catecholamine on heart. Class III drugs are very good for the management of reentrant arrhythmias, and their action is by prolonging the effective refractory period of cardiac action potential without decreasing conduction velocity. Class IV drugs block the calcium entry.

10.7.17.1 Lidocaine

It is a class I anti-arrhythmic drug which depresses automaticity in the Purkinje fibers. It produces dramatic decrease in conduction velocity in hypoxic cardiac tissues. It is a popular anti-arrhythmic drug in intensive care units. It shortens Q-T interval without affecting P-R interval. The drug is indicated in the management of ventricular arrhythmias. The lidocaine is used at 2–4 mg/kg by slow intravenous injection and can be repeated as per need up to a total dose of 8 mg/kg. The drug is ineffective in the management of supraventricular arrhythmias and is contraindicated in the dogs with sick sinus syndrome, second- or third-degree heart block, and idioventricular rhythm with no evidence of "P" wave.

10.7.17.2 Procainamide

It is also a class I anti-arrhythmic drug having little or no effect on the SA or AV node and pace maker current. In rational dose it produces insignificant prolongation of the QRS and Q-T intervals. It is used at the dose rate of 6–8 mg/kg intravenously over a period of 5 min. For sustained effect, initial dose should be followed by intermittent intramuscular or oral routes. Procainamide is recommended in the management of acute atrial flutter, atrial fibrillation, and sick sinus syndrome. The drug may cause hypotension that can be attended with intravenous fluids and dopamine.

10.7.17.3 Quinidine

It is also a class I anti-arrhythmic drug. It depresses impulse conduction in the heart and repolarization of the working myocardium and Purkinje fibers. It prolongs the P-R and Q-T intervals and widens QRS. It is used in the clinical management of ventricular premature complexes, ventricular tachycardia, acute atrial fibrillation, and refractory supraventricular tachycardia. Dose rate of quinidine is 6–16 mg/kg. Quinidine is frequently used orally and occasionally intramuscularly and in special situations intravenously.

10.7.17.4 Beta Adrenergic Blockers

Propranolol, atenolol, and metoprolol are the beta adrenergic blocking drugs commonly used for the management of canine arrhythmias. Propranolol is a class II anti-arrhythmic drug that is relatively nonselective. It acts through stabilizing effects like quinidine. The drug is effective in reducing heart rate caused by increased sympathetic tone or by thyrotoxicosis. But sinus tachycardia caused by hypokalemia, fever, or abnormal form of automaticity does not respond favorably to propranolol. Propranolol can be used in the management of supraventricular arrhythmias. It is used at the dose rate of 0.04–0.06 mg/kg intravenously slowly or 0.2–1.0 mg/kg orally thrice daily.

10.7.17.5 Phenytoin

Its effect is similar to lidocaine but is more effective in the management of supraventricular tachycardia. The drug is also effective in digitalis-induced arrhythmias. Because of erratic absorption, its systemic bioavailability is only 40%, hence generally not preferred in the management of arrhythmias.

10.7.17.6 Calcium Channel Blockers

Verapamil, calcium channel blocking drug, falls under the category of class IV anti-arrhythmic drugs. It inhibits the entry of calcium ion into arterial smooth muscles as well as myocytes and conduction tissues. Verapamil is the drug of choice for the management of paroxysmal supraventricular tachycardia, atrial flutter, atrial fibrillation, and extrasystole. It is used at 0.05–0.15 mg/kg intravenously. The dose is titrated by administrating 0.05 mg/kg IV every 5–30 min. In dogs with congestive heart failure, it can be combined with digoxin. The drug has also been used to control atrial arrhythmias in cats with hypertrophic cardiomyopathy. The most common side effect of verapamil is hypotension. It should not be used in dogs with hypersensitivity to verapamil and in pregnant bitches.

10.7.17.7 Vagolytic Drugs

Atropine and glycopyrrolate are commonly used as vagolytic drugs in the management of sinus bradycardia, SA arrest, and incomplete AV block. Atropine is used at the dose rate of 0.02 mg/kg intramuscularly or intravenously to block excessive vagal tone. After the effect is established, atropine may be given orally. Its prolonged use is associated with side effects. Glycopyrrolate (0.005–0.01 mg/kg intravenously or intramuscularly), isopropamide (2.5–5.0 mg orally twice or thrice daily), and isoproterenol (0.4 mg in 250 ml DNS through slow intravenous drip to effect) are the other vagolytic drugs advocated in the management of SA arrest, AV block, or bradycardia.

10.7.17.8 New Drugs

- Amiodarone—It has been found effective for both atrial and ventricular arrhythmias in man and has also been tried in the management of ventricular premature complexes, ventricular tachycardia, paroxysmal atrial fibrillation, and Wolff-Parkinson-White syndrome. Its initial trial in dogs has been disappointing.
- Mexiletine—It is a class IB anti-arrhythmic drug similar to lidocaine in action. It is good for continued anti-arrhythmic therapy and is effective in the clinical management of ventricular premature complexes at 0.25–8.0 mg/kg orally twice or thrice daily.
- Tocainide—It is class I type anti-arrhythmic drug and is synthetic analogue of lidocaine. It is of not much use in canine.

10.7.18 Homeopathic Drugs in the Management of Canine Arrhythmias

Homeopathic drugs are being used in human and veterinary practice as an alternative/or complementary medicine for the management of many diseases. A few reports have described the beneficial effects of digitalis and Abies nigra in the management of arrhythmias in canines.

- Digitalis—It has been considered a drug of first choice in the management of diseases where the heart is primarily involved with weak and irregular pulse in humans. Varshney and Chaudhari (2007) reported beneficial effect of homeopathic digitalis 6 C (four drops orally QID for 7 days) in two adult dogs with atrial paroxysmal tachycardia.
- Abies nigra—It is reported to be a unique drug that can be used in the management of both tachycardia and bradycardia in humans. Changkija and Varshney (2007a, b, c) tried Abies nigra 30 C at four drops QID PO in the management of tachycardia, bradycardia, and sick sinus syndrome in a few dogs and found it effective in restoring heart rate.

Table 10.3 Anti-arrhythmic drugs, their indications, and doses at a glance

Drugs	Indications	Dose	Commonly used preparations/brand names
Amiodarone	V. arrhythmias, V. fibrillation	10–15 mg/kg orally BID for 7 days, then 5–7.5 mg/kg orally BID for 14 days, and then 7.5 mg/kg orally OD as maintenance	Inj. Aldarone, inj. Amiodarone, inj. Eurythmic, inj. Duron (50 mg/ml), tab. Duron, Amiodar, Eurythmic (100/200 mg)
Atenolol	Supraventricular premature complex Tachycardia Atrial fibrillation VPC	0.4–1.0 mg/kg orally q 24 hourly	Beta, B-bloc, Aten, Altol, Aloten, Betacard, Telol, Atcom, Atekind (25, 50, 100 mg tab.)
Atropine sulfate	Sinus bradycardia Sinoatrial arrest Incomplete AV block	0.01–0.02 mg/kg IM, IV 0.02–0.04 mg/kg SQ IM	Inj. Atro (0.6 mg/ml) Inj. Atropine sulph Inj. Tropine
Bretylium	Refractory V. tachycardias Recurrent V. fibrillation in humans but not in dogs	2.0–6.0 mg/kg IV subsequent dose 1–2 h interval	Inj. Bretylium Inj. Bretylol Inj. Bretylate 50 mg/ml
Calcium gluconate	Ventricular asystole	10% sol. 1 ml/2.5 kg IV	Calcium Sandoz (10%) 10 ml injection
Digoxin	Supraventricular premature complex Supraventricular tachycardia, atrial fibrillation/flutter	0.01–0.03 mg/kg IV half dose IV, 1/4th dose after a wait of 30–60 min, rest dose again after 30 min Oral maintenance 0.01–0.02 mg/kg in two divided doses	Digox, Cardioxin, Lanoxin (0.25 mg tab. or 0.25 mg/ml inj.)
Diltiazem	Supraventicular arrhythmias	0.4–1.4 mg/kg orally TID	Dilcal, Channel, Dili, Dilgard, Iski, DTM (30, 60 mg tab. SR tabs of 90, 120 mg)
Disopyramide	V. arrhythmias	Dog >18 kg: 100 mg orally TID or QID	Norpace, Regubeat (100, 150 mg cap.)
Esmolol	Supraventricular Premature comp. Tachycardia Atrial fibrillation VPC	200–500 μg/kg IV over 1 min (loading dose) followed by 25–200 μg/kg/min	Esocard, Miniblock (100 mg or 250 mg/10 ml inj.)

(continued)

Table 10.3 (continued)

Drugs	Indications	Dose	Commonly used preparations/brand names
Glycopyrrolate	V. escape complex Sinus bradycardia Second-degree heart block SA arrest	0.005–0.01 mg/kg IV, IM 0.01–0.02 mg/kg SQ	Glyco-P, Pyrolate (0.2 mg/ml inj.)
Isopropamide iodide	Sinus bradycardia SA block AV block	0.05–0.3 mg /kg orally BID or TID	Gastabid Stelbid (5 mg tab.)
Isoproterenol	Bradycardia AV block Cardiac arrest	0.04–0.08 μg/kg/min IV infusion or 0.1–0.2 mg SQ IM 4 hourly or 0.4 mg in 250 ml 5% D/W IV slowly	Not available in market in India
Lignocaine	Life-threatening arrhythmias	2.0–8.0 mg/kg slow IV over 2 min Initial dose 40–80 μg/kg CRI (follow-up)	Gesicain Gesicard Xylocaine Xylocard (2% sol. inj.)
Metoprolol	Atrial fibrillation	0.5–1.0 mg/kg orally TID	Betaloc, Lopresor, Metolar (50 or 100 mg tab., 5 ml inj. of 1 mg/ml)
Mexiletine hydrochloride	Symptomatic V. arrhythmia	2.0–8.0 mg/kg orally BID or TID	Mexitil (50 or 150 mg cap., or 250 mg inj.)
Morphine sulfate	Supraventricular premature beats	0.2 mg/kg IM, SQ as needed	Morcontin (CR), Relimorf SR (10, 30, 60, 100 mg tab.), morphine sulfate (inj. 10 mg/ml)
Nadolol	-------do-------	Initial—0.2 mg/kg orally TID or BID (max. up to 1.0 mg/kg)	Not available
Phenytoin	V. arrhythmias caused by myocardial ischemia or digitalis	2.0–4.0 mg/kg IV in increments max. 10 mg/kg	Dilantin, Phenytoin Epsolin (50 mg/ml inj.)
Propranolol	Supraventricular premature com. Tachycardia, atrial fibrillation, ventricular premature complex	0.04–0.06 mg/kg IV slow 0.2–1.0 mg/kg orally TID	Ciplar, Propal, Inderal, Beta bloc, Lol, Lol-SR, Propal (10, 20, 40 mg tab.)

(continued)

Table 10.3 (continued)

Drugs	Indications	Dose	Commonly used preparations/brand names
Procainamide	VPC, V. tachycardia	6.0–20.0 mg/kg QID/TID (SR form) 6.0–8.0 mg/kg IV over 5 min 500–1000 mg in 500 ml 5% D/W, slow IV infusion at 10–40 μg/kg/min or to effect	Pronestyl caps. or injection
Propantheline	Sinus bradycardia	0.5–1.0 mg/kg orally TID	Pro-Banthine (15 mg tab.)
Quinidine	VPC, V. tachycardia	6.0–20 mg/kg orally TID or QID until loading dose controls arrhythmia	Quinidine Natcardine (100, 200 mg tab.)
Sotalol	V. arrhythmias	1.0–2.0 mg/kg orally BID	Sotagard (40 or 80 mg tab).
Tocainide	Ventricular arrhythmias	15–30 mg/kg orally BID or TID	Tonocard Tocolol Mexitil Tocain
Verapamil	Supraventricular tachycardia specially preexcitation type	1.0–4.0 mg/kg orally TID or BID 0.1–0.3 mg/kg slow IV 2–10 μg/kg/min IV infusion	Calaptin, VPL Isoptin (40, 80 mg tab. or 2.5 mg/ml inj.)

References

Changkija B, Varshney JP (2007a) Clinical management of tachycardia in dogs with abies nigra. In: Varshney JP, Swaminarayan S (eds) Research findings homeopathic bioefficacy and management of animal health. Sintex International Limited, Kalol, pp 163–166

Changkija B, Varshney JP (2007b) Clinical management of bradycardia in a nondescript dog with abies nigra—a case report. In: Varshney JP, Swaminarayan S (eds) Research findings homeopathic bioefficacy and management of animal health. Sintex International Limited, Kalol, pp 167–170

Changkija B, Varshney JP (2007c) Sick sinus syndrome in a spitz dog and its management with abies nigra. In: Varshney JP, Swaminarayan S (eds) Research findings homeopathic bioefficacy and management of animal health. Sintex International Limited, Kalol, pp 171–174

Engel TR, Meister SG, Frankl WS (1978) The "R-on-T" phenomenon: an update and critical review. Ann Intern Med 88:221–225

Smirk FH (1949) R waves interrupting T waves. Br Heart J 11:23–36

Varshney JP, Chaudhuri S (2007) Atrial paroxysmal tachycardia in dogs and its management with homeopathic digitalis—two case report. Homeopathy 96:270–272

Varshney JP, Sutaria P, Deshmukh VV, Chaudhary PS (2013) Prospective study of cardiac arrhythmias—a survey of 20000 canines. Intas Polivet 14:129–136

Willium VM (1970) Classification of anti-arrhythmic drugs. In: Symposium on cardiac arrhythmias. Astra, Elsinore

Wright KN, Connor CE, Irvin HM, Knilans TK, Webber D, Kass PH (2018) Atrioventricular accessory pathway in 89 dogs. Clinical features and outcome after radiofrequency catheter ablation. J Vet Int Med 32:1517–1529

Further Reading

Allen DG (1991) Small animal medicine. J.B. Lippincott Company, Philadelphia, PA

Birchard SJ, Sherding RG (2000) Saunders manual of small animal practice, 2nd edn. W.B. Saunders Company, Philadelphia, PA

Boericke W (2001) Pocket manual of homeopathic materia medica, 9th edn. India Books and Periodicals Publishers, New Delhi

Bolton GR (1975) Handbook of canine electrocardiography. W.B. Saunders Company, Philadelphia, PA

Ettinger SJ, Feldman EC (2000) Textbook of veterinary internal medicine. Diseases of the dog and cat, 5th edn. W.B. Saunders Company, Philadelphia, PA

Kumar A, Varshney JP (2005) Atrial standstill in a German Shepherd dog. Indian Vet Med J 29:225–226

Morgan RV (1992) Handbook of small animal practice, 2nd edn. W.B. Saunders Company, Philadelphia, PA

Nelson RW, Couto CG (1998) Small animal internal medicine, 2nd edn. C.G. Mosby, St. Louis, MO

Tilley LP (1985) Essentials of canine and feline electrocardiography, 2nd edn. Lea and Febiger, Philadelphia, PA

Electrocardiographic Findings in Cardiac and Non-cardiac Diseases

11

11.1 Electrocardiographic Findings in Cardiac Diseases

Diseases	ECG findings
Aortic stenosis	Arrhythmias
	ECG changes are related to left ventricular hypertrophy
Subaortic stenosis	ECG may be normal
	In severe cases "R" wave amplitude is increased (more than normal limits)
	There is left axis deviation (MEA < 40°)
	S-T segment may be depressed
	"T" wave changes are also conspicuous
Pulmonic stenosis	Axis is deviated toward right (MEA > 104° on frontal plane)
	ECG changes are related to right ventricular hypertrophy
	(There is large S wave in leads I, II, III)
	"S" wave is deep (in left precordial chest leads)
Mitral valve stenosis	"P" wave may be tall (>0.4 mV) and broad (>0.04 s)
	Supraventricular premature complexes may be seen
	Tachycardia may be conspicuous
	Atrial fibrillation/flutter may occur
	ECG changes related to right ventricular enlargement may be seen
Atrial septal defect	ECG may show right heart enlargement pattern
Ventricular septal defect	Biventricular enlargement pattern may be seen in ECG
Patent ductus arteriosus (left to right shunting)	ECG may show left ventricular enlargement pattern with a normal QRS axis on frontal plane
	"R" wave voltage is increased in leads II, III, aVF, V2, and V4
	Presence of broad "P" wave (>0.4 s) is suggestive of left atrial enlargement

J. P. Varshney, *Electrocardiography in Veterinary Medicine*,
https://doi.org/10.1007/978-981-15-3699-1_11

Diseases	ECG findings
Patent ductus arteriosus (with pulmonary hypertension)	ECG may show right ventricular enlargement pattern
	There may be right axis deviation (MEA > 104°)
	Deep "S" wave may be seen in leads I, II, and III
Chronic mitral insufficiency (CMI)	ECG may be normal or abnormal
	"P" wave may be broad (>0.04 s), tall (>0.4 mV), and notched (lead II)
	"R" wave amplitude is increased in lead II (more than 2.5 mV)
	CV6LU (more than 3.0 mV), and CV5RL (more than 0.5 mV)
	"Q" and "S" wave amplitudes are increased (more than 0.5 mV) in leads I, II, and III
	"QRS" is broad
	"T" wave depression may be evident
	S-T segment abnormalities are common
	Supraventricular premature beats may occur
	Ventricular premature complexes may be seen
	Atrial fibrillation may be evident
	Paroxysmal supraventricular tachycardia may occur
Tricuspid insufficiency (TI)	Lead II may show broad "P" with/without increase in amplitude
	P-R interval is increased (more than 0.14 s)
	"Q" and "S" waves show increased depth in leads II, III, aVF, and CV6LU
	"QRS" is broad (more than 0.06 s)
	"R" wave amplitude is decreased in leads II, III, aVF, and CV6LU
	"R" wave amplitude is increased in leads aVR and CV5RL
	Arrhythmias are not very common
Aortic insufficiency (AI)	Changes similar to chronic mitral insufficiency (CMI)
Pulmonary valve insufficiency (PI)	Lead II may show broad "P" with/without change in amplitude
	P-R interval is increased (more than 0.14 s)
	"R" wave amplitude is decreased in leads II,III, aVF, and CV6LU
	"S" wave is deep in leads II, III, aVF, and CV6LU
	"QRS" is broad
	Some dogs with congenital PI may show normal ECG
	Arrhythmias are not very common
Bacterial endocarditis	ECG may be normal or abnormal
	There may be sinus tachycardia

Diseases	ECG findings
	S-T segment abnormalities (elevation or depression) are common
	"Q" wave in lead II is deep
	ECG may show premature ventricular complexes/ventricular
	Tachycardia or AV blocks
	"R" wave amplitude is increased in leads II, avF, CV6LL, and CV6LU
	"QRS" is broad
	"P" wave duration is increased
Dilated cardiomyopathy (DCM)	Rhythm is sinus
	"P" may be broad and tall
	Arrhythmias (atrial fibrillation, ventricular premature complexes, ventricular tachycardia, "R"-alternans) may be present
	"QRS" is broad
	S-T segment may show slurring
Hypertrophic cardiomyopathy (HCM)	Atrioventricular blocks are seen
	Bundle branch blocks may be present
	Left atrium is enlarged ($P > 0.04$ s)
	Left ventricle is hypertrophied
Secondary myocarditis	S-T may show slurring, depression, or elevation in leads II, III, aVF, V2 and V4
	Atrioventricular (AV) block may be present
	Bundle branch block may be present
	Ventricular arrhythmias such as ventricular premature complexes or ventricular tachycardia may be present
Infective myocarditis	S-T segment abnormalities are common
	Conduction disturbances are common
Pericardial effusions	"R" wave amplitude is decreased in all leads including chest leads (<1.0 mV)
	Electrical alternans (varying amplitude of "R" wave) is commonly seen
	Rhythm may be sinus or arrhythmic (tachycardia)
Constrictive pericarditis	Low-voltage complexes are common
	"P" wave may be broad (more than 0.04 s)
	Rhythm may be sinus or arrhythmic
	Supraventricular arrhythmias may be seen
Restrictive cardio-myopathy	Atrial fibrillation may be seen
	Ventricular arrhythmias may be seen

11.2 Electrocardiographic Findings in Non-cardiac Diseases

11.2.1 Non-cardiac Diseases

Diseased conditions	Electrocardiographic changes
Acute renal failure	Progressive bradycardia (progressively decreasing heartbeat) may lead to cardiac arrest
	Ventricular tachycardia is also seen
American	Tachycardia
Trypanosomiasis (*T. cruzi*)	"T" wave inversion
	Ventricular premature complexes
	Atrioventricular block
	Prolongation of P-R interval
	Decreased "P" wave amplitude
	MEA is shifted
	Enlargement of right heart (tall "P" wave and presence of "S" and "Q" waves with right axis deviation suggestive of right heart enlargement)
Babesiosis (*B. gibsoni*)	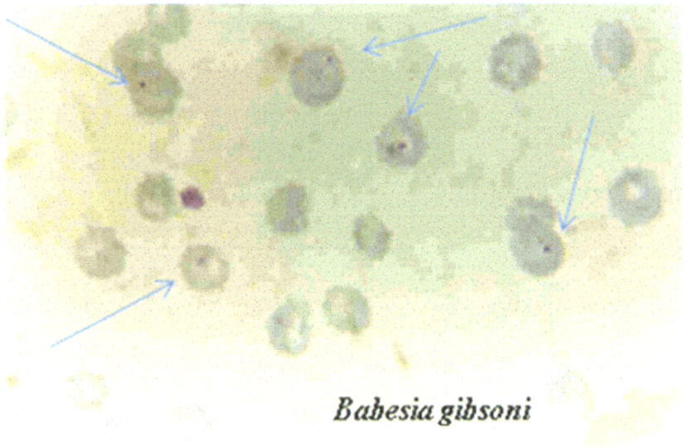
	Sinus arrhythmia
	Tachycardia
	Tachyarrhythmia
	Bradycardia
	Bradyarrhythmia
	Atrial fibrillation
	Mobitz type II heart blocks
	Low-voltage complexes
	Ventricular premature complex
	S-T elevation
	S-T depression
	S-T coving
	S-T coving and broad QRS (left bundle branch block)
	Atrial fibrillation and S-T depression
	Tachycardia and S-T depression
	Sinus arrest and S-T depression

Diseased conditions	Electrocardiographic changes
	Increased R amplitude and broad QRS
	Enlarged T
Canine distemper	Atrioventricular blocks
	Ventricular arrhythmias
	Abnormalities of S-T segment
Canine parvo	ECG changes depend on the form and stage of the disease
	Changes ranging from sinus rhythm to heart enlargement have been reported. Some of the changes are illustrated below
	Sinus rhythm
	Sinus arrhythmia (varying R-R interval)
	Tachycardia (heart rate > 180 bpm)
	Low-voltage "R" wave (<0.5 mV)
	Multiple premature ventricular complexes
	Repeated bursts of paroxysmal ventricular tachycardia
	Left ventricular enlargement
	Right axis deviation
	Electrical alternans
	Abnormalities of S-T segment
	Wandering pace maker
	Atrial fibrillation
	Wide "QRS" (>0.06 s)
	"QRS" notching
Diabetes mellitus	DM has been observed as significant promoter of cardiac arrhythmias in man (Kannel et al. 1998)
	Diabetic humans have greater risk of developing atrial fibrillation (Huxley et al. 2011)
Dirofilariasis (*D. immitis*)	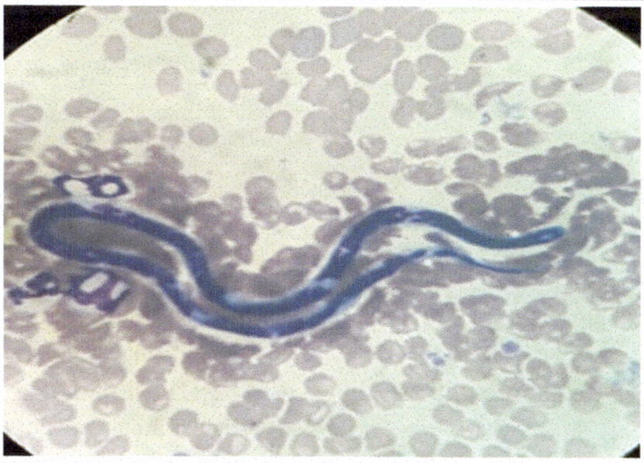
	ECG pattern in most of the dogs with dirofilariasis is normal
	Changes become apparent only when chronic severe pulmonary hypertension is also associated with dirofilariasis
	MEA > 104° on frontal plane or > 105° on transverse plane
	"S" wave in leads I, II, and III
	"QR" or "RSR" pattern in lead aVR

Diseased conditions	Electrocardiographic changes
	"S" > 0.8 mV in lead V2 and > 0.7 mV in lead V4
	R/S ratio < 1.0 in lead CV6LU
	Positive T in lead V10 in dogs other than Chihuahua
Ehrlichiosis (*E. canis*)	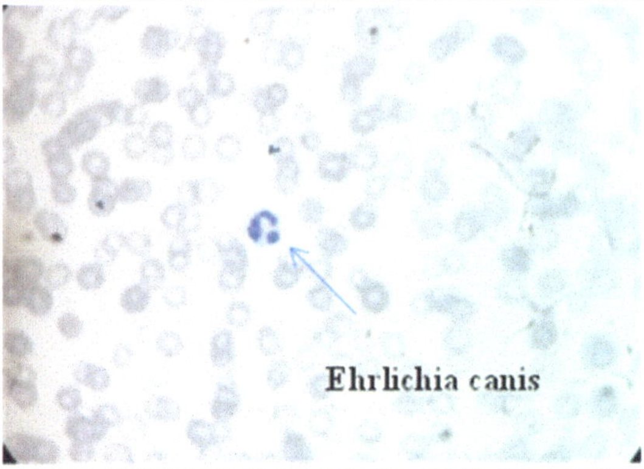
	Tachycardia
	Wandering pace maker
	Supraventricular tachycardia
	Atrial standstill
	Sinus arrhythmia
	Sinus tachycardia
	Atrial fibrillation
	Atrial Premature complexes
	Second-degree Mobitz type B heart block
	Right bundle branch blocks
	S-T elevation
	S-T depression
	Sinus arrest and low-voltage complexes
	Sinus arrest and ventricular premature complexes
	Sinus arrest, low-voltage complex and S-T elevation
	Ventricular premature complex, broad QRS, and tachycardia
Electric injury	Ventricular arrhythmias
	Cardiac arrest
Infectious canine	Low-voltage complexes
Hepatitis with pericardial effusion	Absence of "P" wave
Hypoparathyroidism	Prolonged Q-T interval
	Ventricular premature complexes
	Deep and wide "T" wave
	Bradycardia
Hyperparathyroidism	Shortened Q-T interval
	Atrioventricular block
	Ventricular premature complexes
Hypothyroidism	Low-voltage complexes particularly "R" wave
	Inverted "T" wave

Diseased conditions	Electrocardiographic changes
	Bradycardia
	Atrial fibrillation
	Arrhythmias
Hyperthyroidism	Sinus tachycardia
	Increased "R" wave amplitude
	Supraventricular and ventricular arrhythmias
	Intraventricular conduction abnormalities.
Hypoadrenocorticism (Addison's disease)	Bradycardia
	Widening and flattening of "P" wave
	Prolongation of P-R and Q-T intervals
	Wide QRS complex
	S-T segment alterations
	Tall "T" wave
Hyperadrenocorticism (Cushing syndrome)	ECG changes suggestive of left ventricular enlargement (occasional)
Hypoglycemia	S-T segment depression
	Flattened "T" wave
	Prolonged Q-T interval
	Arrhythmias
Hypoglycemia (pregnancy)	Sinus arrhythmia (heart rate within range but highly variable R-R interval)
	Bradycardia
	Bradyarrhythmia
Hypoglycemia (juvenile)	Slow heart rate
	Decreased "R" wave amplitude with descending notch
	Arrhythmia (R-R and P-R intervals are varying)
	Tall "T" wave
	Absence of "P" wave
Heat stroke	Tachycardia
	Abnormalities of S-T segment
	Abnormality of "T" wave
	Ventricular arrhythmia
	Third-degree heart block
Hypothermia	Bradycardia
	"J" wave
Obesity	Sinus rhythm
	Low voltage of "R" wave
	Tachycardia
	Arrhythmias
Pancreatitis (acute and severe)	Ventricular arrhythmias
Pulmonary thromboembolism	Abnormalities of S-T segment
	Sinus bradycardia

dilatation-volvulus, acute pancreatitis, and shock have been associated with microscopic myocardial necrosis. Myocardial damage has also been recorded in excessive sympathetic stimulation, hyperthermia, irradiation, and electric shock.

12.1.1.2 Diagnostic Profile

1. Clinical signs

   Clinical manifestations in dilated cardiomyopathy are highly variable and are related to heart failure. The clinical signs are enumerated below.
   - Weakness.
   - Lethargy.
   - Exercise intolerance.
   - Coughing.
   - Anorexia.
   - Chronic muscle wasting.
   - Abdominal distension/ascites.
   - Muddy colored mucus membranes.
   - Weak femoral pulse.
   - Jugular distension/pulse.
   - Tachypnea or dyspnea.
2. Physical Examination

   Physical examination in cases of dilated cardiomyopathy provides very useful information with regard to the heart and lung. The following information may be obtained on physical examination in dogs with dilated cardiomyopathy.
   - Nature of the pulse. Pulse deficit can be detected.
   - Nature of cardiac rhythm. It may be chaotic on auscultation if serious arrhythmias such as atrial fibrillation, atrial, or ventricular premature complexes are associated.
   - Murmurs. Various grades of systolic murmurs (grades 1–3) may be detected.
   - Variations (increase or decrease) in lung sound.
   - Pulmonary crackles/rales. These may be heard in dorso-caudal region of the lungs if pulmonary edema has set in due to left heart failure.
   - Changed character of heart sound. Heart sounds are muffled if there are effusions in pericardium or pleura.
   - Nature of cardiac apex beat. It is very pronounced.
3. Electrocardiography

   Electrocardiographic examination is not very specific to diagnose cardiomyopathy. It provides information regarding heart rhythm and rate, conduction disturbances, enlargement of the atrium and ventricles, variations in segments and intervals, etc. The following changes in electrocardiogram are generally seen in cases of dilated cardiomyopathy.
   - Electrocardiogram may be normal.
   - Tall and/or broad "P" wave with or without morphological changes.
   - Atrial fibrillation (Fig. 12.1). Commonly seen in giant breeds.
   - Ventricular tachycardia.
   - "R" wave alternans.

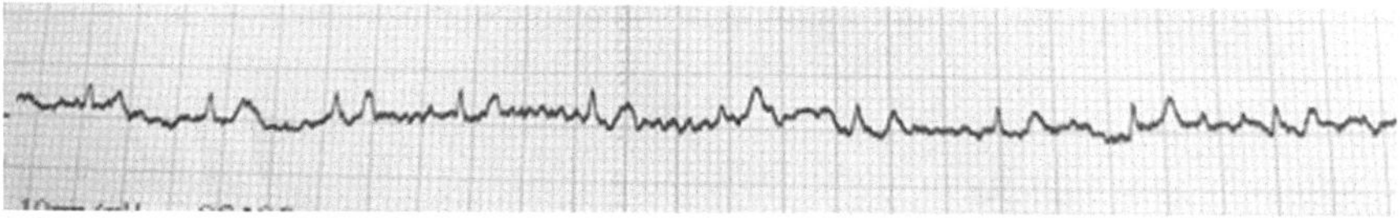

Fig. 12.1 Electrocardiogram (lead, II, sensitivity 1, speed 25 mm/s) of a 6-year-old female Labrador with exercise intolerance, dyspnea for 4 months showing ventricular heart rate as 120 bpm, atrial fibrillations, low-voltage R (0.2–0.25 mV) in all leads. Biochemical investigations revealed normal liver (SAP 96 U/L, ALT 44 IU/L, total bilirubin 1.2 mg%) and kidney (creatinine 1.4 mg%, BUN 26 mg%) functions and increased level of cTn-I (0.96 ng/mL) suggesting cardiomyopathy

- Increased amplitude of "R" wave.
- Decreased amplitude of "R" wave.
- Low-voltage complex (Fig.12.1).
- Broad and slurred "QRS."
- Ventricular premature complexes. Commonly seen in large breeds of dog and Dobermans.
- Paroxysmal ventricular tachycardia.
- Depression of S-T segment.
- Notched QRS complex if associated with microscopic intramural myocardial infarction (MIMI).

4. Radiography

   Radiographic examination of thorax provides valuable information regarding size of the heart, shape of the heart, lung congestion, position of trachea, and size of the large vessels. The following changes in radiographs are generally seen in dogs with cardiomyopathy:
   - Generalized cardiac enlargement.
   - Reduced angle of trachea with spine.
   - Left ventricle touching diaphragm.
   - Right ventricle touching sternum.
   - Increased size of pulmonary veins.
   - Interstitial pulmonary or alveolar edema (Fig.12.2a, b).
   - Left atrial enlargement (left heart failure).
   - Changed cardiac silhouette due to more pleural fluid.
   - Enlarged caudal vena cava.
   - Ascites and hepatomegaly (right heart failure).
5. Echocardiography

   Echocardiographic examination is very valuable. It provides vital information about the size of the chambers, wall thickness, status of various valves, and large vessels. This is the only diagnostic tool that can differentiate between dilated and hypertrophic cardiomyopathy. Therefore confirmatory diagnosis of dilated cardiomyopathy can only be made on echocardiographic examination. The following changes in echocardiogram are generally seen in cases of dilated cardiomyopathy.

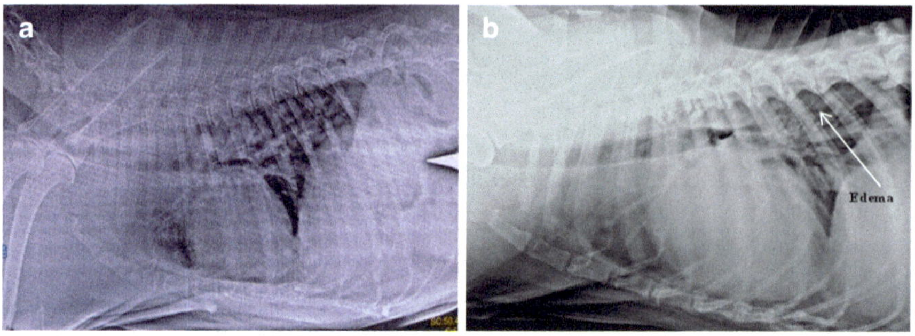

Fig. 12.2 (**a**) Right lateral radiograph of a 6-year-old female Labrador with exercise intolerance, dyspnea for 4 months showing edema in dorso-caudal region of the lung suggesting left heart failure. (**b**) Right lateral radiographs of a 10-year-old German Shepherd with severe dyspnea, panting, and exercise intolerance showing edema in dorso-caudal region of the lung suggesting left heart failure

- Heart enlargement (left atrium, left ventricle, right ventricle) (Fig. 12.3).
- Decreased left ventricular shortening fraction (loss of contractility) and ejection fraction.
- Thin left ventricular posterior wall and interventricular septum.
- Irregular motion of mitral valve leaflets.
- Increased E-point septal separation.
- Diminished excursion of the aortic wall.

6. Biochemical Examination

   Examination of various biochemical indices provide additional information regarding the other associated changes.

   - Slight increase in the values of blood urea nitrogen and serum creatinine is suggestive of prerenal azotemia.
   - Serum albumin is decreased in chronic heart failure with ascites.
   - Moderate increase in serum alkaline phosphatase and alanine aminotransferase activities is suggestive of hepatic congestion and hypoperfusion.
   - Fall in arterial PO_2 is indicative of increasing pulmonary edema.
   - Fall in venous PO_2 is suggestive of diminished cardiac output.
   - Lower values of plasma carnitine may be seen in giant breeds and Boxers with cardiomyopathy.
   - Lower values of plasma taurine may be seen in Cocker Spaniels with cardiomyopathy.
   - Hyponatremia and hypochloremia in dogs with congestive heart failure may indicate poor prognosis.

7. Cardiac Biomarkers

   Recently some cardiac biomarkers are being advocated for detecting myocardial damage.

   - Increase in levels of cTn-I and NT-pro BNP (levels depend on extent of damage to cardiac muscle and heart failure) are suggestive of cardiac muscle damage.

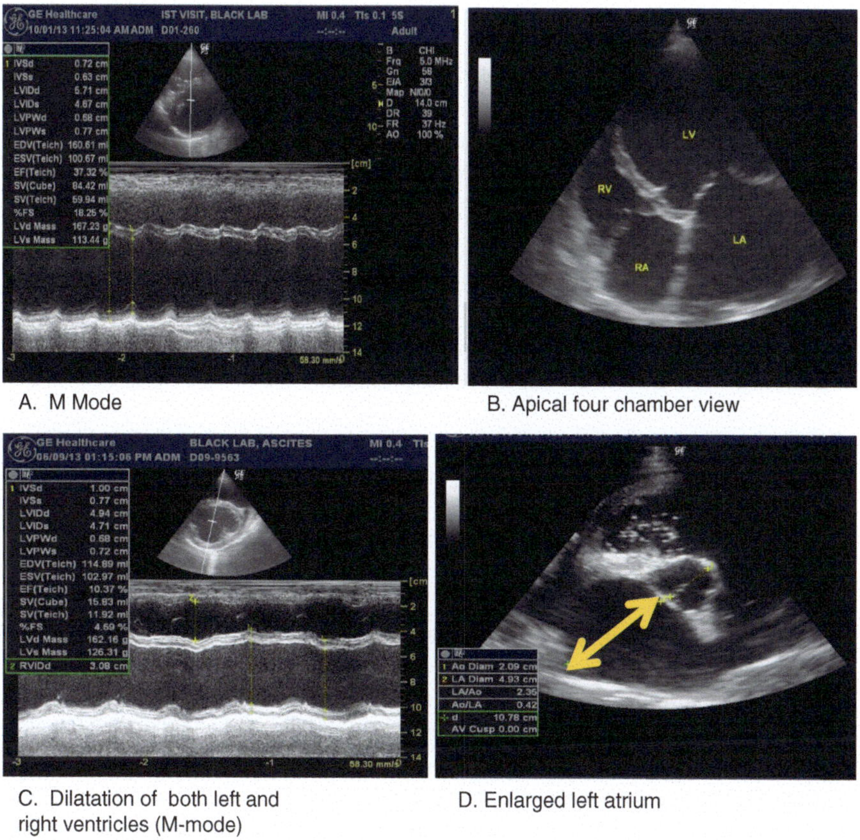

Fig. 12.3 (**a–d**) Echocardiogram of a dog with dilated cardiomyopathy showing apical four-chamber view, enlarged left atrium, and dilatation of both ventricles in M-mode. (These figures are with the courtesy of Dr. Neetu Saini, Associate Professor, Department of Medicine, Guru Angad Dev Veterinary and Animal Sciences University, Ludhiana)

12.1.1.3 Therapy

The aim of treatment is to increase cardiac output by increasing contractility, to control signs of congestive heart failure and manage arrhythmias. Details of the treatment are given under heart failure and arrhythmias. Nevertheless some outline is being given below.

Mild Heart Failure

- Restrict exercise.
- Give low-salt diet.
- Use diuretics.

Moderate to Severe Heart Failure

- Restrict exercise.
- Milrinone or digitalis glycoside may be used.

- Control edema with furosemide. Monitor dehydration with furosemide therapy.
- Give low-salt diet.
- In refractory cases consider the use of milrinone or digitalis or other vasodilators.

Life-Threatening Heart Failure

- Cage rest.
- Oxygen therapy.
- Furosemide (every hour until diuresis is induced).
- Nitroprusside IV with monitoring of blood pressure.
- Oral hydralazine and topical nitroglycerine.
- Manage arrhythmias with anti-arrhythmic drugs.
- Pimobendan to deal with congestive heart failure.

12.1.1.4 Prognosis

Prognosis of the cases of dilated cardiomyopathy is very much variable. In established cases of dilated cardiomyopathy, survival rate varies from 2–3 months to 10–12 months. Survival rate in giant and large breed is low.

12.1.2 Hypertrophic Cardiomyopathy (HCM)

Hypertrophic cardiomyopathy is characterized by hypertrophy of the left ventricle. The hypertrophy of ventricular septum is not in proportion to the left ventricular free wall. The condition is more common in cats. However it has been reported in German Shepherds. Its etiology is uncertain. It is clinically manifested as development of cardiac failure leading to sudden death. Sometimes coughing, dyspnea, left apical systolic murmurs, syncope, or collapse are observed.

12.1.2.1 Diagnostic Profile

The diagnosis of hypertrophic cardiomyopathy requires a comprehensive approach right from clinical examination to echocardiography. Neither clinical manifestation nor radiography and electrocardiography are of confirmatory value. The distinction between dilated and hypertrophic cardiomyopathy can only be done by echocardiography. Important observations with respect to clinical signs, electrocardiography, radiography, echocardiography, etc. in hypertrophic cardiomyopathy are enumerated as follows:

1. Clinical Signs

   Clinical signs in hypertrophic cardiomyopathy are not different from dilated cardiomyopathy. The signs are related to heart failure as are seen in other heart diseases.

   - Coughing.
   - Dyspnea with mild exercise.
   - Weak pulse.
   - Pale mucus membrane.
   - Murmurs.
   - Syncope/collapse.

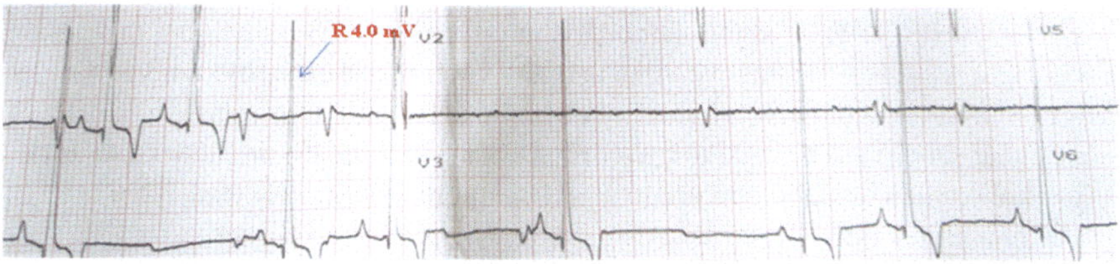

Fig. 12.4 Electrocardiogram (lead II, sensitivity 1, speed 25 mm/s) of an 8-year-old male Dachshund dog showing heart rate as 60 bpm, broad P (0.1–0.3 mV, 0.06 s), increased amplitude of R wave (4.0 mV), broad QRS (0.06 s), negative T (1.0 mV, 0.08 s), and variable R-R interval (0.6–1.44 s) suggesting left ventricular enlargement/hypertrophy

2. Electrocardiographic Changes
   Electrocardiograms may show changes related to left heart enlargement and conduction disturbance.
   - AV block.
   - Bundle branch block.
   - Left atrial enlargement.
   - Left ventricular hypertrophy (Fig. 12.4).
3. Radiography
   Established case of hypertrophic cardiomyopathy may reveal changes in the radiographs suggesting left heart enlargement. But it cannot be said with certainty that the changes are definitely due hypertrophic cardiomyopathy. Changes commonly observed in radiographs are as follows:
   - Left heart enlargement.
   - Tracheal elevation.
   - Pulmonary venous enlargement.
   - Pulmonary congestion or edema.
4. Echocardiography
   Echocardiography is the only diagnostic tool that can definitely differentiate between dilate and hypertrophic cardiomyopathy. Dogs with hypertrophic cardiomyopathy may show following important changes:
   - The ratio of thickness of septal to left ventricular wall is increased.
   - E-point septal distance is reduced.
   - Left ventricle shows hyperdynamic motion.
   - Mitral valve shows anterior motion during systole.
   - Aortic valve may show closure during mid systole.
   - Ratio of the size of left atrium and aorta is increased.
5. Pathology
   Pathological examination may provide following information:
   - Asymmetric concentric left ventricular hypertrophy.
   - Thickened mitral and aortic valves.
   - Fibrous endocardium.
   - Myocardial fiber hypertrophy.

12.1.2.2 Therapy

There is no definite treatment for hypertrophic cardiomyopathy. Nevertheless some relief can be provided.

- Digoxin should not be used in hypertrophic cardiomyopathy as it increases myocardial oxygen demand and predisposes to ventricular arrhythmias.
- Pulmonary edema is controlled using diuretics as detailed earlier.
- Beta-blockers (propranolol @ 0.2–1.0 mg/kg PO TID, metoprolol @5–50 mg PO TID, or atenolol@ 20–100 mg PO TID) or calcium entry blockers (diltiazem, verapamil) are effective in lowering heart rate, prolonging ventricular filling time, and reducing ventricular contractility and myocardial oxygen demand.
- Control of fluid buildup and arrhythmias may provide symptomatic improvement.
- No exercise is advocated.

12.1.3 Doberman Pincher Occult Cardiomyopathy

Doberman cardiomyopathy is an insidious, slowly progressive myocardial disease. Healthy Doberman may have occult left ventricular dysfunction. Cardiomyopathy in Doberman Pinschers has been described as a hereditary disease with genetic predisposition.

12.1.3.1 Etiology

- Not properly known.
- Deficiency of taurine and carnitine is suspected.
- Impairment of intracellular energy homeostasis.
- Hereditary. Genetic predisposition likely.

12.1.3.2 Symptoms

- Generally asymptomatic.
- Protracted occult nature of pre-congestive heart failure.
- Signs of progressive left ventricular failure.
- Episodic weakness and syncope.
- Other symptoms are similar to idiopathic dilated cardiomyopathy.

12.1.3.3 Diagnosis

Clinical examination, radiography, electrocardiography, and Holter electrocardiography are valuable aids in the diagnosis of Doberman cardiomyopathy.

1. Symptoms
    - Dobermans with exertion, weakness, reduced exercise intolerance, murmurs, or syncope should invariably be subjected to cardiac evaluation. These signs are nonspecific and arise suspicion of heart disease.
2. Electrocardiography
    - Not sensitive early marker.
    - Detection of ventricular tachyarrhythmia (VPCs) is an important indication of cardiomyopathy in Doberman.

- Prolonged duration of P wave, i.e., wide P (> 0.04 s) and wide QRS (>0.06 s), is seen in advanced cases of cardiomyopathy.
- Many Doberman with cardiomyopathy may also show atrial fibrillation.

3. Holter electrocardiography
   - Ventricular tachyarrhythmia.
   - Ventricular premature complexes >100 VPCs per 24 h.
4. Radiography
   - Not sensitive for mild to moderate cases.
   - No radiographic abnormality is detected in many Dobermans.
   - In advanced and non-treated cases, left atrial enlargement may be seen.

12.1.3.4 Treatment Approach

- Digoxin may exert positive influence but aggravates ventricular arrhythmias.
- Regular furosemide therapy is advocated when left ventricular dysfunction, pulmonary edema, pulmonary lobar vein distension, and gallop rhythm are detected.
- Carvedilol (mild beta-1 selective blocking drug with antioxidant property) may be tried (0.3 mg/kg BID). It may be used alone, with digoxin, or with amiodarone.
- Sotalol (new anti-arrhythmic with beta-blocking activity) may be used (40–80 mg BID PO). Be helpful in preventing sudden deaths.

12.1.3.5 Prognosis

- Guarded to grave as sudden deaths are common in Doberman.

12.1.4 Boxer Cardiomyopathy

It is also known as Boxer arrhythmogenic right ventricular cardiomyopathy. The disease is similar to human arrhythmogenic right ventricular cardiomyopathy. Its definite etiology is not known. It is considered as a genetic disease. There is no ventricular dilatation, but histologically there is replacement of ventricular myocardium with fatty or fibrous fatty tissues. Boxer cardiomyopathy is characterized by ventricular arrhythmias, syncope, and death. Ventricular systolic dysfunction and congestive heart failure are relatively less common in Boxer cardiomyopathy.

12.1.4.1 Diagnostic Profile

1. Clinical Signs
   - Unique to boxers, similar to human arrhythmogenic right.
   - ventricular cardiomyopathy.
   - Adult onset disease.
   - In obscure form the disease is asymptomatic despite having VPCs.
   - In overt form, the disease shows symptoms such as episodic weakness, exercise intolerance, dyspnea, coughing, arrhythmias, and syncope in many cases.

2. Electrocardiography
   - Sometimes ECG is normal. Holter monitoring is recommended.
   - Ventricular premature complexes and ventricular tachycardia are very common. Most of the VPCs may be of right ventricular origin (positive major deflection of VPC in lead II).
   - Supraventricular arrhythmias are not very common.
   - "QRS" complex is broad reflecting LBBB.
3. Radiography
   - Normal in many cases.
   - Cardiomegaly with mild to moderate left atrial enlargement.
   - Pulmonary venous congestion or edema is seen in uncommon form with LV systolic dysfunction.
4. Echocardiography
   - In most of the Boxers, echocardiogram is almost normal.
   - Left ventricular dilatation and systolic dysfunction are seen in some cases.

12.1.4.2 Therapy

Therapy of Boxer cardiomyopathy can be broadly divided into four categories.

1. Boxers with no clinical signs but with ventricular arrhythmia. Anti-arrhythmic therapy is indicated when VPCs are more than 500 per day or evidence of ventricular tachycardia or there is R-on-T phenomenon.
   - Sotalol 1.5–3.5 mg/kg orally twice daily.
   - Or combination of mexiletine (5–8 mg/kg orally thrice daily) with atenolol (0.3–0.6 mg/kg orally twice daily or 12.5 mg/dog orally twice daily).
2. Boxers with syncope showing ventricular arrhythmia but no sign of heart failure.
   - Lidocaine (2 mg/kg intravenously as a bolus followed by 25–75 microgram/kg/min).
3. Boxers showing congestive heart failure and arrhythmias.
   - Therapy as outlined under heart failure.
   - Manage arrhythmias as outlined under arrhythmia.
4. Ideal therapy is implantable cardioverter defibrillator (Nelson et al. 2006).

12.1.4.3 Prognosis

In asymptomatic Boxers prognosis is good. Boxers in second category may survive well with anti-arrhythmic therapy as episodes of syncope are reduced dramatically, but with development of congestive heart failure, their survival rate is reduced. Boxers in third category have low survival rate and may not live long.

12.1.5 Secondary Myocarditis

Secondary myocarditis is due to systemic diseases or other cardiovascular diseases. Though this is an incidental finding in other diseases, it is of prognostic value as the case may collapse if myocarditis is not attended to. Secondary myocarditis may be due to various reasons. The important reasons are ischemia due to MIMI, arterial hypertension, gastric dilatation-volvulus, or vegetative endocarditis; physical factors, viz., heat stroke, trauma, or CNS trauma; metabolic disorders such as uremia, chronic electrolyte imbalance, hypo- or hyperthyroidism, diabetes mellitus, or hyperadrenocorticism; nutritional deficiencies, viz., taurine, carnitine, thiamine, vitamin E, or deficiency of blood; immunological disorder (lupus erythematosus); and toxicities (prolonged use of doxorubicin, halothane anesthesia). Traumatic myocarditis or myocardial injuries due to blunt chest trauma has also been described in dogs. Blunt chest traumas have been described in humans also. Nowadays, chest traumas are commonly seen in dogs met with automobile accidents, animal attacks, or falls from high buildings. These accidents may cause blunt trauma to chest cage and lead to myocardial injury owing to chest compression and/or concussion.

12.1.5.1 Diagnostic Profile

1. Clinical signs
   - Arrhythmias.
   - Cardiac murmurs.
   - Weakness.
   - Anorexia and weight loss.
   - Tachypnea, orthopnea, dyspnea, and cough.
   - Abdominal distension.
   - Tachycardia.
   - Subcutaneous edema.
   - Syncope and sudden death.
2. ECG changes
   - Arrhythmias and conduction disturbances are common.
   - S-T segment abnormalities, viz., slurring, elevation, or depression in different leads (II, III, aVF, CV6LL, and CV6LU), are commonly observed.
   - AV blocks and bundle branch blocks may also occur.
   - Ventricular premature complexes may be seen.
   - Ventricular tachycardia is also common.
   - Bradycardia and silent atrium are also seen in some cases.
   - Dogs on doxorubicin therapy may develop atrial and ventricular enlargement.

   ECG changes in dogs with chest injuries may be delayed up to 48 h. Hence ECG should be repeated frequently (2–24 h. interval).

   In these cases Holter monitoring is most suitable. Premature ventricular complexes (multiform), ventricular tachycardia, and R-on-T phenomenon are very common in traumatic cardiac injury.
3. Echocardiography
   - It should be done in severely injured dogs having normal electrocardiogram and not responding to resuscitative measure.
   - Changes such as increased end-diastolic wall thickness, wall motion abnormality with decreased fractional shortening, increased echogenicity, or localized areas of echolucency may suggest cardiac trauma in dogs met with an accident/fall.

4. Biochemical changes
    - Levels of SGOT and CPK are increased in acute ischemic myocardial disease.
    - Increased level of cTn-I in dogs recently met with an accident/fall is a good indicator of cardiac trauma/insult.

12.1.5.2 Therapy

Treatment is generally similar to cardiomyopathy described earlier. However, for traumatic cardiac injury, following protocol may be followed:

- Primary concern in traumatic cardiac injury is to deal with life-threatening arrhythmia and to restore hemodynamics.
- Stabilize the dog first with fluid and electrolyte therapy and analgesics.
- Anti-arrhythmic treatment is initiated when arrhythmias are associated with hypotension, marked weakness, increased capillary refill time, pale mucus membranes, or syncope.
- Lidocaine is the anti-arrhythmic drug of choice in ventricular arrhythmias (VPC, ventricular tachycardia).
- Beta-blockers (propranolol, sotalol, metoprolol, atenolol) may be considered with caution in cases refractory to lidocaine (even when shock and pain has been managed).
- Side effects of beta-blockers (hypotension, AV block, bronchoconstriction) should always be kept in mind.
- Continuous ECG monitoring (Holter) is needed in cases with traumatic cardiac injuries.

12.1.6 Infectious Myocarditis

Infectious myocarditis is caused by invasion by various microorganism (*Corynebacterium*, *Staphylococcus*, *Streptococcus*, *Pseudomonas*, *E. coli*, and many other organism), virus (parvo, distemper, pseudorabies), protozoa (*Trypanosoma cruzi*, *Toxoplasma gondii*), or fungus (*Aspergillus* sp.). Clinical manifestations are not specific. However, arrhythmias and sudden death are seen in puppies with viral myocarditis. During chronic phase of the disease, signs of heart failure occur.

12.1.6.1 Diagnostic Profile

1. Clinical signs
    - During acute phase fever, depression, lethargy, weakness, dyspnea, cough, and lung congestion are observed on clinical examination.
    - Chronic phase of infectious myocarditis is associated with megaly (hepatic, splenic, and cardiac), ascites, cardiac murmurs, pleural effusion, low-grade cyclic fever, depression, dullness, weakness, anorexia, and cardiac arrhythmias.
2. Blood examination
    - Leukocytosis with neutrophilia is common in bacterial myocarditis.
    - Blood smear examination for blood protozoa.

- Leucopenia may be seen in viral infections.
- Immunological examination for confirmation of parvo or distemper.

3. Blood culture
   - May be attempted in cases with suspected bacterial etiology.
4. ECG changes
   - S-T segment abnormalities.
   - Sinus tachycardia.
   - Premature beats.
   - Paroxysmal tachycardia.
   - Sinoatrial abnormalities.
   - Atrial arrhythmias.
   - AV blocks.
   - Intraventricular blocks.
5. Echocardiographic changes
   - Cardiac dilation.
   - Reduced cardiac contractility.
   - No valvular abnormity.
   - Changed myocardial echogenicity.

12.1.6.2 Therapy

1. Antibiotic therapy. Antibiotics should be administered as per culture report for 3–6 months. The following antibiotics are generally used in the management of myocarditis of bacterial origin.
   - Sodium/potassium penicillin's (initial 60,000 units/kg IV and then 40,000 units/kg four to six times daily).
   - Amoxicillin @ 20 mg/kg IV, PO, BID.
   - Cephalothin @ 20–40 mg/kg IV, IM, PO BID-TID.
   - Gentamicin @ 2.2 mg/kg IV, IM TID.
   - Amikacin @ 5 mg/kg IV, IM TID.
2. Anti-protozoal drugs in case of protozoal myocarditis.
3. Anti-arrhythmic drugs for the management of arrhythmias.
4. Parenteral fluid therapy to restore hemodynamics.
5. Drugs for the management of congestive heart failure.
6. Use of aspirin (5–10 mg/kg PO SID-QID) or heparin (40–80 mg/kg SQ, BID-TID) is controversial.
7. Diuretics are used to control fluid buildup.
8. Low-sodium diet should be given.

12.2 Bacterial Endocarditis

Endocarditis in dogs is mostly infectious due to invasion of cardiac valves and endocardium by organisms. Many microorganisms have been associated with bacterial endocarditis in dogs. *Staphylococcus aureus*, *E.coli*, beta-hemolytic streptococci, *Aerobacter aerogenes*, *Pseudomonas aeruginosa*, *Erysipelothrix rhusiopathiae*,

Pasteurella multocida and *Pasteurella pneumotropica*, *Corynebacterium*, *Enterococcus*, *Klebsiella*, and *Proteus* spp. are the common microbes found associated with endocarditis in dogs. The invasion of microbes causes inflammation and thickening of the valves. Clinical manifestations are characterized by fever, toxemia, generalized sepsis, cardiac arrhythmias, congestive heart failure, and systemic emboli.

12.2.1 Diagnostic Profile

1. Clinical signs. Commonly observed clinical signs are fever, weakness, lethargy, vomiting, anorexia, shifting and intermittent lameness, posterior limb paresis/paralysis, pulse deficit, toxemia, arrhythmia, muscular pain, hemorrhages, uveitis, and retinal hemorrhages.
2. Laboratory findings. Laboratory investigations may reveal leukocytosis and neutrophilia with left shift; increased CPK, aspartate aminotransferase, and LDH; decreased blood glucose and serum albumin; increased BUN, serum creatinine, amylase, and lipase; and proteinuria. Blood culture examination may identify organism.
3. Electrocardiography. Electrocardiographic changes are not specific to bacterial endocarditis. The changes are related to arrhythmias, conduction disturbances, and change in chamber size. The commonly observed changes in an electrocardiogram are sinus tachycardia, S-T segment changes (during early stage), increased depth of "Q" wave in lead II if associated with myocardial infarction, premature ventricular complexes, ventricular tachycardia, AV blocks, bundle branch blocks, broad "P," tall "R" wave in different leads (II, avF, CV6LL, and CV6LU), and broad "QRS."
4. Echocardiography. It may reveal thickening of cardiac valves, dilation of left ventricle, increased movement of septum and free wall during systole, and changes in AV valve leaflets as per the heart structures involved.
5. Doppler echocardiography. Turbulence in blood flow or regurgitation may be detected.
6. Radiography. Radiographic examination may reveal cardiomegaly and interstitial/alveolar densities of the lung.

12.2.2 Therapy

- Antibiotic therapy based on sensitivity for 3–6 months.
- Anti-arrhythmic therapy to control arrhythmias.
- Management of circulatory heart failure as described under heart failure.
- No corticosteroids.
- Use of anticoagulants (aspirin or heparin) is also controversial.

References

Nelson OA, Lehmers S, Schneider T, Thompson P (2006) The use of an implantable cardioverter defibrillator in a boxer to control clinical signs of arrhythmogenic right ventricular cardiomyopathy. J Vet Intern Med 20:1232–1237

Further Reading

Allen DG (1991) Small animal medicine. J.B. Lippincott Company, Philadelphia, PA

Birchard SJ, Sherding RG (2000) Saunders manual of small animal practice, 2nd edn. W.B. Saunders Company, Philadelphia, PA

Bolton GR (1975) Handbook of canine electrocardiography. W.B. Saunders Company, Philadelphia, PA

Ettinger SJ, Feldman EC (2000) Textbook of veterinary internal medicine. Diseases of the dog and cat, 5th edn. W.B. Saunders Company, Philadelphia, PA

Meurs KM (2004) Boxer cardiomyopathy. An update. Vet Clin North Am Small Anim Pract 34:1235–1244

Meurs KM, Fox PR, Norgard M (2007a) A prospective genetic evaluation of familial dilated cardiomyopathy in the Doberman Pinscher. J Vet Intern Med 21:1016–1020

Meurs KM, Spier AW, Millers MW (1999) Familial ventricular arrhythmias in boxers. J Vet Intern Med 13:437–439

Morgan RV (1992) Handbook of small animal practice, 2nd edn. W.B. Saunders Company, Philadelphia, PA

Nelson RW, Couto CG (1998) Small animal internal medicine, 2nd edn. C.G. Mosby, St. Louis, MO

Tilley LP (1985) Essentials of canine and feline electrocardiography, 2nd edn. Lea and Febiger, Philadelphia, PA

Valvular Insufficiency 13

13.1 Chronic Mitral Insufficiency (CMI)

Chronic mitral valve insufficiency (CMI) is the most common cause of congestive heart failure in small breeds of dogs. It may lead to mitral regurgitation (MR) in long-standing cases. CMI is characterized by back flow of blood from the left ventricle to left atrium during ventricular systole leading to forward failure, backward failure, and volume overload. Its clinical severity is classified as functional heart failure class I, II, III, and IV. Clinically CMI is characterized by various grades of cardiac murmurs, fatigue, polydipsia, tachycardia, orthopnea, tachypnea, infrequent cyanosis, jugular pulse, hepatojugular reflex, arrhythmias, and dyspnea.

Etiology of CMI varies from genetic predisposition (especially in small and medium breeds of dog), arrhythmias, dilated cardiomyopathy, chest trauma to tumors. Diagnosis of CMI poses challenge in routine practice and requires a systematic approach employing detailed clinical examination, radiological examination, electrocardiographic examination, and echocardiographic examination.

13.1.1 Diagnostic Profile

1. Murmurs of various grades are detected on chest auscultation. Soft, blowing and an early systolic murmurs over mitral valve in the left chest may arise suspicion of mitral valve insufficiency.
2. Clinical manifestations are not very much specific. These are of general nature such as marked exertion, weakness, polydipsia, tachypnea, orthopnea, tachycardia, or cyanosis depending on the class of functional heart failure. In many cases there is a chronic productive or nonproductive cough. Cough is easily elucidated by tracheal palpation if tracheal collapse is coexisting. Therefore clinical signs may arise clinical suspicion especially in aged dogs of small breeds. The dogs with clinical suspicion of CMI should be thoroughly investigated.

J. P. Varshney, *Electrocardiography in Veterinary Medicine*,
https://doi.org/10.1007/978-981-15-3699-1_13

3. Echocardiography is of vital importance in the diagnosis of CMI as status, and functioning of the heart valves can only be evaluated by echocardiography. The echocardiographic changes in the mitral valve, left atrium, or left ventricle depend on the class of functional heart failure. Echocardiography in dogs with CMI may reveal mitral value thickening (Fig. 13.1) and change in its shape, increase in chamber size (atrium, ventricle) and/or contractility force of ventricles, etc.
4. Doppler echocardiography can detect retrograde flow of blood (from the left ventricle to left atrium during systole). Echo- and Doppler echocardiography are the only means to confirm chronic mitral insufficiency.
5. Electrocardiography is not of diagnostic value in confirming CMI. But it may detect some changes associated with different class of heart failure (class II, III, and IV functional heart failure). The following changes have generally been observed in cases of CMI:

   - P wave in lead II may be broad (duration more than 0.04 s) and notched, and its amplitude may remain between 0.35 and 0.4 mV suggesting left atrial enlargement.
   - Amplitude of R wave may increase in different leads. It may be more than 2.5 mV in lead II, more than 3.0 mV in lead CV6LU, and 1.0 mV in lead CV5RL.
   - Amplitude of S wave may be more than 0.5 mV in standard bipolar leads (lead I, II, III).
   - QRS may be broad (more than 0.06 s).
   - There may be frequent depression of T wave.
   - There may be changes in S-T segment (depression or elevation).
   - Atrial fibrillation, supraventricular premature beats, ventricular premature complexes, and ventricular tachyarrhythmia may be detected in electrocardiogram.
   - In class IV heart failure, low-voltage complexes may be evident.

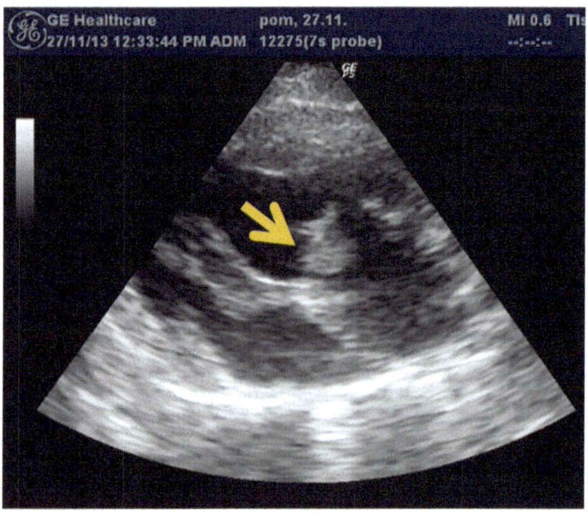

Fig. 13.1 Echocardiogram of a dog showing thickened mitral valve (This figure is with the courtesy of Dr. Neetu Saini, Associate Professor, Department of Medicine, Guru Angad Dev Veterinary and Animal Sciences University, Ludhiana)

6. Radiography may be helpful in detecting the changes in the size of left atrium. Enlargement of the left atrium (Fig. 13.2) and/or left ventricle is a common radiographic finding in cases of CMI. In advanced cases pulmonary edema in dorsocaudal hilar region, enlargement of pulmonary vasculature, and dorsal displacement of trachea (suggesting left heart failure) are also evident. When mitral valve insufficiency is complicated with pulmonary hypertension, hepatomegaly and ascites (suggesting right heart failure) can also be visualized in radiographs.
7. Cardiac biomarkers- Levels of cardiac biomarkers (Cardiac troponin-I, NT pro-BNP) have been found elevated in cases of chronic mitral valve insufficiency.

13.1.2 Therapy

No medical treatment provides complete cure. Nevertheless, treatment of chronic mitral valve insufficiency is directed as per degree of heart failure. It is outlined below:

Class I Heart Failure

Dogs with CMI showing class I heart failure generally do not require any treatment. But their excitability can be controlled with phenobarbital (@ 1–2 mg/kg given orally twice or thrice daily) and diazepam (@ .5–2.2 mg/kg orally). Isosorbide dinitrate can be given @1–2 mg/kg twice daily orally for reducing heart load and improving coronary circulation.

Class II Heart Failure

Dogs with CMI showing class II heart failure should be treated to prevent deterioration. The purpose of using phenobarbital, diazepam, and isosorbide is same as described under class I heart failure. Use of propranolol (@ 0.3–1.0 mg/kg orally twice or thrice daily) is to control associated tachycardia. To prevent or delay the

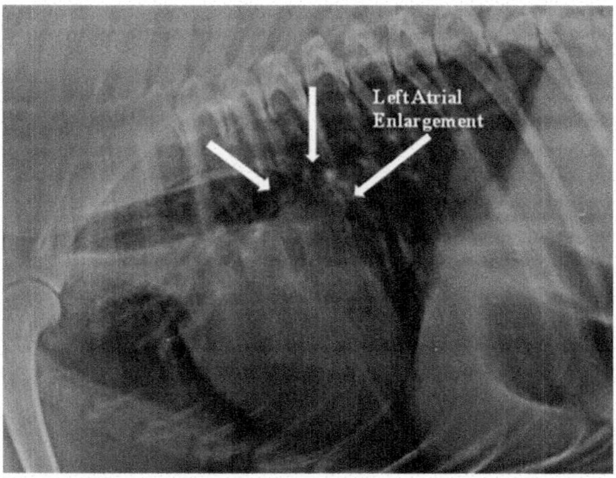

Fig. 13.2 Right lateral radiograph of a 5-year-old male German Shepherd with murmurs at left apex showing enlargement of the left atrium suggestive of chronic mitral valve insufficiency. Electrocardiogram of this dog showed broad P wave (0.6 s) with normal amplitude (0.4 mV) confirming left atrial enlargement

onset of congestion, use of captopril (@ 0.25–0.5 mg/kg orally twice or thrice daily) is recommended. Enalapril (@ 0.25–0.5 mg/kg orally twice or thrice daily) can also be used in place of captopril. Furosemide (@ 2–4 mg/kg orally twice daily) is recommended as a diuretic to control congestion. Its dose should be reduced/discontinued when the dog is stabilized.

Class III Heart Failure

Dogs with CMI showing class III heart failure can be managed with the use of furosemide, nitroglycerine, isosorbide dinitrate, hydralazine, enalapril/captopril/prazosin, and digoxin as per clinical need. Furosemide (@ 4–5 mg/kg orally, IV, IM, twice or thrice daily until pulmonary edema is reduced followed by once daily dose) and spironolactone (4–5 mg/kg orally twice daily) are recommended for reducing the pulmonary edema. Nitroglycerine ointment (2%) is applied on skin area (0.5–3.0 cm) in groin region or on inner ear flap of dogs having problem in oral administration of diuretics. Isosorbide dinitrate can be given @1–2 mg/kg twice daily orally for reducing heart load and improving coronary circulation. If furosemide is less effective, its efficacy may be improved by using hydralazine (@ 1–3 mg/kg orally twice daily). Enalapril (0.25–0.5 mg/kg orally twice daily) or prazosin (1.0 mg/15 kg orally twice or thrice daily) is used to delay congestion and manage arrhythmias. ACE inhibitors like enalapril and diuretics like spironolactones should not be used together as electrolyte imbalance may occur. Digoxin (@ 0.005–0.008 mg/kg PO BID) is recommended in the management of supraventricular tachycardia.

Class IV Heart Failure

- Dogs with CMI showing class IV heart failure need aggressive management.
- Cage rest with low-sodium diet and oxygen therapy is the first priority. The following drugs are used as per the need of the ailing dog.
- Morphine sulfate @ 0.2 mg/kg IM, SQ. It is given for its euphoric effect and to reduce pulmonary edema.
- Furosemide—as above.
- Nitroglycerine ointment—as above.
- Dobutamine hydrochloride @ 5–20 μg/kg/min in 5% dextrose IV. It is an inotropic agent.
- Dopamine hydrochloride @ 2–10 μg /kg/min in distilled water IV. It is an inotropic agent.
- Nitroprusside @ 1–10 μg/kg/min IV to maintain arterial blood pressure. It is an arterial and venous dilator.
- Pimobendan 0.25 mg/kg PO BID (empty stomach). It is given to manage congestive heart failure.
- Theophylline (long-acting bronchodilator) can be used to facilitate easy.
- respiration (15–20 mg/kg orally twice daily).
- Sildenafil is known to reduce pulmonary hypertension. Pulmonary hypertension has been seen in cases of heart failure also. It may be used @ 0.5–2.0 mg/kg orally twice daily to improve general activity.

- Hydralazine is a vasodilator and reduces hypertension. It can be used @ 0.5–2.0 mg/kg orally twice daily.

13.2 Tricuspid Insufficiency (TI)

Tricuspid insufficiency (TI) is characterized by back flow of blood from the right ventricle to right atrium during ventricular systole leading to forward failure, backward failure, and volume overload of the right heart. It usually occurs with CMI. In mild form TI is asymptomatic. Right heart forward failure is characterized by tachypnea, tachycardia, orthopnea, exercise intolerance, and cyanosis. While right-sided backward failure is manifested by jugular pulse, hepatojugular reflux, peripheral venous engorgement, hepatomegaly, splenomegaly, ascites, peripheral and dependent edema, and polyuria/polydipsia.

Diagnosis of TI poses challenge in routine practice and requires a systematic approach employing detailed clinical examination, radiological examination, electrocardiographic examination, and echocardiographic examination.

13.2.1 Diagnostic Profile

1. Murmurs—Grade II–III/Vi systolic to holosystolic murmurs can be heard over right apical thrust.
2. Thoracic radiography—It may show enlargement of the right heart and caudal vena cava; pericardium and/or pleural effusions; ground glass appearance of the abdomen; and splenomegaly and hepatomegaly in dogs with tricuspid valve insufficiency.
3. ECG changes—Electrocardiographic changes are not of confirmatory diagnosis. The following changes in an electrocardiogram may be observed in dogs suffering with TI.
   - P wave may be tall in lead II (amplitude more than 0.4 mV).
   - P-R interval may be prolonged (more than 0.14 s).
   - R wave amplitude may be decreased in different leads (II, III, aVF, CV6LU).
   - R wave amplitude may increase in aVR and CV5RL leads.
   - Amplitude of Q and S waves may increase in leads II, III, aVF, and CV6LU.
   - QRS is broad (more than 0.06 s).
   - There may be right axis deviation.
4. Echocardiography—It is of great assistance in confirmatory diagnosis of TI. The following findings may be observed in echocardiography.
   - Right atrium and ventricular dilation.
   - Abnormal tricuspid valve leaflets.
   - Septum encroaching left ventricular chamber.
5. Doppler echocardiography—It may reveal retrograde flow of blood from the right ventricle to right atrium during ventricular systole.

13.2.2 Therapy

No medical treatment provides complete cure. Nevertheless palliative treatment may be undertaken as per clinical needs.

- Diuretics—Furosemide or spironolactone or combination.
- Low-sodium diet.
- Antiarrhythmic drugs as per need.
- Isosorbide dinitrate and nitroglycerine ointment to relieve pulmonary hypertension.

13.3 Mitral Stenosis (MS)

Stenosis of mitral valve interferes with blood flow from the left atrium to the left ventricle during ventricular diastole and atrial diastole/systole. It is clinically recognized by signs related to underlying diseases (mitral valve atresia, isolated mitral valve stenosis, bacterial endocarditis, enlargement of the left heart without simultaneous enlargement of atrioventricular ring, mitral valve insufficiency, dilated cardiomyopathy), decreased exercise tolerance, cough, episodic weakness, tachypnea/dyspnea, wheezes, hepatomegaly and/or splenomegaly, jugular pulse, hepatojugular reflex, and ascites.

13.3.1 Diagnostic Profile

Diagnosis of mitral stenosis requires a comprehensive approach involving clinical examination, radiography, electrocardiography, echocardiography, and Doppler echocardiography.

1. Early diastolic murmurs may be detected over left apical thrust on chest auscultation. There may be presystolic accentuation.
2. Clinical signs such as decreased exercise tolerance, cough, episodic weakness, tachypnea/dyspnea, wheezes, hepatomegaly and/or splenomegaly, jugular pulse, hepatojugular reflex, and ascites are observed.
3. Thoracic radiography. It may reveal left atrial enlargement, pulmonary venous congestion, pulmonary edema, and prominence of main pulmonary artery.
4. Electrocardiography. Electrocardiogram of a dog with mitral stenosis may show following changes:

   - Increased amplitude (more than 0.4 mV) and prolonged duration (more than 0.04 s) "P" wave suggesting biatrial enlargement.
   - Arrhythmias (tachycardia, supraventricular premature complexes, atrial fibrillation, atrial flutter) are common.
   - Changes suggesting right ventricular enlargement may be seen.

5. Echocardiography. The following changes may be observed:
    - Change in amplitude and E-F slope of the septal mitral valve leaflet.
    - Enlargement of the left atrium.
    - Dilation of the right ventricle.
6. Doppler echocardiography
    - Turbulent flow from the left atrium to ventricle during diastole.

13.3.2 Therapy

- No effective treatment.
- Restrict exercise.
- Try treatment detailed under CMI.
- Valvulotomy or valve replacement.

13.4 Aortic Insufficiency (AI)

Aortic valve insufficiency is characterized by regurgitant flow of the blood from the aorta to the left ventricle during ventricular diastole. Clinical signs are related to primary congenital anomaly (ventricular septal defect, aortic stenosis, malformation of the aortic valve, rupture of the aortic sinus, Marfan's syndrome, or ruptured chordae tendineae). General clinical signs are similar to CMI.

13.4.1 Diagnostic Profile

1. Early to mid-diastolic murmurs or holosystolic murmurs may be detected.
2. Peripheral pulse is prominent and bounding.
3. Thoracic radiography. The following changes may be observed on radiographic examination of the thorax.
    (a) Left ventricular enlargement.
    (b) Tracheal elevation in lateral view.
    (c) Increased prominence of apex.
    (d) Left atrial enlargement.
    (e) Pulmonary edema.
4. Electrocardiography. ECG changes are similar to CMI.
5. Echocardiography. The following changes are generally observed in cases of AI:
    - Left ventricle may be dilated.
    - Septal and left ventricular free wall thickness may be normal.
    - Left atrium may be dilated.

- Systolic motion of septal and left ventricular free wall is exaggerated.
- Aortic valve leaflets may be thickened.

6. Doppler echocardiography

- Turbulence may be detected below the aortic valve during ventricular diastole.

13.4.2 Therapy

- Medical treatment is generally not rewarding.
- Valve replacement is the only remedial measure to correct insufficiency in severe cases.
- Nevertheless arteriolar dilators, cardiac glycoside, and other treatment outlined under CMI may be adopted to provide relief. Exercise and excitement should be avoided.

13.5 Pulmonic Insufficiency (PI)

Pulmonic insufficiency is characterized by regurgitant blood flow from the pulmonary artery to right ventricle during ventricular diastole due to inability of the pulmonary semilunar valve to close. The insufficiency of pulmonary semilunar valve is ascribed to congenital malformation, dirofilariasis, pulmonary hypertension, tricuspid insufficiency, dilation of the pulmonary valvular sinus, or bacterial endocarditis. It is clinically manifested by decreased exercise tolerance, weight loss, ascites, hepatomegaly, splenomegaly, jugular pulse, hepatojugular reflux, and diastolic murmurs over the left cranial ventral chest referred to right chest.

13.5.1 Diagnostic Profile

1. Diastolic murmurs are heard over the left thorax in cranioventral area. These murmurs are also referred to the right ventral thorax.
2. Thoracic radiography may reveal following changes:

   - Right ventricle enlargement.
   - A bulge of pulmonary artery segment (PAS) at 1 o'clock position in dorsoventral radiographs.
   - Enlargement of the caudal vena cava.

3. Electrocardiography may reveal following alteration in an electrocardiogram:

   - Broad "P" (>0.04 s) with or without change in its normal amplitude in lead II.
   - Increase in P-R interval (>than normal).
   - Decrease in "R" wave amplitude in lead II, III, aVF, and CV6LU.
   - Increase in "R" wave amplitude in lead aVR and CV5RL.

- Increase in "S" wave amplitude in lead II, III, aVF, and CV6LU.
- Broad "QRS" (> than normal range).

4. Echocardiography
   - Dilation of the main pulmonary artery.
   - Dilation of the right ventricle.
   - Pulmonary valve is not fully opened during diastole.
   - Absurd septal motion.
5. Doppler echocardiography
   - Ventricular diastole is associated with turbulence near pulmonary semilunar valve.

13.5.2 Therapy

- Medical treatment is not very much rewarding.
- Diuretics (furosemide, spironolactone, or combinations) are advised to reduce fluid buildups.
- Low-sodium diet is a usual requirement of diseases of the heart.
- Pulmonary hypertension can be attended with isosorbide dinitrate and nitroglycerine ointment.
- Surgery is the only rational approach. The following surgical interventions are taken to repair insufficiency of pulmonary valve:
  - Patch graft procedure.
  - Valvulotomy via MPA.
  - Valvuloplasty.
  - Implantation of a valved or non-valved conduit.
  - Balloon valvuloplasty.

Further Reading

Abbott JA (2008) Acquired valvular disease. In: Tilley LP, Smith FWK, Oyama MA, Sleeper MM (eds) Manual of canine and feline cardiology, 4th edn. Saunders Co., Philadelphia, PA

Allen DG (1991) Small animal medicine. J.B. Lippincott Company, Philadelphia, PA

Birchard SJ, Sherding RG (2000) Saunders manual of small animal practice, 2nd edn. W.B. Saunders Company, Philadelphia, PA

Ettinger SJ, Feldman EC (2000) Textbook of veterinary internal medicine. Diseases of the dog and cat, 5th edn. W.B. Saunders Company, Philadelphia, PA

Morgan RV (1992) Handbook of small animal practice, 2nd edn. W.B. Saunders Company, Philadelphia, PA

Nelson RW, Couto CG (1998) Small animal internal medicine, 2nd edn. C.G. Mosby, St. Louis, MO

Pericardial Effusion

14

Pericardial effusion (PE) is a common disease of the pericardium in dogs and cats and is characterized by accumulation of increased amount of fluid in the pericardial sac. PE is classified on the basis of physical and cytological characteristics of the pericardial fluid. Pericardial effusion is commonly diagnosed in German Shepherds, Boxer, English bulldog, and Boston Terrier due to various reasons. Clinical signs depend on the rate and degree of cardiac compensation. Moderate cardiac compression causes signs related to right heart failure such as weakness, lethargy, exertion, dyspnea, ascites, and syncope on exertion, while signs of low cardiac output such as marked weakness or collapse, dyspnea/tachypnea, and death develop rapidly in case of severe compression of the heart. Pericardial effusion in dogs is of idiopathic or neoplastic origin (hemangiosarcoma is common). Pericardial disorders are of two main types, i.e., congenital or acquired. Congenital pericardial disorders may be due to diaphragmatic hernia, pericardial cyst, or pericardial defects, while acquired pericardial disorders are further subcategorized as pericardial effusion, constrictive pericarditis, and pericardial mass with or without effusion or fibrosis.

14.1 Diagnostic Profile

For the diagnosis of pericardial effusion, a comprehensive approach involving examination of clinical signs, electrocardiographic evaluation, echocardiographic evaluation, radiographic evaluation, routine laboratory investigations, and analysis of pericardial fluid is necessary.

1. Clinical signs—The following clinical signs may be observed in different combination:

   Jugular pulsation or distension.

   Weak arterial pulse.

   Pulsus paradoxus (decrease in arterial pulse pressure during inspiration).

   Muffled heart sound.

   Arrhythmias.

J. P. Varshney, *Electrocardiography in Veterinary Medicine*,
https://doi.org/10.1007/978-981-15-3699-1_14

Pericardial fractional rub.
Ascites.
Diminished lung sounds.
Restlessness at night.
Tachypnea.
Prolonged capillary refill time.
Pallor mucus membrane.
A triad of muffled heart sound, jugular venous distension, and weakened arterial pulse is very much suggestive of pericardial effusion.

2. Electrocardiography—The following electrocardiographic abnormalities may be observed in dogs with pericardial effusion:
   Sinus tachycardia.
   Supraventricular or ventricular arrhythmia.
   S-T segment deviation (elevation).
   Low-voltage complexes (<1.0 mV) in limb and thoracic leads (Fig. 14.1).
   Electrical alternans of R (Fig. 14.2) or T waves (Fig. 14.3).
   Normal ECG also does not rule out pericardial effusion.
   Electrocardiography is not of definite diagnostic value in cases of pericardial effusion.
3. Echocardiography—This is the technique that provides confirmatory diagnosis of pericardial effusion. The following changes may be observed in echocardiography of dogs with pericardial effusion.
   Echo-free pericardial fluid space surrounding the heart is commonly seen (Fig. 14.4a, b).
   There is left to right swinging of the heart.
   Dimensions of ventricular chamber are reduced.
   In case of cardiac tamponade, diastolic collapse of the right atrium and right ventricles is seen.
   It provides confirmatory diagnosis of pericardial effusions.
4. Radiography—Radiographic examination of dogs with suspected pericardial effusion is very valuable as it can provide information about cardiac size and its silhouette, size of the vena cava, and tracheal position. The following changes in the radiographs are commonly observed in dogs affected with pericardial effusion.

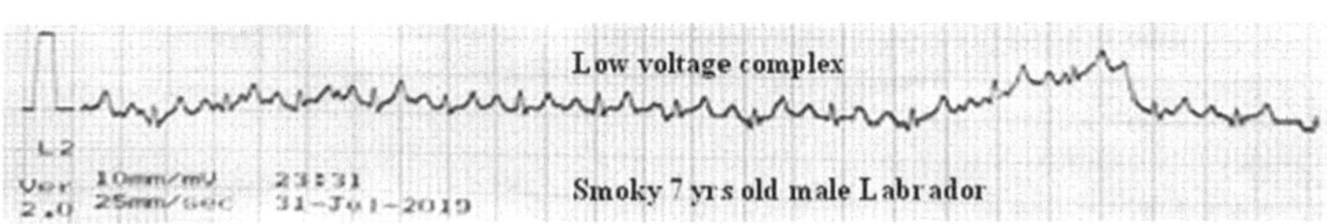

Fig. 14.1 Electrocardiogram (lead II, sensitivity 1, speed 25 mm/s) of a male, 7-year -old Labrador dog (Smoky) with jugular distension, muffled heart sound, recumbency, osteoarthritis of both hips and left knee, globoid heart (chest X-ray Fig. 14.5c), and minor increase in cTn-I (0.16 ng/mL) showing low-voltage complex (0.20 mV) in all leads and sinus tachycardia (heart rate 200 bpm) suggesting pericardial effusion

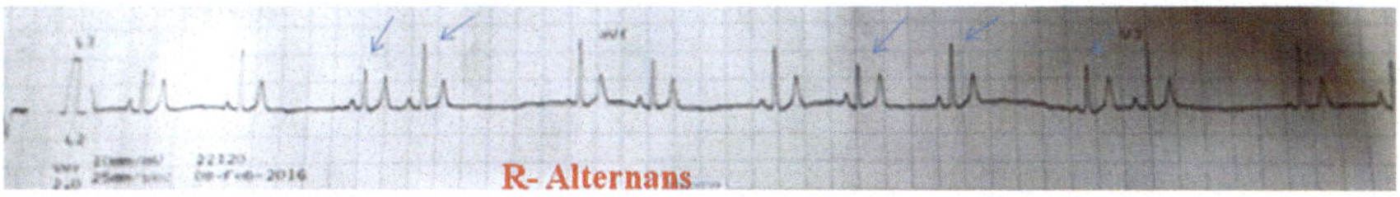

Fig. 14.2 Electrocardiogram (lead II, sensitivity 1, speed 25 mm/s) of a dog with globoid heart on chest X-ray showing R-alternans (amplitude of R wavevarying) suggestive of pericardial effusion

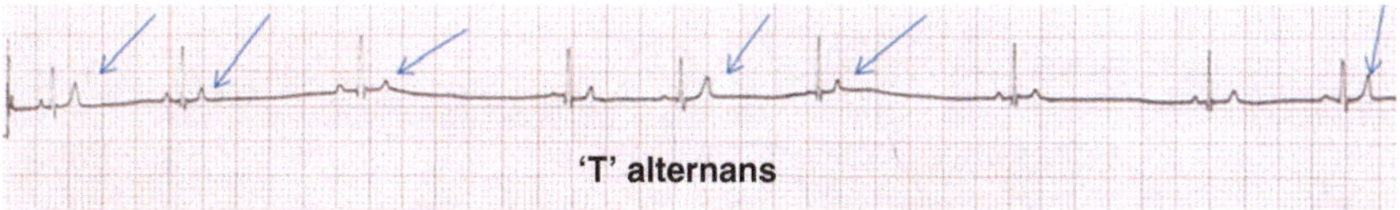

Fig. 14.3 Electrocardiogram (lead II, sensitivity 1, speed 25 mm/s) of a dog showing T-alternans (amplitude of T wave varying) suggestive of pericardial effusion

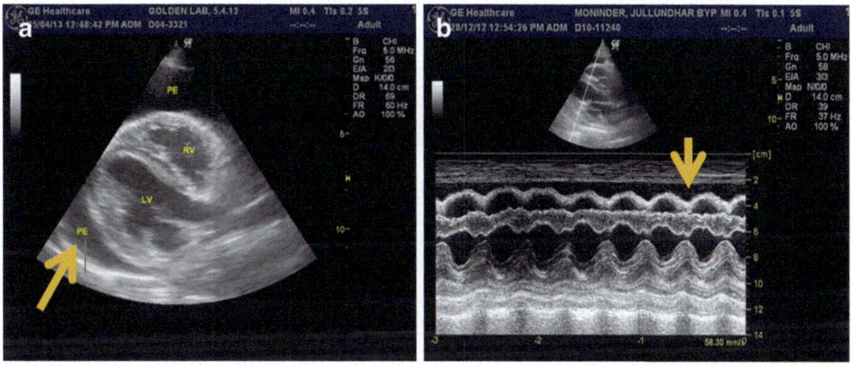

2-D image M-Mode image

Fig. 14.4 (**a**, **b**) Echocardiogram of a dog showing pericardial effusion (PE)
(This figure is with the courtesy of Dr. Neetu Saini, Associate Professor, Department of Medicine, Guru Angad Dev Veterinary and Animal Sciences University, Ludhiana)

Enlargement of the heart (Fig. 14.5a, b).
Changed heart shape. Cardiac silhouette becomes rounded or globoid (Fig. 14.5c).
Enlargement of vena cava (Fig. 14.5d).
Pleural effusions.
Tracheal elevation.

5. Laboratory findings—Pericardial fluid analysis is helpful in revealing the nature of effusions. For example the presence of erythrocytes, increased nucleated cell count, and high protein (more than 3.0 g/dL) may suggest hemangiosarcoma. Increased pH of pericardial fluid (more than 7.3) is seen in dogs with neoplasia. Hemogram may show nucleated erythrocytes, schistocytes, regenerative changes, and almost normal packed cell volume in hemangiosarcoma.

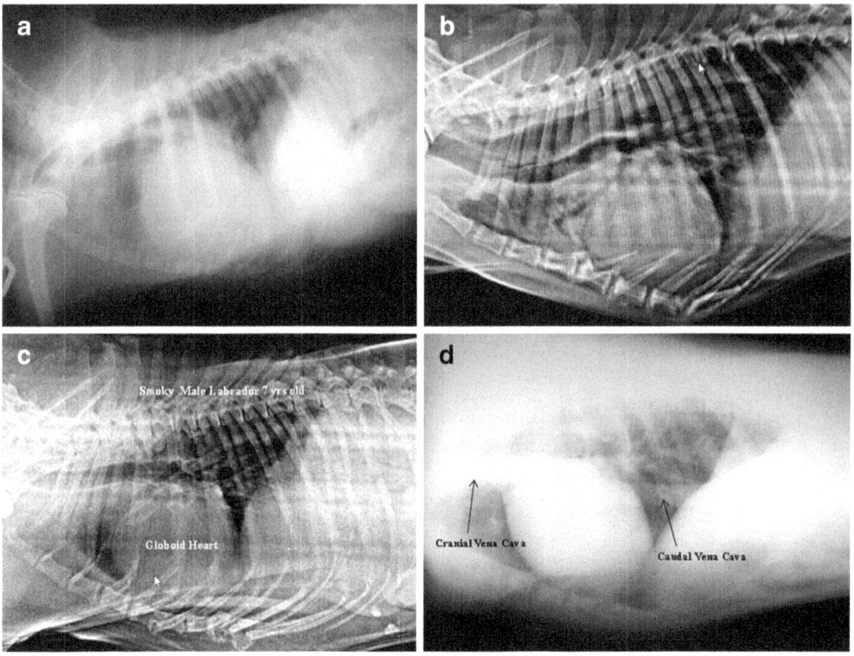

Fig. 14.5 (**a–d**) Radiographs of dogs showing globoid heart silhouette and distended vena cava suggestive of pericardial effusion

14.2 Therapy

Medical management of pericardial effusion may not always be rewarding. It needs pericardiocentesis repeatedly. The following protocol may be adopted:

- Pericardiocentesis. Repeated pericardiocentesis may be required.
- Prednisolone @ 0.5 mg/kg orally twice daily tapering over 7–14 days. It may reduce recurrence in idiopathic cases.
- Heart failure is to be managed with diuretics and pericardiocentesis.
- Treatment with digitalis, vasodilators, and angiotensin-converting enzyme inhibitors (ACEI) is contraindicated in dogs with pericardial effusions.
- Surgical removal of the pericardial sac is a better treatment option to provide quality life.

14.3 Prognosis

- Prognosis in cases of PE is good only when the pericardial sac is surgically removed.
- Repeated pericardiocentesis is needed.
- Pericardial effusions of neoplastic origin have poor prognosis.

Further Reading

Allen DG (1991) Small animal medicine. J.B. Lippincott Company, Philadelphia, PA

Birchard SJ, Sherding RG (2000) Saunders manual of small animal practice, 2nd edn. W.B. Saunders Company, Philadelphia, PA

Ettinger SJ, Feldman EC (2000) Textbook of veterinary internal medicine. Diseases of the dog and cat, 5th edn. W.B. Saunders Company, Philadelphia, PA

Morgan RV (1992) Handbook of small animal practice, 2nd edn. W.B. Saunders Company, Philadelphia, PA

Nelson RW, Couto CG (1998) Small animal internal medicine, 2nd edn. C.G. Mosby, St. Louis, MO

Heart Failure, Cardiopulmonary Arrest, and Cardiogenic Shock

15

15.1 Heart Failure

Sometimes the terms heart failure, heart attack, and heart disease are loosely used. In fact heart attack and heart failure are due to heart diseases, while all heart diseases may not have heart failure or heart attack. Heart attack, most common in humans, is an acute condition when heart functioning is stopped suddenly. It is due to severe arrhythmias or blockage of the coronary arteries. On the other hand, heart failure is a gradual process which is an outcome of severe heart disease and is characterized by inability of the heart to pump blood efficiently to different parts of the body. It is commonly due to mitral valve insufficiency and dilated cardiomyopathy. Chronic mitral valve insufficiency is seen primarily in small breeds of dog, while dilated cardiomyopathy is seen most commonly in adults of large and medium breeds such as Doberman, Great Dane, Labrador, Irish wolfhound, German Shepherd, Scottish Dearhound, etc. Timely diagnosis is of great importance in the management of the cases with heart failure as delay may be catastrophic. Compromised heart in heart failure leads to systolic (forward failure) and/or diastolic (backward failure) dysfunctions causing inadequate perfusion of tissues and/or edema of dependent parts with varying clinical manifestations. Various cardiac and systemic compensatory responses are also triggered in the body in response to failing heart.

Backward failure—It is a diastolic dysfunction. It leads to tissue congestion and edema due to increased capillary hydrostatic pressure or increased arterial pressure.

(a) Right-sided heart backward failure—The backward failure of the right heart during diastole leads to hepatic congestion, ascites, and hydrothorax.
(b) Left-sided heart backward failure—The backward failure of the left heart during diastole leads to pulmonary congestion and edema.

Forward failure—The heart is unable to pump the blood effectively to the different organs of the body. It is a systolic dysfunction that leads to a poor cardiac output clinically characterized by exhaustion and exercise intolerance.

J. P. Varshney, *Electrocardiography in Veterinary Medicine*,
https://doi.org/10.1007/978-981-15-3699-1_15

(a) Right-sided heart forward failure—The forward failure of the right heart during systole leads to inadequate systemic blood flow.
(b) Left-sided heart forward failure—The forward failure of the left heart during systole leads to inadequate tissue perfusion.

15.1.1 Common Causes of Heart Failure

- Poor contraction of myocardium.
- Chamber enlargement.

15.1.2 Diagnostic Approach in Heart Failure

Diagnosis of heart failure in dogs in early stage is sometimes very illusive where characteristic features are absent. Therefore a systematic approach right from history, detailed clinical examination, use of noninvasive modern diagnostic technology (electrocardiography, echocardiography, and radiography), to the use of cardiac biomarkers is to be taken in each suspected case of heart failure.

15.1.2.1 History

- History in cases of heart failure is quite variable except marked weakness.
- Some owners report disruption of sleep due to intense, persistent nocturnal coughing and dyspnea.
- Dogs show marked weakness and easy exertion.
- Dogs show panting on exercise.
- Dogs are uncomfortable with exercise and show no interest physical activities.
- Dogs are reluctant to lie down. Panting increases in recumbency (orthopnea).
- Distension of abdomen and/or edema of hind limbs is also reported.

  The history of persistent cough may indicate left-sided heart involvement. But solely it may not be diagnostic as it is also seen in cases with tracheal collapse, pulmonary edema, dirofilariasis, compression of the left main stem bronchus, and chronic pulmonary disease. Development of dyspnea in dogs with heart failure is a late sign, while in humans it is an early sign of failing heart. All these complaints are nonspecific but definitely create suspicion of heart failure and suggest further investigations to exclude or include other causes.

15.1.2.2 Clinical Manifestations

Clinical manifestations of heart failure relate to arrhythmias, forward failure, and backward failure (Fig. 15.1). In very early stage, nonspecific signs such as marked weakness and laziness may only be evident. With a period of time, symptoms may appear in various permutations and combinations.

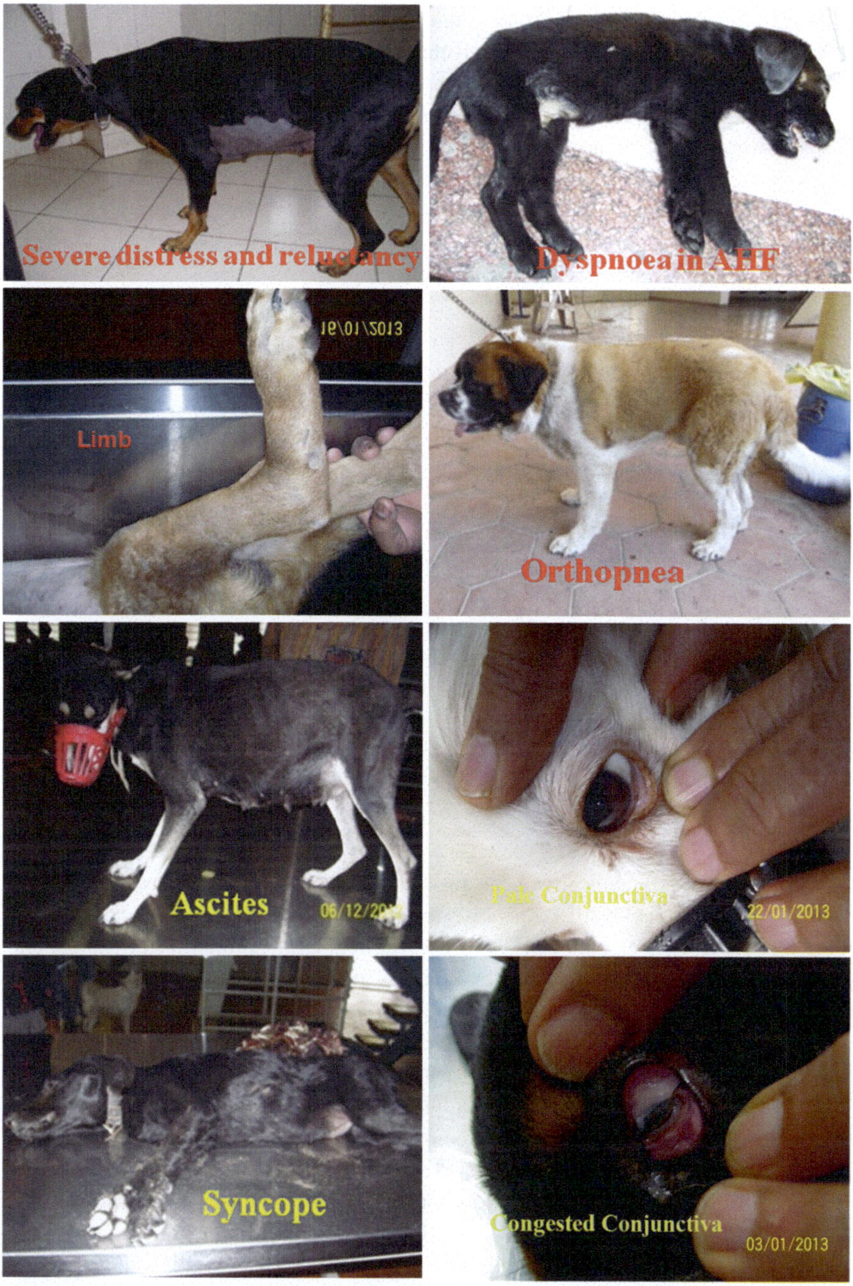

Fig. 15.1 Dogs with heart failure showing different clinical manifestations

(a) Heart and lung abnormalities.
- Arrhythmia on auscultation of the heart.
- Gallop sounds (in severe left ventricular volume overload).
- Rales and crackling sound on thorax auscultation in dorso-caudal area of the lung.

(b) Signs related to low output/forward failure.
- Weakness/lethargy/dullness/depression.
- Exercise intolerance.
- Sudden unconsciousness/syncope/collapse.
- Hypothermia/cool extremities.
- Dehydration/prolonged capillary refill time.
- Weak femoral pulse and pulse deficit.

(c) Signs related to backward failure.
- Persistent unproductive coughing.
- Dyspnea/orthopnea/panting.
- Reluctance to lie down.
- Jugular vein distension (hepatojugular reflux).
- Abdominal distension (ascites).
- Hydrothorax.
- Peripheral edema.
- Muddy or cyanotic mucus membrane.

(d) Other signs.
- Anorexia/decreased appetite.
- Weight loss.
- Anxiety/anxious look.
- Sudden death.

Classical cases of congestive heart failure are clinically characterized by signs of both forward and backward failure. Respiratory diseases also exhibit dyspnea and rales; therefore their origin needs to be ascertained. Clinical manifestations depend on the side of the heart involved. Nevertheless, decreased stamina, difficult breathing, sleepiness, decreased appetite, fainting, and coughing are commonly associated with heart failure due to either mitral valve insufficiency or dilated cardiomyopathy. Right-sided heart failure is associated with poor venous return causing leakage of some blood into the right atrium through the tricuspid valve backing the blood into circulation and consequently causing congestion. The fluid is build up in the abdomen (ascites) and/or in limbs (peripheral edema).

In case of left heart failure, due to poor contractility, whole blood is not pushed into circulation, and some blood is leaked into the left atrium through mitral valve and then back into the lungs causing lung edema leading to coughing and dyspnea.

Classification of Functional Heart Failure

New York Heart Association (NYHA) has modified and suggested functional classification of heart failure into four categories. The classification is based on the development of clinical sign in relation to different grades of activity.

Class I. There are no symptoms in class I heart failure. Clinical signs do not develop even on exercise.

Class II. Clinical signs such as fatigue, dyspnea, or coughing develop on exercise in class II heart lure.

Class III. Clinical signs such as fatigue, dyspnea, or coughing develop with minimum activity.

Class IV. Clinical signs such as dyspnea and coughing are seen even at rest. These signs are exaggerated with minimum physical activity making the dog highly uncomfortable.

15.1.2.3 Cardiac Biomarkers and Biochemical Indices

CBC, liver and kidney profile, thyroid profile, and sodium and potassium estimations are generally done to create basic profile to exclude or include liver, kidney, ions, or thyroid involvement. But none of these is specific for heart failure. Recently cardiac biomarkers such as cTn-I and NT pro-BNP are being estimated to detect myocardial damage even in early stage when signs of heart failure are not apparent. These biomarkers indicate myocardial damage. Lowered vitamin D_3 level (4.2–28.9 ng/mL, mean 13.55 ± 1.67 ng/mL) and increased cTn-I level (mean 5.19 ± 1.78 ng/mL, range 0.1–28.97 ng/mL) have recently been observed in dogs with left heart failure at this hospital (Varshney et al. 2019). Some reports indicate that lower level of vitamin D_3 predisposes to heart failure. Oxygen tension (PvO_2) is also lowered in dogs with heart failure.

15.1.2.4 Radiographic Findings in Heart Failure

All cases suspected with heart failure showing clinical signs of dyspnea, orthopnea, panting, coughing, syncope, seizures, or collapse should be subjected to chest radiography. X-ray is an important diagnostic modality of great assistance in differentiating respiratory sign of cardiac and pulmonary origin. Right lateral and ventro-dorsal radiographs should be taken in all cases of suspected heart failure. Radiographs may reveal following changes:

- Pulmonary venous and capillary congestion.
- Change in cardiac silhouette (Fig. 15.2a).
- Interstitial and/or alveolar edema (Fig. 15.2b).
- Hepatomegaly/splenomegaly (Fig. 15.3).
- Pleural effusions.
- Ground glass appearance of the abdomen suggesting ascites (Fig. 15.4).
- Compression of main stem bronchi (Fig. 15.5).

Pulmonary edema in dorso-caudal hilar region gives strong indication of left heart failure. Cases showing these radiographic changes should invariably be subjected to electrocardiography.

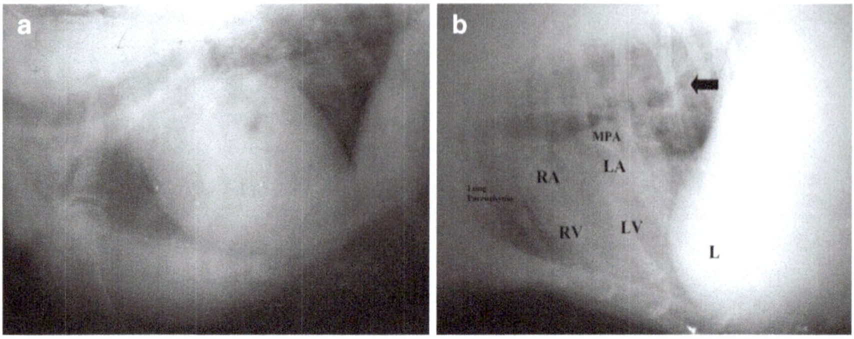

Fig. 15.2 Radiographs of dogs with congestive heart failure. (**a**) Showing changed cardiac silhouette with increased sternal contact. (**b**) Showing pulmonary edema in dorso-caudal hilar region of the lung

Fig. 15.3 Radiograph of a dog with congestive heart failure showing hepato- and splenomegaly

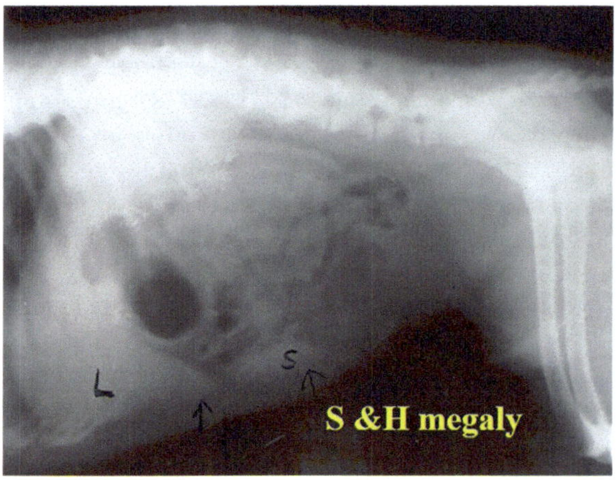

Fig. 15.4 Radiograph of a dog with congestive heart failure showing ground glass appearance suggestive of ascites

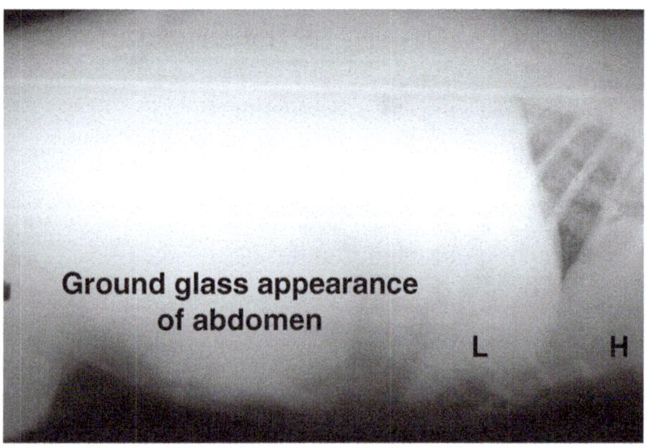

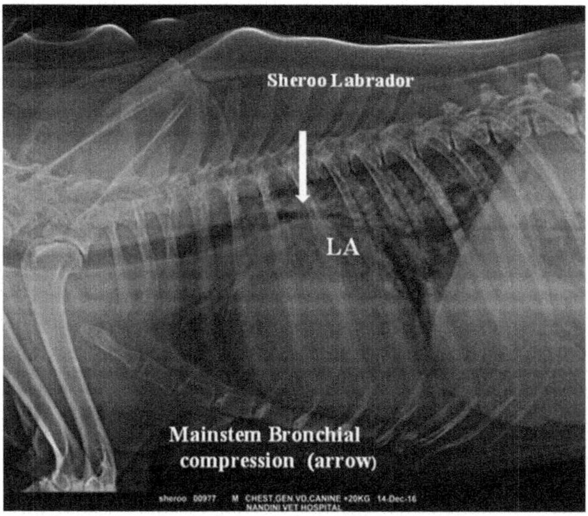

Fig. 15.5 Right lateral radiograph of an adult male Labrador with congestive heart failure showing main stem bronchial compression

15.1.2.5 Electrocardiographic Findings in Heart Failure

Electrocardiogram in heart failure is variable depending upon underlying cause. Any of the following electrocardiographic changes may be associated with heart failure. Electrocardiographic features of these abnormalities have been described in detail in previous chapters.

(a) Sinus rhythm.
(b) Sinus tachycardia.
(c) Sinus bradycardia.
(d) Sinus arrest.
(e) Atrial flutter/fibrillations.
(f) Atrial premature complexes.
(g) Atrial tachycardia.
(h) "P" mitrale (broad P wave)
(i) Ta wave (right atrial enlargement).
(j) Tall "R" (increased amplitude of R wave).
(k) Broad "QRS".
(l) Low-voltage complexes.
(m) Ventricular escape complex.
(n) Ventricular premature complex.
(o) Ventricular tachycardia.
(p) S-T segment changes.
(q) Sinus arrest.
(r) Prolonged Q-T interval.
(s) Change in polarity of "T" wave.
(t) Bundle branch blocks.
(u) Alternans of "R".

15.1.2.6 Echocardiographic Findings in Heart Failure

Echocardiography is another important noninvasive diagnostic modality for imaging the heart and its surrounding structures. It is one step ahead of electrocardiography and provides important information about size of the heart chambers, thickness of the wall, wall motion, valve configuration and motion, the proximal great vessels, and detection of the pericardial and pleural effusions which is not possible by electrocardiography. Identification of mass lesion within and adjacent to the heart is also possible by echocardiographic examination. Like other diagnostic modalities, its interpretation should be viewed within the context of a thorough history, clinical picture, complete cardiovascular examination, and other tests.

15.1.3 Treatment of Heart Failure

Since congestive heart failure is a progressive condition with physical changes in the valves and weakened ventricular wall, the primary aim of the treatment is to relieve distressing symptoms and make the life of the dog comfortable. Specific treatment of congestive heart failure depends on the underlying heart disease and its severity. Nevertheless, the main goal of the treatment is to reduce fluid buildup and to increase contractibility of the heart to improve the quality and length of pet's life. Nowadays a variety of medications and specific diets are available to help in achieving this target. Cough suppressant (butorphanol or hydrocodone), restricting sodium in the diet (to reduce water holding), cardiac diet, providing rest, oxygen therapy, and medications such as diuretics, ACE inhibitors, vasodilators, and positive inotropes are the integral part of the management of congestive heart failure to relieve fluid accumulation, to maintain or increase cardiac output, to minimize myocardial oxygen demand, and to control rhythm disturbances.

Relieving Fluid Accumulation (Congestion and Edema)—Fluid buildup is reduced with the use of low-sodium diet and diuretics.

- Low-sodium diets. Dogs with heart failure should be given diets low in sodium. Potassium chloride may be added to improve the flavor of the diet.
- Supplementation of L-carnitine (50 mg/kg orally twice or thrice daily) may have an added advantage.

  Use of diuretics—Diuretics are the main part of therapy in the management of heart failure.

  Furosemide is used @ 2–4 mg/kg BID-TID PO or 2–8 mg/kg IV up to every hour. If furosemide alone is not working satisfactorily, triamterene (2–4 mg/kg PO daily) can be considered. Spironolactone (2–4 mg/kg per day PO), hydrochlorothiazide (2–4 mg/kg BID PO), and chlorothiazide (20–40 mg/kg BID-TID PO) are the other diuretics of choice.

Increasing Myocardial Contractility—Contractility of the myocardium is increased by using following drugs:

- Digoxin is used @ 0.22 mg/m^2 or 0.005–0.01 mg/kg PO BID. This dose is followed by the maintenance dose of 0.01–0.02 mg/kg orally.
- Dobutamine is advocated @5–20 μg/kg/min as a constant rate infusion.
- Dopamine is recommended @ 1–10 μg/kg/min. as a constant rate infusion.
- Milrinone is recommended @ 0.5–1.0 mg/kg PO.TID-QID. It may cause ventricular arrhythmia. Hence ECG monitoring is must whenever the drug milrinone is used.
- Pimobendan is also recommended for increasing myocardial contractility. It is given @ 0.25 mg/kg PO BID on empty stomach.
- Carvedilol can also serve this purpose and is used @ 0.5 mg/kg PO BID.

Decreasing Cardiac Workload—It is equally important to reduce work load of the failing heart. Cardiac workload can be reduced by using following drugs:

- Arteriolar dilators—Hydralazine, captopril, enalapril, and prazosin are used for this purpose. Hydralazine is given @ 1–3 mg/kg BID PO, while captopril is given @ 2.0 mg/kg TID PO. Their efficacy is to be assessed with blood pressure monitoring. Enalapril @0.5 mg/kg OD or prazosin @ 1 mg BID-TID PO can also be used.
- Venodilators—Nitroglycerine or isosorbide dinitrate can be used for this purpose. Nitroglycerine ointment is applied on the skin @ 0.25–2 in. TID-QID as a cutaneous application. Isosorbide dinitrate is used @1.0–1.5 mg/kg TID-QID PO).
- Complete rest—Cage rest may be required with life-threatening heart attack.

Improving Oxygen Supply—Availability of oxygen at tissue level is a very important factor. Oxygen supply in cases of heart failure can be improved by taking following measures:

- Check airways, remove froth. Nebulization using 20% ethanol may reduce airways foaming.
- Oxygen (40–50%) through oxygen mask @6–10 L/min.
- Bronchial dilators—Aminophylline can be used for this purpose as a slow intravenous or intramuscular injection. It can also be given subcutaneously. Aminophylline may be given TID-QID. Its dose rate is 10–20% higher than theophylline. Theophylline is also given to dilate bronchi @ 9 mg/kg TID-QID.

Arrhythmia Management—Use of antiarrhythmic drugs as per nature of arrhythmia is recommended. Details of antiarrhythmic therapy are given in the chapter of arrhythmias.

2. Laboratory investigations—Detailed laboratory investigations are needed in these cases because of comorbidities.
   - Hemogram to rule in or rule out anemia and blood protozoan/rickettsial infections.
   - Serum electrolyte (sodium, potassium, chloride, calcium) for ion imbalance, if any.
   - Thyroid hormones for thyroid disorders, if any.
   - Cortisol level for adrenal disorders, if any.
   - Liver function tests to detect liver failure if any.
   - Kidney function tests to detect renal failure if any.
   - Other routine biochemical examination.
3. Blood pressure monitoring—It is needed to detect blood pressure changes.
4. Thoracic radiographs—Chest X-rays should be taken for detecting coexisting respiratory disease, if any.
5. Complete urine analysis—Complete urine examination including culture examination should be conducted in such cases. It is also needed to monitor compliance with dietary restrictions and use of diuretics.
6. Electrocardiography—It is necessary in all cases suspected of heart failure. It assists in detecting arrhythmias, heart blocks, or changes in chamber size, if any.
7. Echocardiography—It is needed for detecting valvular defect and distinguishing dilated and hypertrophic cardiomyopathy.

15.2.2 Therapy

Treatment protocol has already been outlined under heart failure. But doses have to be increased to achieve the goal.

1. Refractory Heart Failure.
   (a) Angiotensin-converting enzyme inhibitors (ACE Inhibitors) and vasodilators.
   - Enalapril or lisinopril is given @ 0.5 mg/kg BID. Dose may be increased to 0.75 mg/kg if response to 0.5 mg/kg is poor.
   - Hydralazine is given @ 0.5 mg/kg BID as an initial dose. If response is poor, it can be gradually increased to a maximum of 1.5 mg/kg till clinical signs are eliminated.
   - Nitroglycerine is used as 2% ointment on inner surface of pinna (1/4–3/4 in. every 6–8 h).
   - Isosorbide nitrate (@0.5–2.0 mg/kg TID/QID) is used when ACE inhibitors are not tolerated well.

   (b) Diuretics.
   - Furosemide is used in higher doses (4.4 mg/kg IV) as continuous infusion in refractory cases.
   - Combination of thiazide- like drug with furosemide is a better choice than using high doses of furosemide. Chlorothiazide may be added @20–40 mg/kg OD or BID. Hydrochlorothiazide (@2–4 mg/kg) is an another alternative to be combined with furosemide.

(c) Positive inotropic drugs.
  - Dobutamine—It is sometimes effective in dogs with dilated cardiomyopathy and intractable heart failure. It is infused @ 2–5 μg/kg per min for 12–24 h. Infusion is repeated at 2–6 weeks interval.

(d) General measures.
  - Restrict exercise.
  - Low-sodium diet.
  - Complete rest.

2. Uncontrollable Congestive Heart Failure.
   Uncontrollable congestive heart failure is characterized by severe pulmonary edema of cardiac origin with or without pleural and/or peritoneal effusions. The dog with uncontrollable congestive heart failure remains in great stress requiring absolute cage rest and oxygen therapy as a first line of therapy followed by other measures as outlined below:
   - Complete cage rest. No physical activity.
   - Oxygen supplementation (50%) @ 6–10 L per minute.
   - Keep airways patent by sucking froth.
   - Tracheal intubation.
   - Thoracocentesis is desirable if pleural effusions have developed.
   - Furosemide in higher doses (2–5 mg/kg IV or IM 2–4 hourly) until respiration rate is decreased. Then reduce the dose to BID or TID.
   - Nitroglycerine ointment 2% 0.5–1.5 in. cutaneously TID-QID.
   - Nitroprusside @ 0.5–1 μg/kg/minute as a constant rate infusion (CRI).
   - Morphine @ 0.05–0.1 mg/kg IV boluses.
   - Aminophylline @6–10 mg/kg slow IV, IM, SC, PO TID-QID.
   - Diazepam@ 5–10 mg IV to reduce anxiety.
   - Enalapril @0.5 mg/kg PO OD-BID or hydralazine @ 0.5–1.0 mg/kg PO repeated at 2–4 h and then BID.
   - Dobutamine @1–10 μg/kg/min CRI or dopamine 1–10 μg/kg/min CRI or amrinone 1–3 mg/kg IV.
   - Restricted sodium in diet.
   - Pimobendan @ 0.25 mg/kg PO BID.

15.3 Common Drugs for Heart Failure: At a Glance

The following group of the drugs are used in the management of the heart failure.

1. **Diuretics**—The purpose of diuretics is to reduce fluid overload and thus provide relief in lung edema, ascites, and pleural effusion.
   Furosemide (Lasix, Ridema, etc. injectable and tablets)—It is a loop diuretic and also has venodilation properties. It may cause hypokalemia in anorectic dogs. Cats are more prone.
   Dog—2–5 mg/kg BID/TID IV, IM, PO or 1–2 mg/kg/h. IV CRI (constant rate infusion).
   Cat—1–2 mg/kg BID/TID IV, IM, PO or 0.25–1.0 mg/kg/h IV CRI.

Thiazides (tablets Aquazide, hydride, thiazide, Bpzide, Esidrex 12.5/25 mg)—It reduces the dose of furosemide by 50% when used with furosemide and augment dieresis in cases where furosemide alone is ineffective.
Dogs and cats—2–4 mg/kg PO BID (alone), 1–2 mg/kg PO BID (with furosemide at 50% dose rate).
Spironolactone (Aldactone Tables 25/100 mg with furosemide, Fruselac, Lasilactone, Spiromide tabs)—It is a potassium-sparing weak diuretic. Used alone or in combination with furosemide.
Dogs and cats 1–2 mg/kg PO BID

2. **Venodilators**—These drugs reduce preload by trapping blood in venous circulation and thus reduce pulmonary edema and consumption of myocardial oxygen. The disadvantage of venodilators is hypotension.
   Nitroglycerine 2% ointment (Myovin ointment, Nitroderm patches)—It is used topically.
   Small dogs—1/4 to 1/2 in. topically TID.
   Large dogs—1/4 to 2 in. topically TID.
   Cats—1/8 to 1/4 in. topically TID.
   Isosorbide dinitrate (sorbitrate Tables 5 and 10 mg, Isordil 10 mg tab, Ditrate Table 5 or 10 mg).
   Dogs—0.5 to 2.0 mg/kg PO BID.
   Isosorbide mononitrate (Angitab, Anginex, Isotec, Monosorbitrate, Monopark tabs of 10–60 mg).
   Dogs—0.25 to 2.0 mg/kg PO BID
3. **Arteriolar Dilators**—These drugs reduce afterload, increase stroke volume, and reduce myocardial oxygen consumption. Arteriolar dilators also cause hypotension and may reflex tachycardia. The drugs are indicated in cases of chronic mitral valve insufficiency or dilated cardiomyopathy and are contraindicated in valvular stenosis, hypertrophic cardiomyopathy, hypotension, and cardiac tamponade.
   (a) Calcium channel blockers—amlodipine (AM-5, Amcard, Amodep, Amtas, Amlol, Amdepin, Amlosafe of 2.5–10 mg tabs).
   Dogs—0.05 to 0.2 mg/Kg PO SID.
   Cats—0.625 mg/cat PO SID
   (b) Arterial smooth muscle relaxants—Hydralazine (Apresoline, Hydrila 25 mg tablets).
   Dogs—0.5 to 3.0 mg /kg PO BID.
   Cats—2.5 mg/cat PO SID.
   The drug causes reflex increase in heart rate and fluid retention. So it is not used alone. It is usually prescribed with beta-blockers (propranolol) and diuretic.
   (c) Alpha receptor blocker—Prazosin (Prazopress 1 and 2 mg tab, Minipress XL 2.5 and 5.0 mg tab).
   (d) ACE inhibitors—Enalapril (Bol, Enapril, Dilvas, Minipril, Encardil, Tenam Tables 2.5/5.0/10 mg).
   Dogs.0.5 mg/kg PO SID-BID.

Cats 0.5 mg/kg PO SID.
Benazepril (Benace 5/10/20 mg FC tabs).
Dogs 0.5 mg/kg PO BID.
Cats 0.25–0.5 mg/kg PO SID.
Imidapril (Tanatril 5/10 mg tabs).
Ramipril (Ramicard, Ramic, Ramichek, Zipril, Ramil, Ramipress 1.25/2.5/5 mg tabs)

4. **Positive Inotropes**—These drugs increase contractility and are suitable in systolic dysfunction. Not recommended in hypertrophic cardiomyopathy.
   (a) Pimobendan—Has been detailed heart failure.
   (b) Digoxin—It inhibits the Na/K ATPase pump and increase cytosolic calcium. It is a weak positive inotrope and drug of choice in cases of sinus tachycardia, atrial fibrillation, and supraventricular arrhythmias. It potentiates ventricular arrhythmias. Therefore, electrocardiographic monitoring should be regularly done when digoxin is used for longer periods.
   Dogs 0.006 mg/kg PO BID.
   Cats 0.03125 mg PO 24–48 h
   (c) Dobutamine—Generally given as CRI (constant rate infusion). It may cause tachycardia and may promote ventricular arrhythmias. It is recommended to be used under intensive care unit (ICU) settings. The drug should not be used in cases of hypertrophic cardiomyopathy.
   Dogs 5–15 μg/kg/min. CRI.
   Cats 1–2 μg/kg/min CRI
5. **Negative Inotropes**—Their action is via negative chronotropic properties. The drugs are suitable in hypertrophic cardiomyopathy; supraventricular arrhythmias; and stenosis of the aortic, subaortic, or pulmonic valves. These drugs reduce myocardial oxygen demand and heart rate.
   Beta-blockers—There are three types of beta-blockers (selective, nonselective, or third-generation beta-blockers). These drugs can protect against sudden arrhythmic death. Beta-blockers may cause hypotension, bradycardia, and bronchospasm.
   Selective beta-blockers:
   - Atenolol.
     Dogs @5–12.5 mg PO SID-BID or 0.25–1.0 mg/kg PO SID-BID.
     Cats @5–12.5 mg PO SID-BID
   - Metoprolol.
     Dogs and cats @ 0.25–1.0 mg/kg PO BID-TID.

   Nonselective
   - Propranolol.
     Dogs and cats @ 0.2–1.0 mg/kg PO BID-TID.
     Third-generation nonselective beta- and alpha blocker
   - Carvedilol.
     Dogs @0.2–1.5 mg /kg PO BID. Low doses are preferred.
     Its use in cats is not known.

15.4 Cardiopulmonary Arrest (CPA)

CPA is a sudden arrest of cardiac function adversely affecting the circulation and functional ventilation.

15.4.1 Etiology

There are many causes of cardiopulmonary arrest. Some of the common causes responsible for CPA are enumerated below.

- Respiratory failure.
- Circulatory failure.
- Severe anemia.
- Anesthesia.
- Myocardial failure.
- Toxemia.
- Electric shock.
- Trauma of the central nervous system.
- Electrolyte imbalance.
- Acid-base disturbances.
- Severe arrhythmias.

15.4.2 Diagnostic Profile

1. Clinical Signs.
    - Bradycardia.
    - Weak or absence of pulse.
    - No audible/very feeble heart sound.
    - Slow respiration.
    - Unconsciousness or deteriorating consciousness.
    - Cyanotic mucus membranes.
    - Dyspnea or gasping or irregular breathing.
    - Dilated pupil.
2. Electrocardiography.
    - Bradycardia or ventricular asystole.
    - Ventricular flutter or fibrillation.
    - Ectopic ventricular depolarization with bizarre QRS complex.
    - Progressive "T" wave enlargement.
    - S-T segment changes.

15.4.3 Treatment

1. Endotracheal intubation.
2. Oxygen therapy (100% oxygen) (Fig. 15.6).

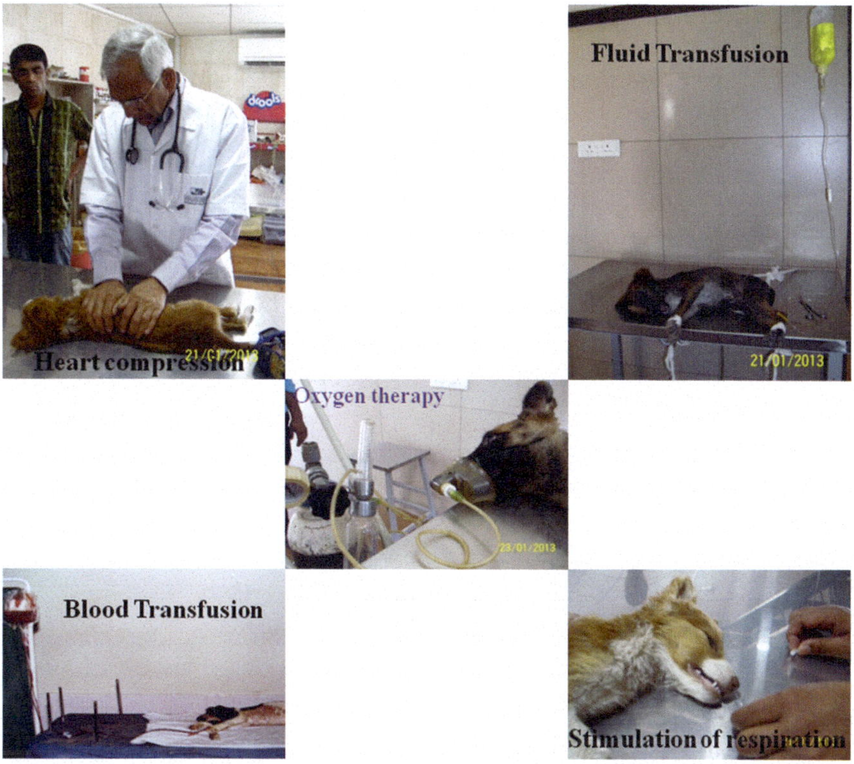

Fig. 15.6 Resuscitation measures in cardiopulmonary arrest

3. External cardiac massage. Keep the dog in right lateral recumbency and apply compression over fifth rib interspace (Fig. 15.6).
4. Stimulate respiration (Fig. 15.6).
5. Medical management.
   - Epinephrine (1:1000) @ 0.2 mg/kg IV or intratracheal.
     It can be repeated in lower doses every 5–10 min during prolonged resuscitation.
   - Fluid therapy (Fig. 15.6).
   - Sodium bicarbonate 1 mEq /kg IV if acidosis is suspected.
   - Blood transfusion in cases of severe anemia (Fig. 15.6).
   - Atropine sulfate IV or glycopyrrolate in case of bradycardia.
   - Calcium (10% calcium chloride 0.1 mL/kg or 10% calcium gluconate 0.4 mL/kg IV) in case of hyperkalemia and hypocalcaemia.
   - Use defibrillator and epinephrine in dogs with ventricular fibrillation.
   - Lidocaine (@ 2–4 mg/kg IV) is used to manage ventricular tachycardia.
   - Continuous ECG monitoring is desirable in such cases.

15.5 Cardiogenic Shock

Shock is common clinical manifestation in seriously ailing dogs demanding aggressive but careful therapeutic approach. Cardiogenic shock is seen in dogs with cardiac disease. Acute circulatory failure of severe nature, irrespective of etiology, results into cardiogenic shock leading to failure of supply of nutrients and oxygen to vital organs and removal of accumulated metabolites. It is mainly due to forward heart failure owing to failure of heart pump. Broadly cardiogenic shock can be defined as inadequate and compromised cellular metabolism as a result of cardiac dysfunction despite adequate intravascular volume. Shock leads to multiple organ failure and death if not attended timely and properly. Cardiogenic shock is characterized by a fall in cardiac output due to impaired systolic function (arrhythmias, ruptured chordae tendineae, aortic or pulmonary stasis, primary cardiomyopathy) or secondary to impaired cardiac filling in diastole (cardiac tamponade, tension pneumothorax, positive pressure ventilation, gastric dilation, or volvulus).

15.5.1 Diagnostic Profile

1. Clinical Signs—Following clinical signs are observed:
   - Tachycardia or bradyarrhythmia.
   - Weak arterial pulse.
   - Arterial hypotension.
   - Pale mucus membranes.
   - Cold extremities.
   - Increased capillary refill time.
   - Peripheral cyanosis.
   - Oliguria.
   - Tachypnea.
   - Decreased intensity of heart sounds.
   - Murmurs.
   - Pulmonary edema (congestive heart failure).
   - Systemic venous distension.
   - Hypothermia.
   - Decreased urinary output.
   - Dilated pupils.
   - Sunken eyes.
   - Changed mentation (depression, disorientation, unresponsiveness).
   - Coma.
   - Death.
2. Electrocardiography—Electrocardiogram may reveal arrhythmias and/or blocks.
   - Sustained ventricular tachycardia.
   - Sustained supraventricular tachycardia.
   - Uncontrolled atrial fibrillation or flutters.
   - Atrial standstill.

- Severe sinus bradycardia.
- Complete heart block.

3. Echocardiography—Echocardiographic findings are variable.
   - Echocardiographic findings vary as per underlying cause.
   - Evidence of systolic dysfunction despite an adequate end-systolic volume.
4. Chest Radiography—It may reveal change in heart shape and lung congestion.
   - May be done when dog is stable.
   - Abnormal cardiac silhouette.
   - Signs of congestive heart failure (enlarged pulmonary, veins, alveolar, or interstitial pattern in perihilar region or pleural effusion).
5. Laboratory Investigations—Laboratory investigations are recommended to rule in or rule out many diseases. Following changes may be observed:
   - Increased PCV is suggestive of hemoconcentration and dehydration.
   - Prerenal azotemia.
   - Hypoglycemia.
   - Hypokalemia.
   - Increased liver-specific enzymes.
   - Metabolic acidosis.
   - Compensatory respiratory alkalosis.

15.5.2 Therapy

Treatment of cardiogenic shock is undertaken with following drugs. Their details have been under therapy of heart failure.

- Inotropic support.
- Diuretics.
- Vasodilators.
- Fluid therapy.

References

Varshney JP, Huma ZI, Sharma N (2019) Evaluation of cardiac troponin-I and vitamin D_3 in dogs with left heart failure and its management using pimobendan. Vet Pract 20:74–77

Further Reading

Chu KM, Shieh SM, Hu OY (1995) Pharmacokinetics and pharmacodynamics of enantiomers of pimobendan in patients with dilated cardiomyopathy and congestive heart failure after single and repeated oral dosing. Clin Pharmacol Ther 57:610–621

Deepti BR, Yathiraj S, Ramesh PT, Rangnath L, Narayanaswamy HD (2016) Diagnosis and treatment of congestive heart failure in canines. Intas Polivet 17:129–131

Ettinger SJ, Suter PF (1970) The recognition of cardiac disease and congestive heart failure. In: Canine cardiology. WB Saunders, Philadelphia, PA, p 215

Fuentes VL (2004) Use of pimobendan in the management of heart failure. Vet Clin North Am Small Anim Pract 34:1145–1155

Hagemeijer F (1993) Calcium sensitization with pimobendan: pharmacology, haemodynamic improvement, and sudden death in patients with chronic congestive heart failure. Eur Heart J 14:551–566

Lefbon BK (2005) Sildenafil and novel cardiovascular therapies. Proc Am Coll Vet Intern Med

Moses BL (1992) Cardiac arrhythmias and cardiopulmonary arrest. In: Morgan RV (ed) Handbook of small animal practice, 2nd edn. W.B. Saunders Company, Philadelphia, London, Toronto, Montreal, Sydney, Tokyo, pp 71–91

Sasaki H, Hiroshi A, Fujita M, Takahama H, Wakeno W, Ito S, Ogai A, Asakura M, Kim J, Minamino T, Takashima S, Sanada S, Sugimachi M, Komamua K, Mochizuki N, Kitakaze M (2009) Metformin prevents progression of heart failure in dogs. Circulation 119:2568–2577

Canine Electrocardiograms in Diseases 16

16.1 Amitraz Toxicity (Fig. 16.1)

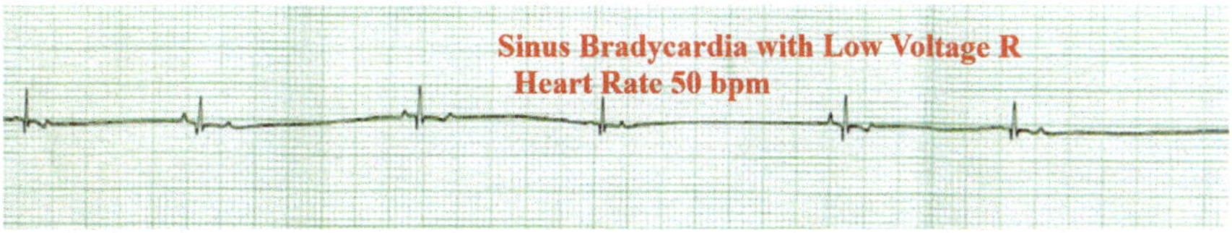

Fig. 16.1 Electrocardiogram (lead II, sensitivity 1, speed 25 mm/s) of a 4-year-old German Shepherd with amitraz toxicity showing sinus bradycardia (heart rate 50 bpm) and low voltage not only for R but for P and T waves also

16.2 Anesthesia (Figs. 16.2 and 16.3)

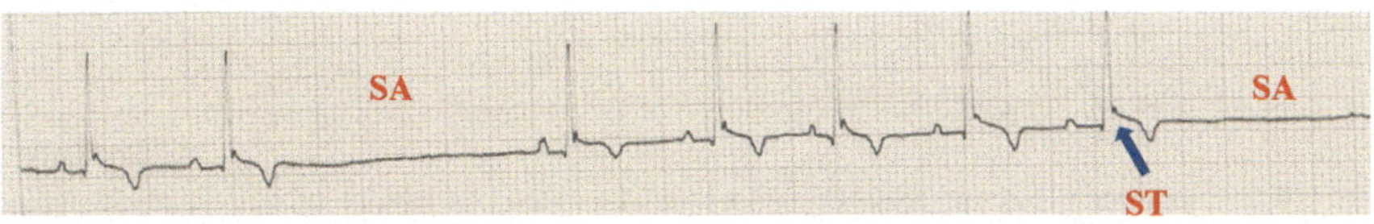

Fig. 16.2 Electrocardiogram (lead II, sensitivity 1, speed 25 mm/s) of an adult male German Shepherd showing ST elevation and sinus arrest during general anesthesia suggesting myocardial hypoxia

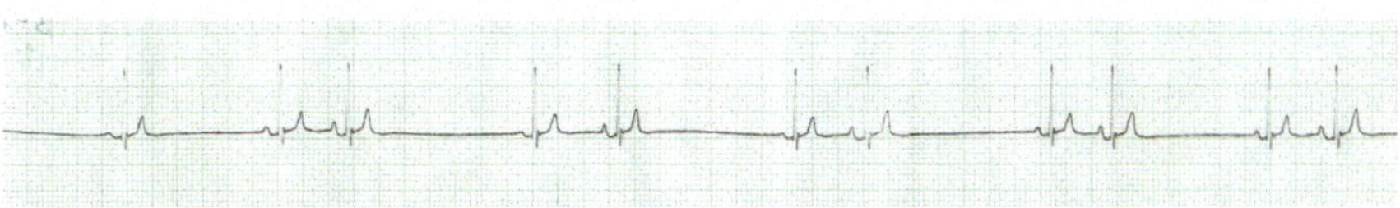

Fig. 16.3 Electrocardiogram (lead II, sensitivity 1, speed 25 mm/s) of an adult nondescript bitch 30 min after xylazine-ketamine anesthesia showing sinus arrest regularly after two sinus beats

J. P. Varshney, *Electrocardiography in Veterinary Medicine*,
https://doi.org/10.1007/978-981-15-3699-1_16

16.3 **Babesiosis** (Figs. 16.4, 16.5, 16.6, 16.7 and 16.8)

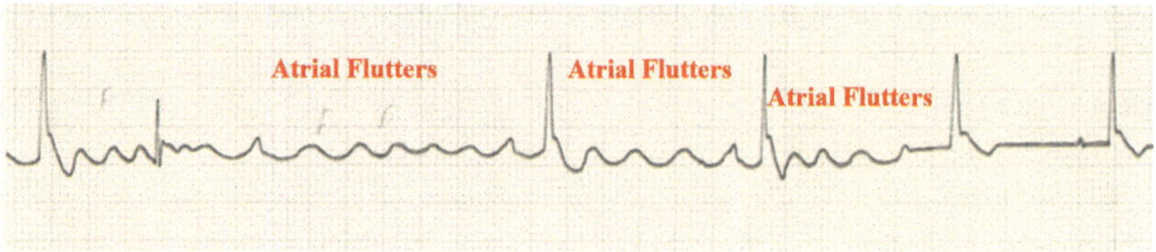

Fig. 16.4 Electrocardiogram (lead II, sensitivity 1, speed 25 mm/s) of a 7-year-old male Great Dane with babesiosis (*B. gibsoni*) and anemia (hemoglobin level 4.6 g/dL) showing atrial flutter

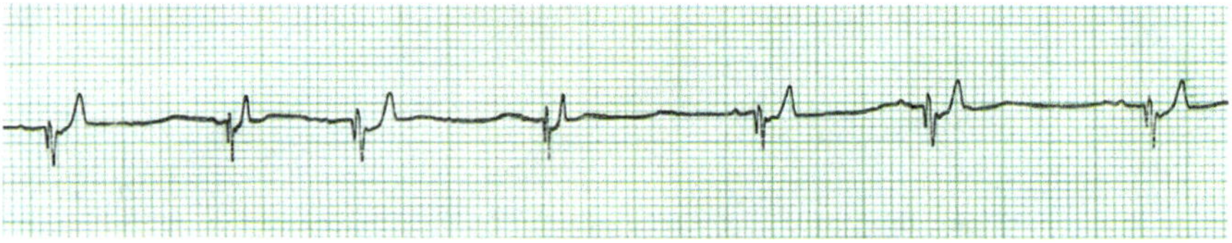

Fig. 16.5 Electrocardiogram (lead II, sensitivity 1, speed 25 mm/s) of a 10-year-old male Pomeranian dog with babesiosis showing low-voltage QRS complex suggestive of pericardial effusion/cardiomyopathy

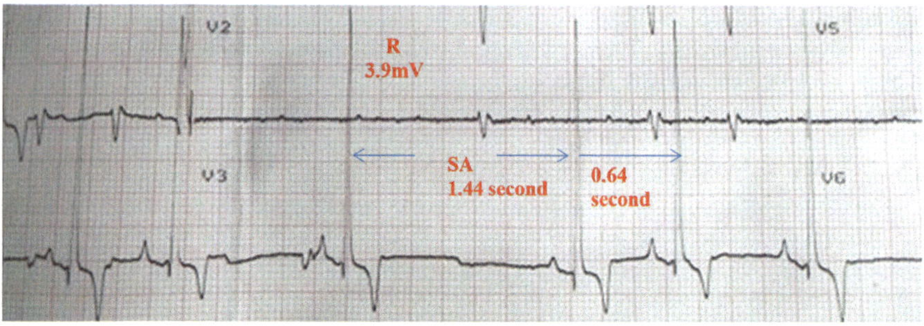

Fig. 16.6 Electrocardiogram (lead II, sensitivity 1, speed 25 mm/s) of a 7-year-old male Dachshund with babesiosis (*B. gibsoni*) showing wandering pace maker (P 0.1–0.3 mV), sinus arrhythmia (heart rate 72 bpm, varying R-R intervals), sinus arrest (blocks are > twice of normal R-R interval), increased amplitude of "R" wave (3.9 mV) and broad QRS (0.08 s) suggesting left ventricular enlargement/hypertrophy with sinus arrest

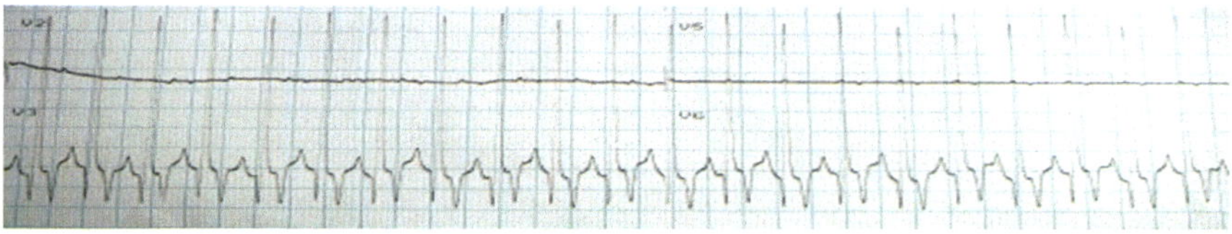

Fig. 16.7 Electrocardiogram (lead II, sensitivity 1, speed 25 mm/s) of a 4.5-year-old Doberman with babesiosis (*B. gibsoni*) showing sinus tachycardia (heart rate 180 bpm, almost constant R-R interval 0.32 s), increased amplitude of Q (0.7–0.8 mV) increased amplitude of R (3.8–3.9 mV), broad QRS (0.07–0.08 s) suggesting biventricular enlargement

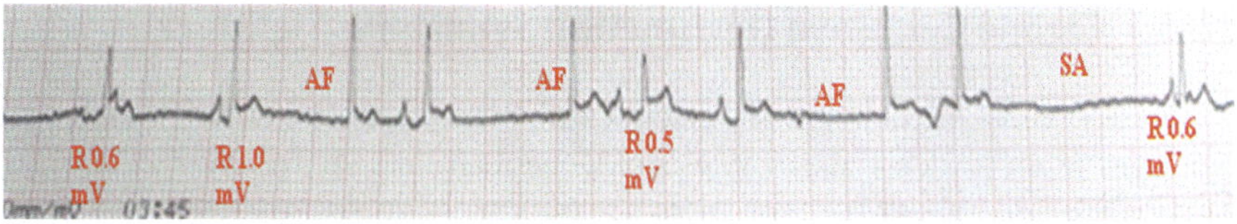

Fig. 16.8 Electrocardiogram (lead II, sensitivity 1, speed 25 mm/s) of a 3-year-old Pomeranian bitch with babesiosis (*B. gibsoni*) showing arrhythmia and sinus arrest (ventricular heart rate 90 bpm, R-R interval 0.4–1.24 s), atrial fibrillation, and electrical alternans of R wave (R amplitude varying from 0.5 to 1.0 mV)

16.4 Carbon Dioxide Pneumoperitoneum (Figs. 16.9 and 16.10)

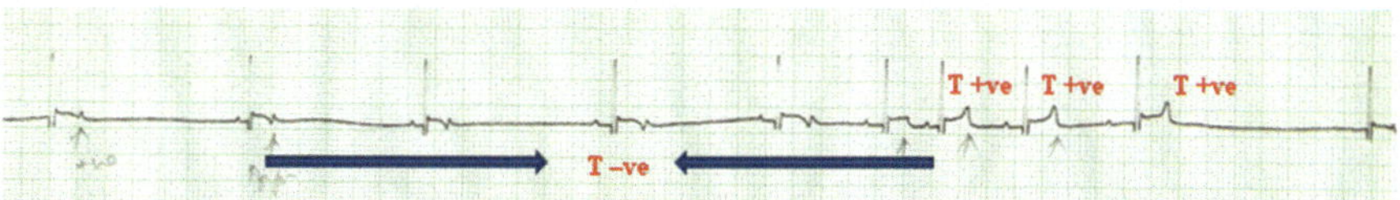

Fig. 16.9 Electrocardiogram (lead II, sensitivity 1, speed 25 mm/s) of a 3-year-old male nondescript dog showing sudden change in polarity of T wave 1 h post carbon dioxide pneumoperitoneum at 6 mm Hg for laparoscopy (Maiti et al. 2013)

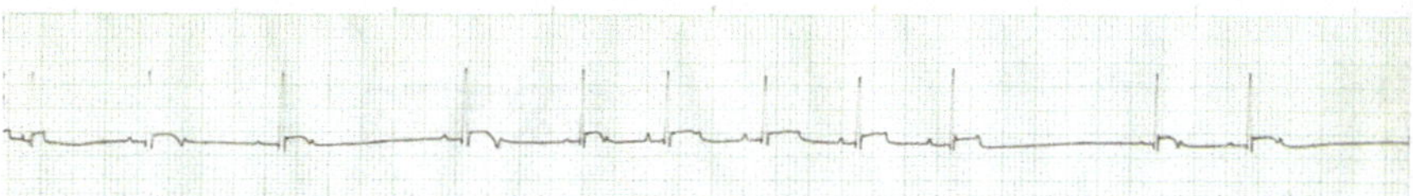

Fig. 16.10 Electrocardiogram (lead II, sensitivity 1, speed 25 mm/s) of a 4-year-old female mongrel dog showing elevation of S-T segment 1 h post carbon dioxide pneumoperitoneum at 6 mm Hg for laparoscopy (Maiti et al. 2013)

16.5 Chocolate Toxicity (Fig. 16.11)

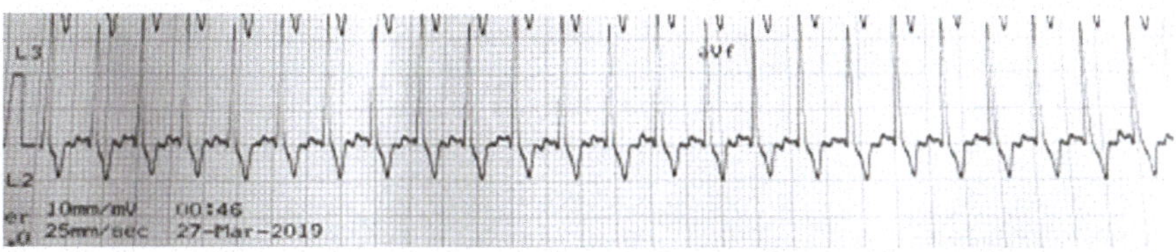

Fig. 16.11 Electrocardiogram (lead II, sensitivity 1, speed 25 mm/s) of a 12-year-old male Cocker Spaniel with chocolate poisoning showing sinus tachycardia (heart rate 220 bpm with almost constant R-R interval of 0.27 s)

16.6 Canine Cognitive Dysfunction Syndrome (Fig. 16.12)

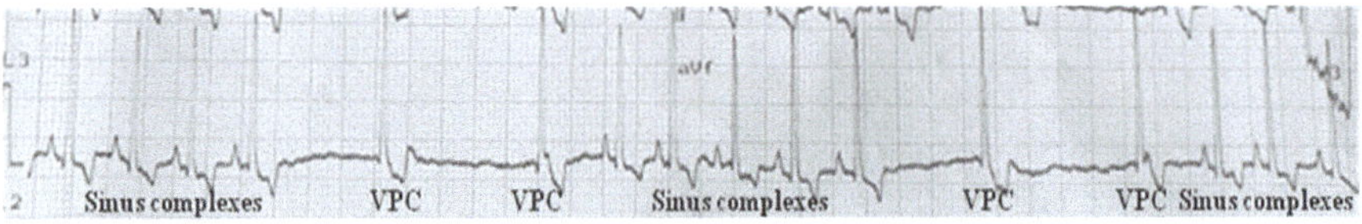

Fig. 16.12 Electrocardiogram (lead II, sensitivity 1, speed 25 mm/s) of a 17-year-old Pomeranian bitch with canine cognitive dysfunction syndrome showing ventricular premature complexes. Since major deflection of VPC is positive, the seat of ectopic focus is in right ventricle

16.7 Diabetic Ketoacidosis (Fig. 16.13)

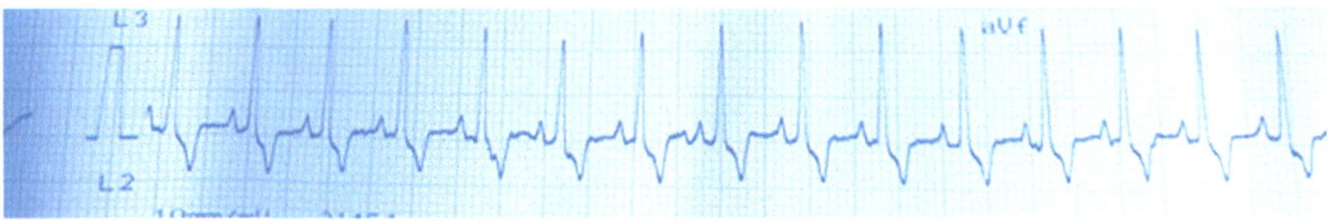

Fig. 16.13 Electrocardiogram (lead II, sensitivity 1, speed 25 mm/s) of a 7-year-old male Pug with diabetic ketoacidosis (blood glucose 595 mg/dL) sowing sinus tachycardia (heart rate 180 bpm) and S-T depression/S-T coving

16.8 Dirofilariasis (Figs. 16.14, 16.15, 16.16, 16.17 and 16.18)

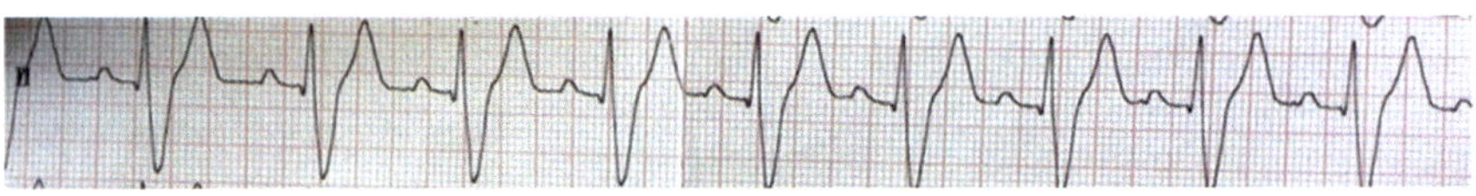

Fig. 16.14 Electrocardiogram (lead II, sensitivity 1,speed 25 mm/s) of a Pomeranian dog with dirofilariasis (*Dirofilaria immitis*) showing sinus rhythm; heart rate 80 bpm; "P" 0.04 s, 0.2 mV; normal "R" 0.8 mV; deep "S" (lead I 0.2, lead II 1.2, lead III 1.0, and avF 1.2 mV); T 0.7 mV, 0.2 s; and axis on frontal plane as −109° (calculated from lead I and lead III) suggesting right ventricular enlargement and right axis deviation (Varshney 2018)

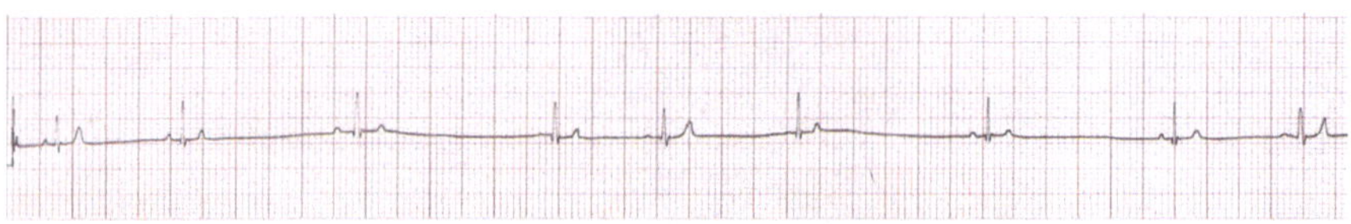

Fig. 16.15 Electrocardiogram (lead II, sensitivity 1, speed 25 mm/s) of a 9-year-old Pomeranian male dog with dirofilariasis (*Dirofilaria immitis*) showing bradyarrhythmia (heart rate 60 bpm, R-R interval varying), wandering pace maker (amplitude of P varying), and T alternans (amplitude of T varying)

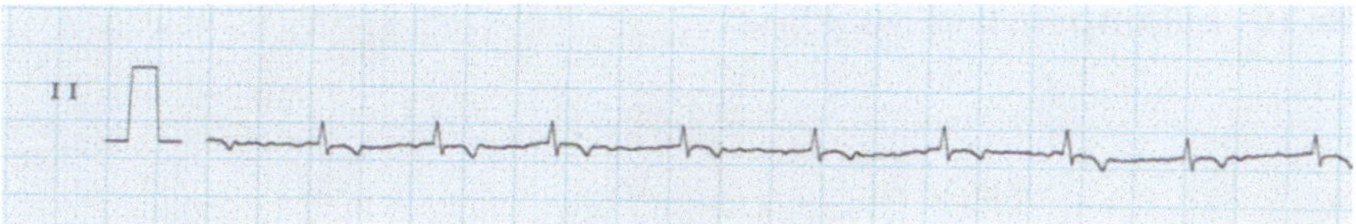

Fig. 16.16 Electrocardiogram (lead II, sensitivity 1, speed 25 mm/s) of a 12-year-old female Pomeranian with dirofilariasis showing atrial fibrillation ("P" wave not recognizable and has been replaced by fine "f" wave) and low voltage "R" wave (0.2 mV). Ventricular heart rate is 140 bpm. There is no coordination between atrial and ventricular contractions

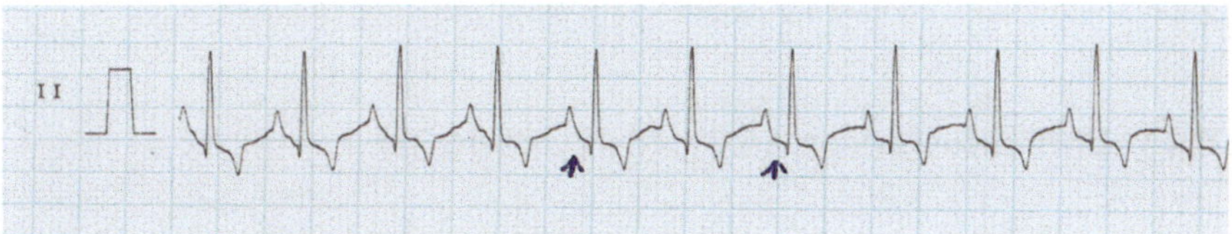

Fig. 16.17 Electrocardiogram (lead II, sensitivity 1, speed 25 mm/s) of a 10-year-old female German Shepherd with dirofilariasis showing "Ta" wave (arrow). The descending arm of "P" wave is long. "Ta" wave is suggestive of right atrial enlargement

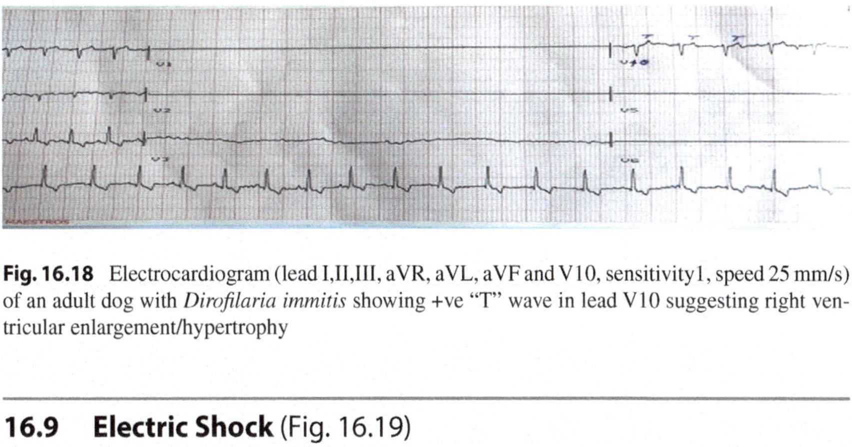

Fig. 16.18 Electrocardiogram (lead I,II,III, aVR, aVL, aVF and V10, sensitivity1, speed 25 mm/s) of an adult dog with *Dirofilaria immitis* showing +ve "T" wave in lead V10 suggesting right ventricular enlargement/hypertrophy

16.9 Electric Shock (Fig. 16.19)

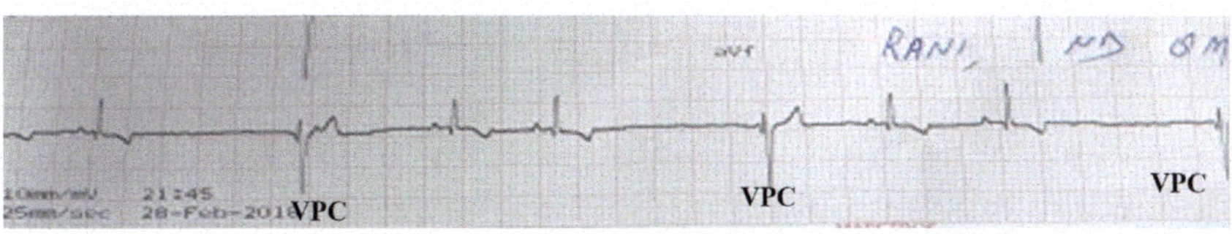

Fig. 16.19 Electrocardiogram (lead II, sensitivity 1, speed 25 mm/s) of an 8-month-old female nondescript dog with electric shock showing ventricular premature complexes after two sinus complexes. Complex 1, 3, 4, 6, and 7 are sinus complex with normal PQRST; and complexes 2, 5, and 8 are ventricular premature complex. Since major deflection of VPC is negative, the seat of ectopic focus is in left ventricle

16.10 Eclampsia (Fig. 16.20)

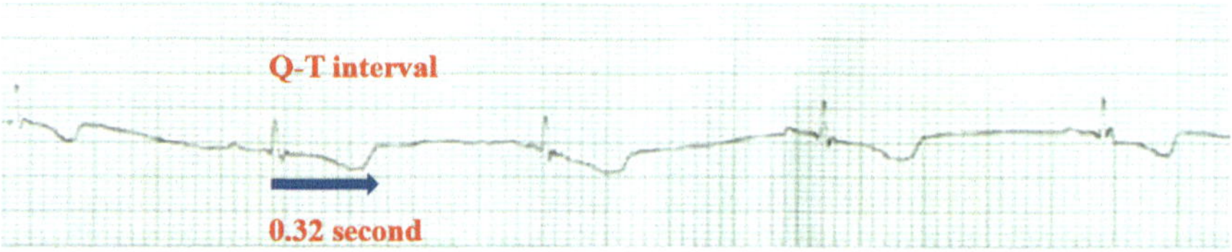

Fig. 16.20 Electrocardiogram (lead II, sensitivity 1, speed 25 mm/s) of a 4-year-old recently whelped Doberman bitch with eclampsia showing prolonged Q-T interval (0.32 s)

16.11 Ehrlichiosis (Figs. 16.21, 16.22 and 16.23)

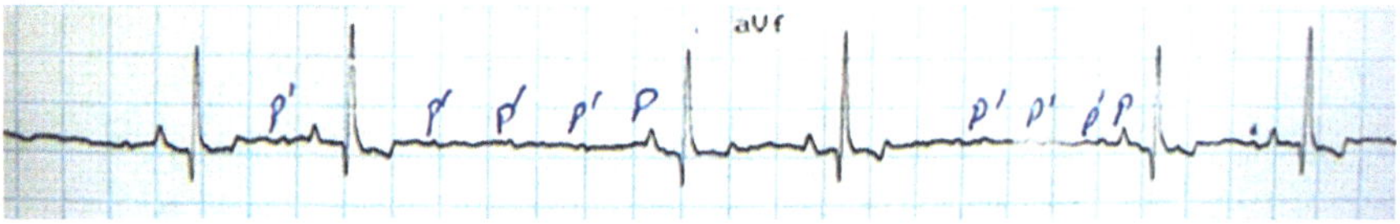

Fig. 16.21 Electrocardiogram (lead II, sensitivity 1, speed 25 mm/s) of a 2-year-old dog with ehrlichiosis (*E. canis*) showing ventricular rate as 52 bpm, atrial rate as 180 bpm, "P" (unconducted), P-R interval 0.17 s (constant in all conducted P waves), R-R interval 0.60–1.4 s, QRS > 0.06 s suggestive of third-degree heart block

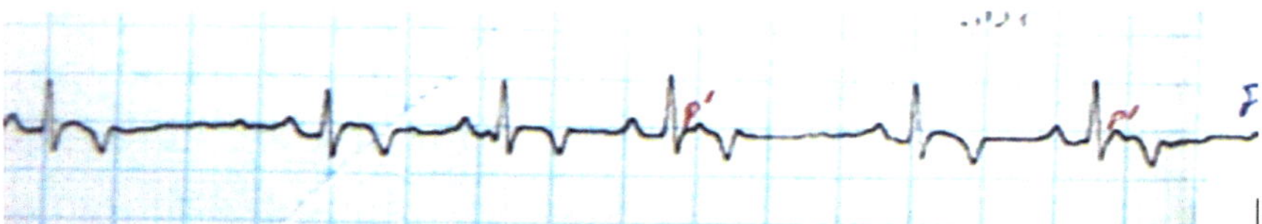

Fig. 16.22 Electrocardiogram (lead II, sensitivity 1, speed 25 mm/s) of a 3-year-old dog with ehrlichiosis showing atrial premature complex (Varshney et al. 2015)

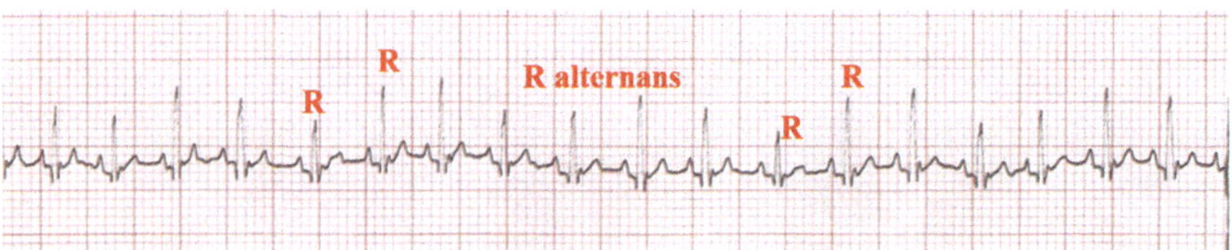

Fig. 16.23 Electrocardiogram (lead II, sensitivity 1, speed 25 mm/s) of a 15-year-old male Dachshund with ehrlichiosis showing electrical alternans of R wave, sinus tachycardia (heart rate 220 bpm) suggesting pericardial effusions

16.12 Heart Failure (Figs. 16.24, 16.25, 16.26, 16.27, 16.28, 16.29, 16.30 and 16.31)

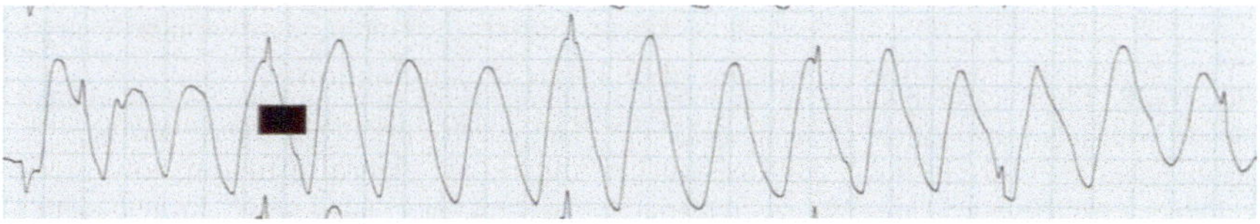

Fig. 16.24 Electrocardiogram (lead II, sensitivity 1, speed 25 mm/s) of 45-day-old Labrador pup with acute heart failure and gasping showing ventricular flutters

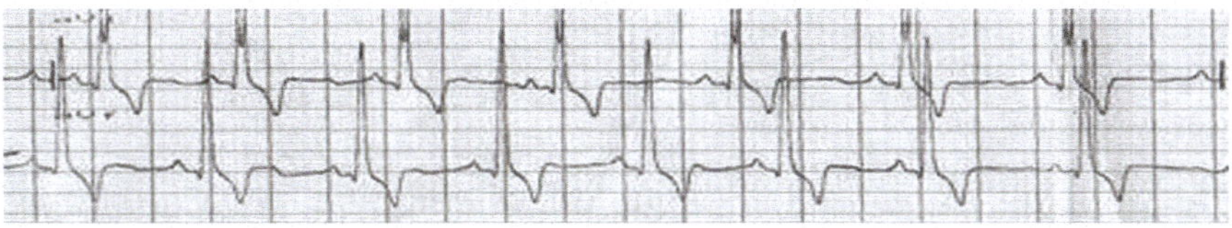

Fig. 16.25 Electrocardiogram (lead II, sensitivity 1, speed 25 mm/s) of a 7-year-old male Great Dane with signs of congestive left heart failure showing broad P (0.06 s), tall R (3.1 mV), broad QRS (0.06–0.07 s), and ST coving suggesting left heart enlargement

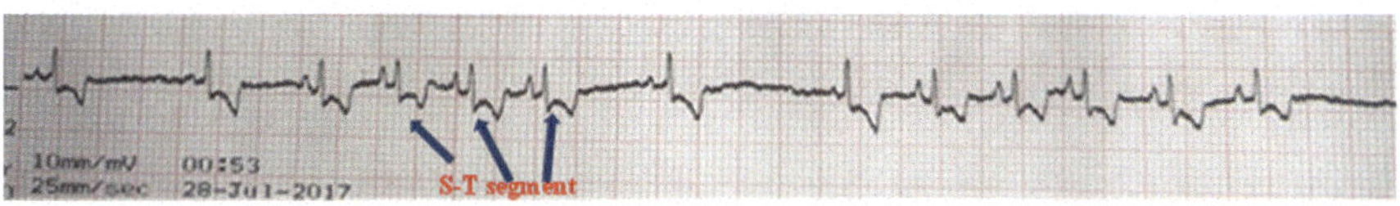

Fig. 16.26 Electrocardiogram (lead II, sensitivity 1, speed 25 mm/s) of a 6-year-old German Shepherd bitch with congestive heart failure showing sinus arrhythmia (heart rate 130 bpm, R-R interval 0.36–0.88 s), SA arrest, low voltage R (0.2–0.3 mV), and S-T segment depression (0.2 mV)

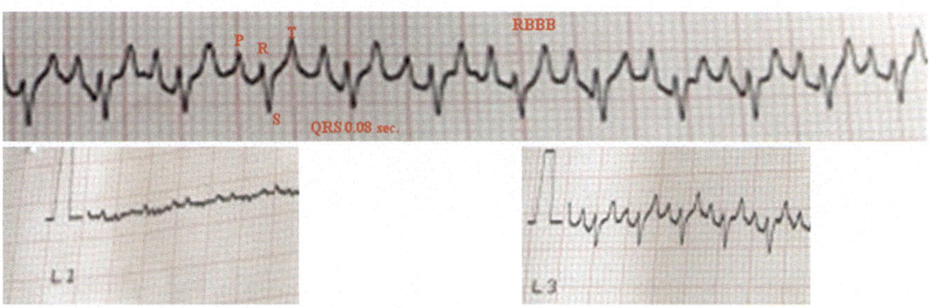

Fig. 16.27 Electrocardiogram (lead I, II, III, sensitivity 1, speed 25 mm/s) of a 12-year-old male Pomeranian with congestive heart failure showing broad QRS (0.08 s) in lead I, II, III, and aVF, small R (0.2 mV lead II), S 0.25 mV, MEA on frontal plane −79 to −81° (calculated from lead I and lead III) suggesting right bundle branch block

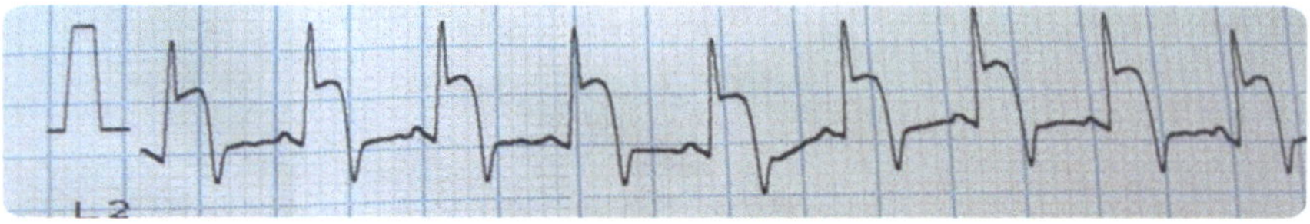

Fig. 16.28 Electrocardiogram (lead II, sensitivity 1, speed 25 mm/s) of an 8-year-old male Mastiff with acute heart failure showing severe S-T elevation (0.6 mV) suggesting severe myocardial hypoxia

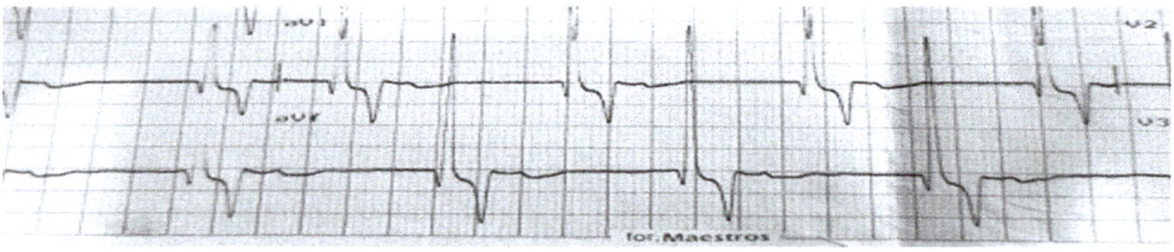

Fig. 16.29 Electrocardiogram (lead II, sensitivity 1, speed 25 mm/s) of a male 6-year-old Labrador suffering from left heart failure showing ventricular heart rate as 60 bpm, increased amplitude of "R" wave (3.2 mV), broad "QRS" (0.10 s) and absence of P wave suggesting left ventricular enlargement with bradycardia and atrial standstill

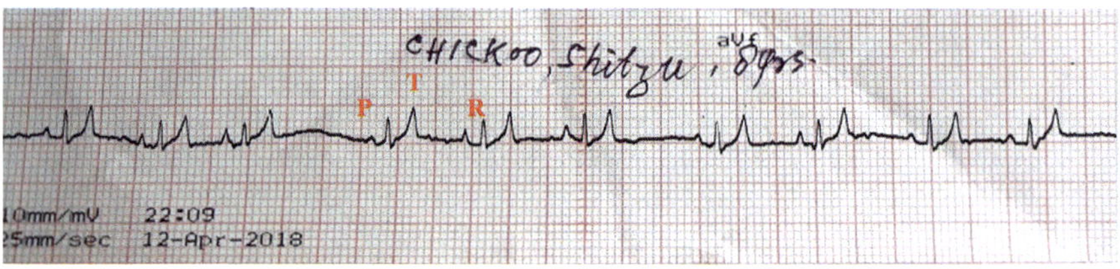

Fig. 16.30 Electrocardiogram (lead II, sensitivity 1, speed 25 mm/s) of an 8-year-old male Shih Tzu dog with heart failure showing large positive T wave (0.4 mV, 0.08 s) as compared to small R (0.3 mV) suggestive of pericardial effusion. T wave changes may be due to pericardial effusions

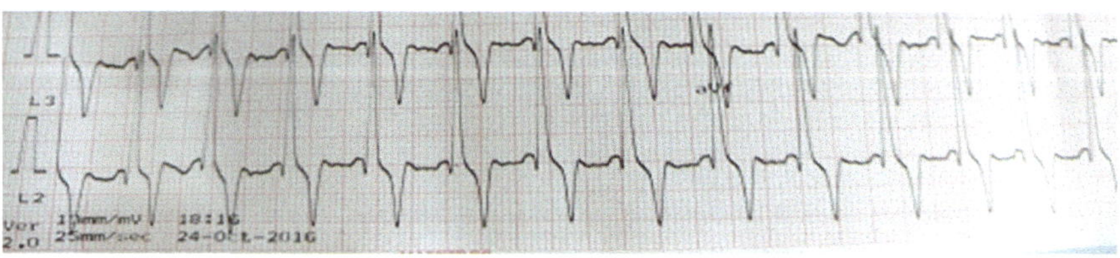

Fig. 16.31 Electrocardiogram (lead II, sensitivity 1, speed 25 mm/s) of a 12-year-old male nondescript dog with congestive heart failure (left) showing atrial fibrillation, ventricular heart rate as 135 bpm, increased amplitude of R (2.6–2.7 mV), broad QRS (0.10 s), and ST slurring at few places suggesting left heart enlargement

16.13 Heat Stroke/Hyperthermia (Figs. 16.32, 16.33, 16.34, 16.35, 16.36, 16.37, and 16.38)

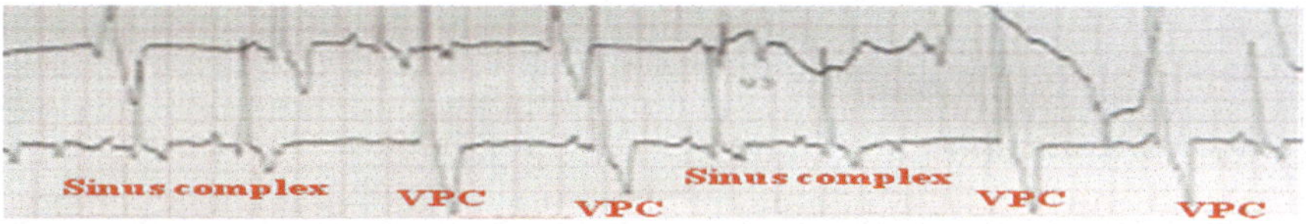

Fig. 16.32 Electrocardiogram (lead II, sensitivity 1, speed 25 mm/s) of a 4-year-old male Golden Retriever with heat stroke (106.8 °F) showing VPC. Two sinus complexes are followed by two ectopic complexes (VPC). Major positive deflection of ectopic QRS indicates that ectopic focus is in right ventricle

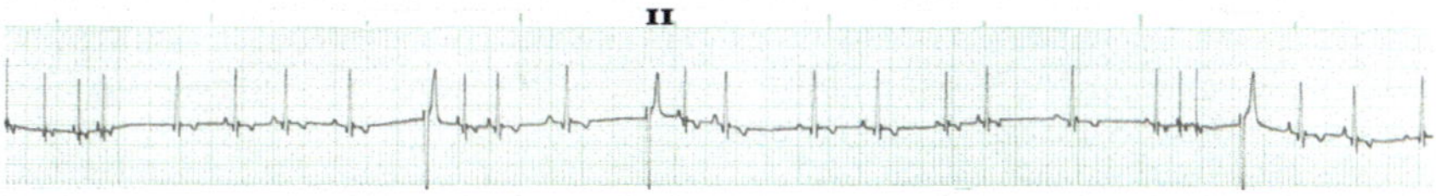

Fig. 16.33 Electrocardiogram (lead II, sensitivity 1, speed 25 mm/s) of a 3.5-year-old male nondescript dog with hyperthermia (106 °F) showing VPCs with major negative deflection and narrowness indicating their origin from one of the proximal intraventricular conduction branches in left ventricle

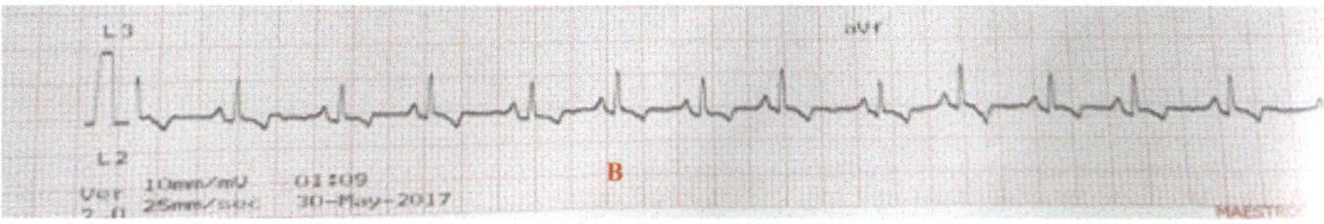

Fig. 16.34 (**a**) Electrocardiogram (sensitivity 1, speed 25 mm/s) of an 8-year and 6-month-old male obese Labrador with heat stroke (108.8 °F) in the month of May (ambient temperature 44.5 °C) showing sinus tachycardia (heart rate 180 bpm), low voltage and changing configuration of complexes. (**b**) Electrocardiogram (lead II, sensitivity 1, speed 25 mm/s) of the above 8-year and 6-month-old male obese Labrador (Fig. 170 A) 6 h post therapy for hyperthermia when temperature came down to 100.2 °F showing uniformity and regularity of the complexes with heart rate of 120 bpm

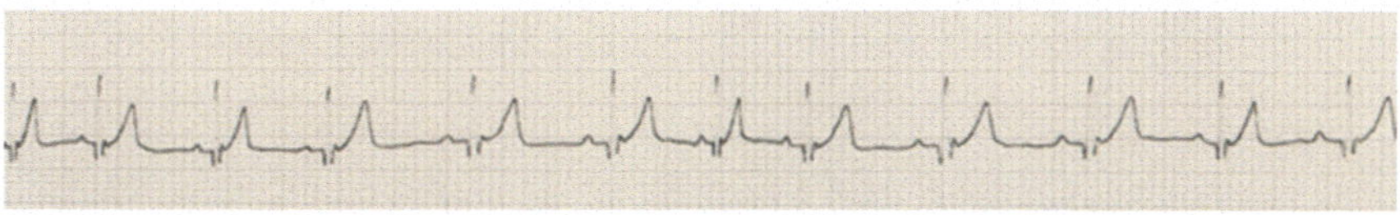

Fig. 16.35 Electrocardiogram (lead II, sensitivity 1, speed 25 mm/s) of a 5-month-old female Labrador with fever (temperature 105.2 °F) showing tall T wave, sinus rhythm with heart rate of 140 bpm and narrow QRS

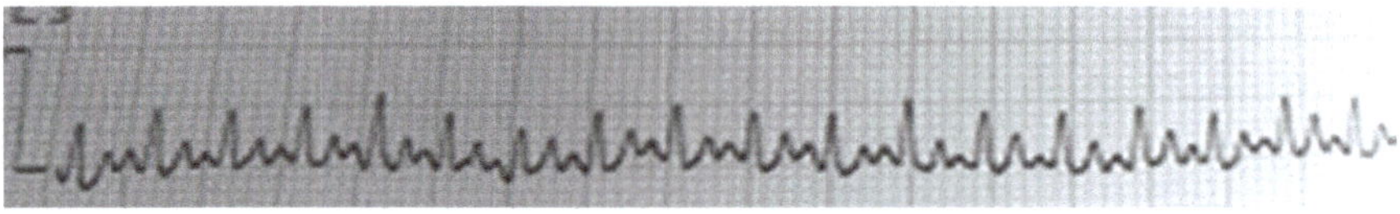

Fig. 16.36 Electrocardiogram (lead II, sensitivity 1, speed 25 mm/s) of a 5-year-old male German Shepherd dog with heat stroke (temperature 109.7 °F) showing sinus tachycardia (heart rate 320 bpm) and low-voltage “R” (0.4–0.5 mV)

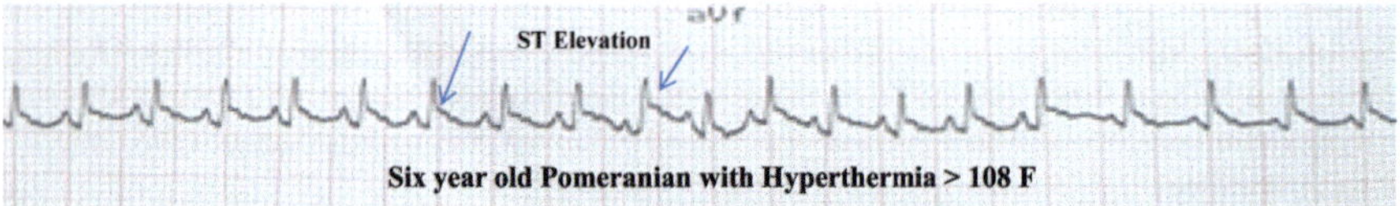

Fig. 16.37 Electrocardiogram (lead II, sensitivity 1, speed 25 mm/s) of a 6-year-old male Pomeranian dog with hyperthermia (>108 °F) showing sinus tachycardia (HR 255 bpm, R-R interval 0.23 s), normal P (0.1 mV, 0.04 s), normal P-R interval (0.08 s), reduced R amplitude (0.6 mV), narrow QRS (0.03 s), S-T elevation (0.2 mV), and very small T wave

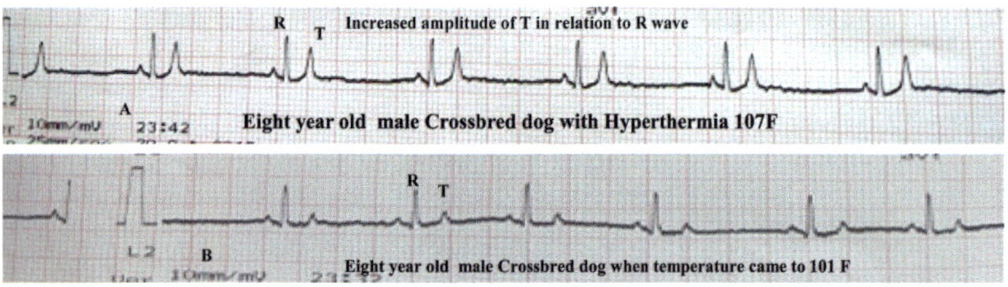

Fig. 16.38 Electrocardiogram (lead II, sensitivity 1, speed 25 mm/s) of an 8-year-old male crossbred dog. (**a**) With hyperthermia (107 °F), ECG is showing sinus rhythm (HR 70 bpm R-R interval 0.85 s), normal P (0.1 mV, 0.04 s), normal P-R interval (0.08 s), normal R (0.8 mV), normal QRS (0.04 s), S-T segment on base line and of 0.12 s duration, increased amplitude of T wave (0.6 mV) in relation to R wave. T wave is 75% of R wave (>25%). R:T ratio is 1:0.75 rather than 1:0.25. (**b**) With normalization of temperature to 101 °F, ECG is showing reduction in T wave amplitude (0.15 mV) and increase in heart rate (HR 100 bpm). Now T wave is about 22% of R wave. R:T ratio is 1.0:0.22

16.14 Hypothermia (Figs.16.39 and 16.40)

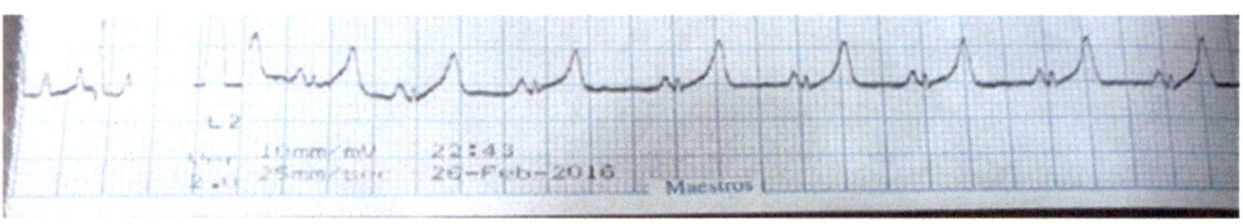

Fig. 16.39 Electrocardiogram (lead II, sensitivity 1, speed 25 mm/s) of a 30-day-old male Golden Retriever pup with hypothermia (95.2 °F) showing heart rate as 100 bpm, very small “r” wave and enlarged and broad T wave (0.6 mV, 0.16 s) suggesting myocardial hypoxia

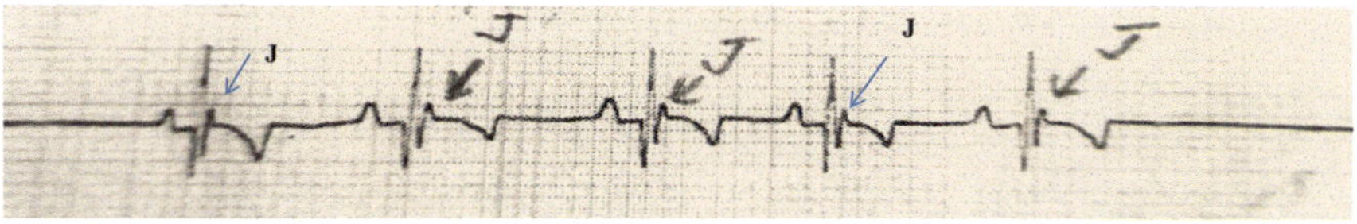

Fig. 16.40 Electrocardiogram (lead II, sensitivity 1, speed 25 mm/s) of a 4.5-month-old male nondescript pup with hypothermia (temperature < 92 °F) showing "J" wave or Osborn wave (Varshney 2016a)

16.15 Liver Diseases (Figs. 16.41, 16.42, 16.43, 16.44 and 16.45)

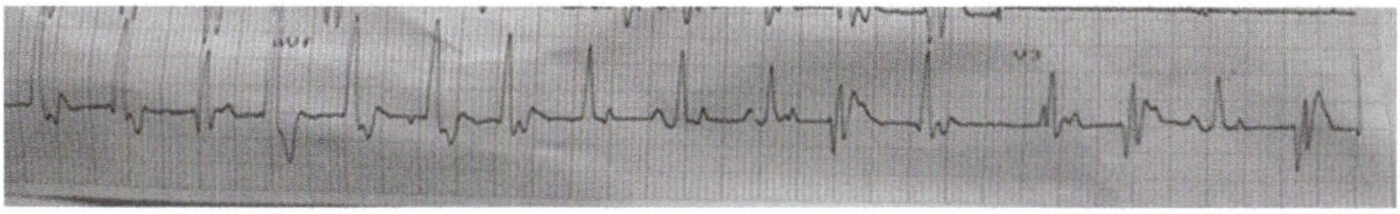

Fig. 16.41 Electrocardiogram (lead II, sensitivity 1, speed 25 mm/s) of an 8-year-old Rottweiler female dog with liver cirrhosis showing multiform ventricular premature complexes

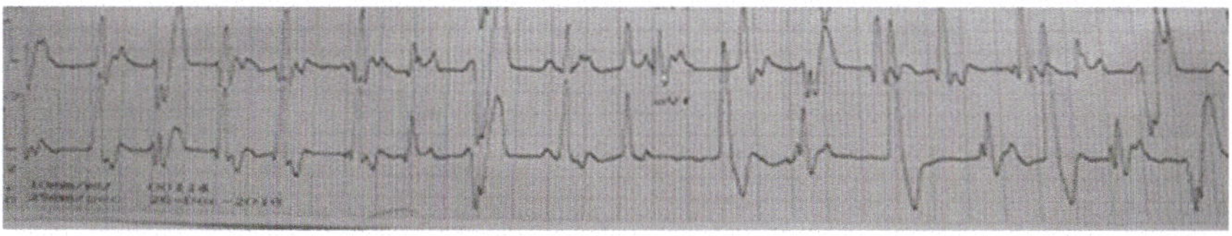

Fig. 16.42 Electrocardiogram (lead II, sensitivity 1, speed 25 mm/s) of an 8-year-old female Rottweiler with hyperechoic liver and ascites showing multiform ventricular tachycardia

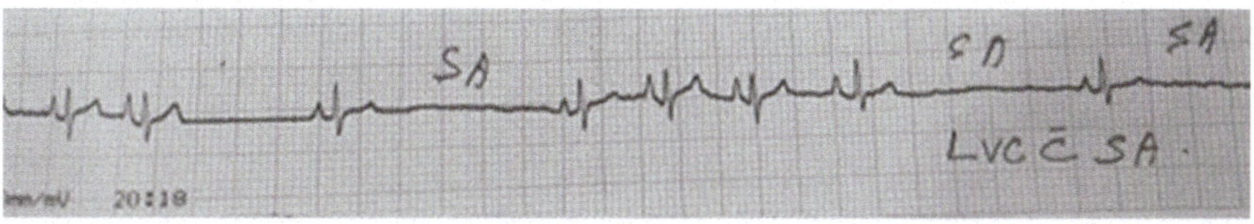

Fig. 16.43 Electrocardiogram (lead II, sensitivity 1, speed 25 mm/s) of a 5-year-old male Labrador with liver cirrhosis showing R alternans and sinus tachycardia (HR 180 bpm)

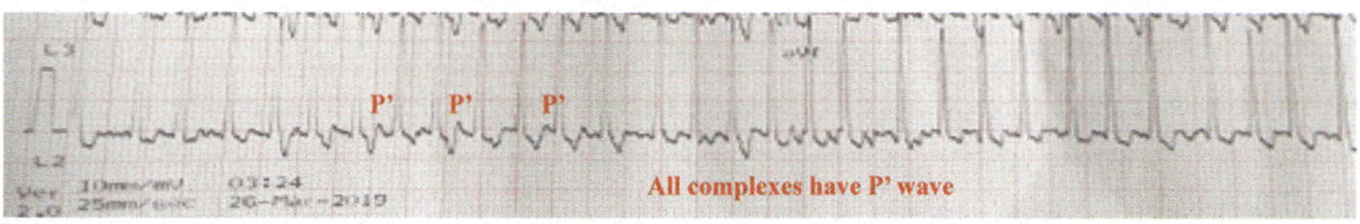

Fig. 16.44 Electrocardiogram (lead II, sensitivity 1, speed 25 mm/s) of a 6-year-old male Pug with hepatic nodular hyperplasia, prostate enlargement, crystalluria, and increased cTn-I (16.1 ng/mL) showing sinus arrest and low-voltage complexes suggesting cardiomyopathy

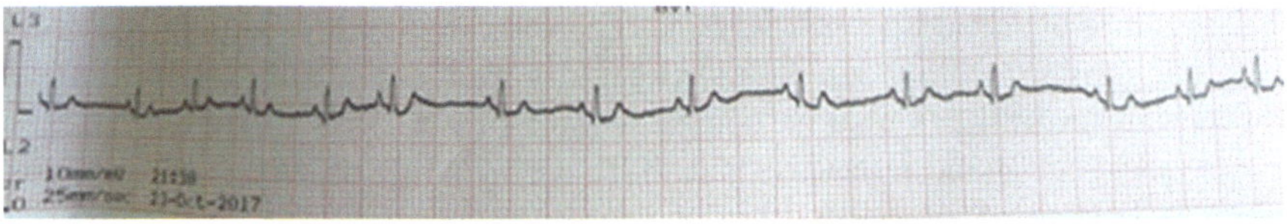

Fig. 16.45 Electrocardiogram (lead II, sensitivity 1, speed 25 mm/s) of a 6-year and 4-month-old Golden Retriever bitch with liver cirrhosis showing continuous atrial tachycardia as P′ wave is seen in previous T wave, and there is no sinus P wave in any complex. Almost all complexes have P′ wave. P′ wave is positive with regular P′-P′ interval. Ventricular heart rate is fast (280 bpm) and P′ waves are of different configurations suggesting multifocal continuous atrial tachycardia

16.16 Obesity (Figs. 16.46 and 16.47)

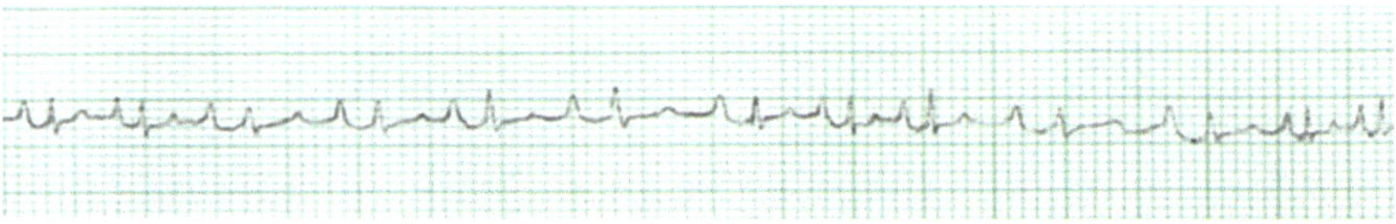

Fig. 16.46 Electrocardiogram (lead II, sensitivity 1, speed 25 mm/s) of a 10-year-old obese female Labrador weighing 59 kg showing low voltage complex (R 0.3 mV). The dog was otherwise healthy

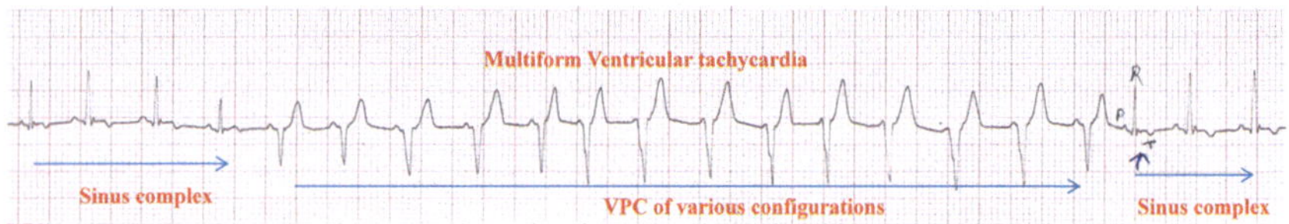

Fig. 16.47 Electrocardiogram (lead II, sensitivity 1, speed 25 mm/s) of an obese female Rottweiler (1.5-year-old) showing low-voltage complexes (R wave 0.2–0.4 mV) sinus tachycardia (heart rate 180 bpm). No other abnormality was detected

16.17 Pancreatitis (Fig. 16.48)

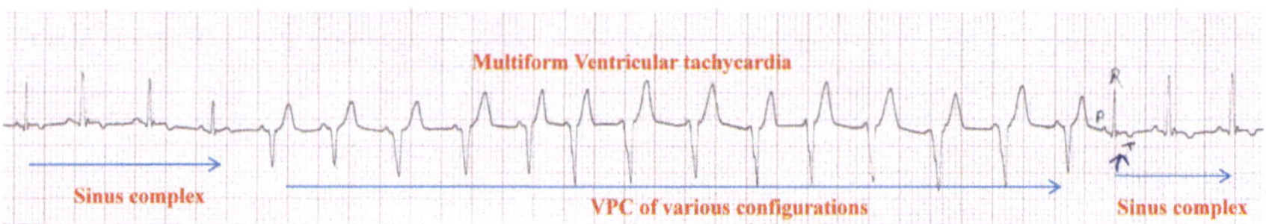

Fig. 16.48 Electrocardiogram (lead II, sensitivity 1, speed 25 mm/s) of a male 9-year-old Doberman Pinschers with severe pancreatitis showing multiform ventricular tachycardia

16.18 **Pneumonia** (Fig. 16.49)

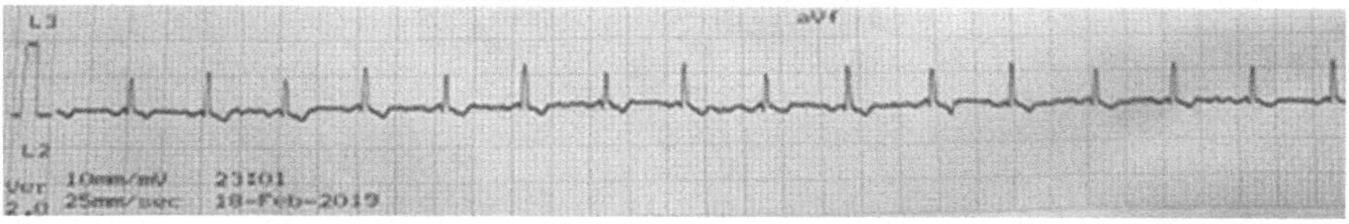

Fig. 16.49 Electrocardiogram (lead II, sensitivity 1, speed 25 mm/s) of a 2-year-old male Pug with severe congestion in cranial lungs masking the silhouette of heart showing atrial fibrillations

16.19 **Renal Failure** (Figs. 16.50, 16.51, 16.52, 16.53 and 16.54)

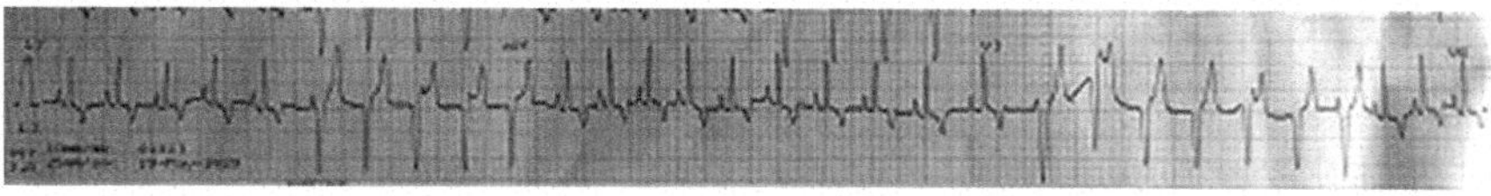

Fig. 16.50 Electrocardiogram (sensitivity 1, speed 25 mm/s) of a 21-year-old female Pomeranian dog with chronic renal failure (serum creatinine 12.8 mg/dL, BUN 43.8 mg/dL, SAP 96 U/L, ALT 14 IU/L, total bilirubin 0.8 mg/dL) showing intermittent ventricular tachycardia (VPCs are in run intermittently)

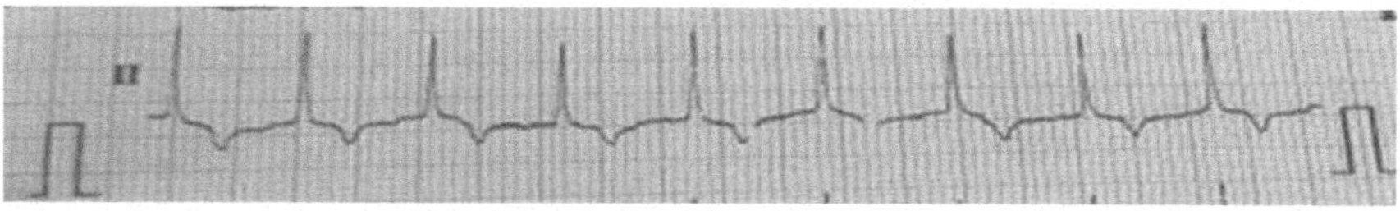

Fig. 16.51 Electrocardiogram (lead II, sensitivity 1, speed 25 mm/s) of an 11-year-old male Labrador with chronic renal failure (serum creatinine 23 mg/dL, BUN 380 mg/dL) showing absence of P wave (atrial standstill)

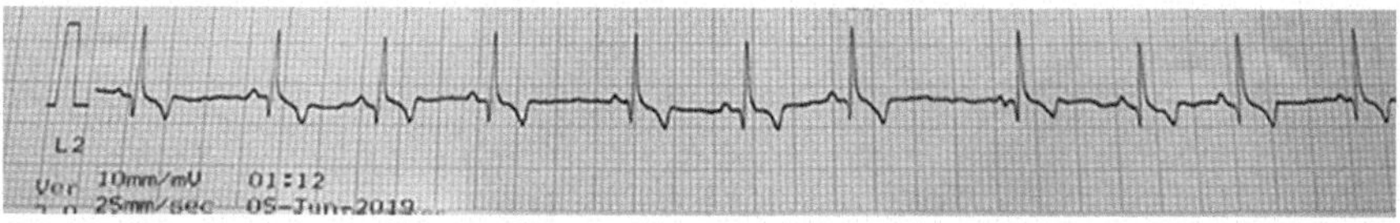

Fig. 16.52 Electrocardiogram (lead II, sensitivity 1, speed 25 mm/s) of a 13-year-old male German Shepherd with grade 4 renal failure (serum creatinine 5.4 mg%, BUN 36.7 mg%), anemia (hemoglobin 8.9 g%, erythrocytes 4.44 million/mm^3, PCV 25.6%) showing sinus arrhythmia (heart rate 100 bpm, R-R variable from 0.52 to 0.8 s), small Q (0.2 mV), broad QRS (0.8 s), and negative T(0.4 mV) suggestive of left bundle branch block (LBBB)

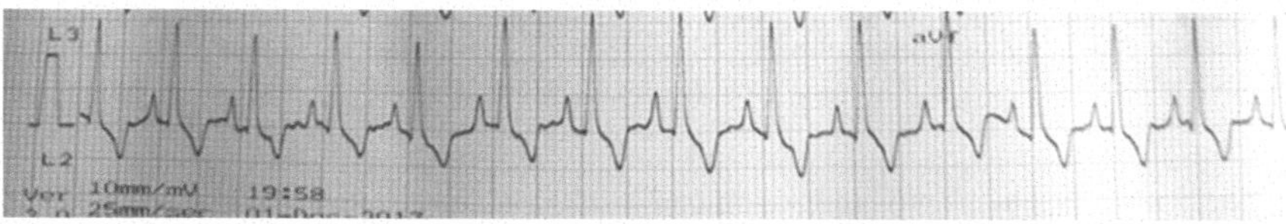

Fig. 16.53 Electrocardiogram (lead II, sensitivity 1, speed 25 mm/s) of a 10-year-old male Pomeranian with renal failure (serum creatinine 10.2 mg%) showing broad QRS (0.08 s) intermittently in all leads suggestive of intermittent left bundle branch block

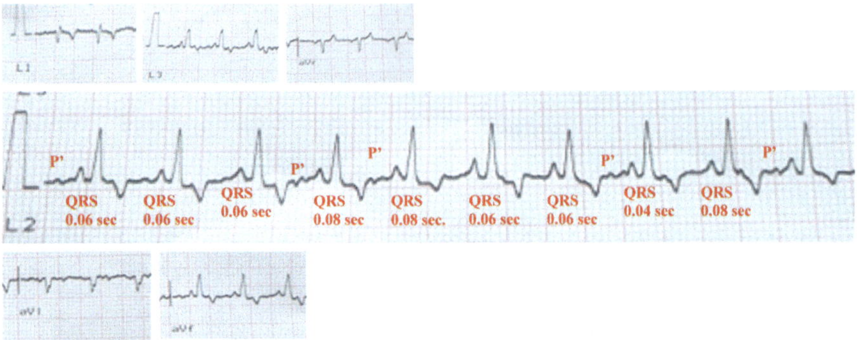

Fig. 16.54 Electrocardiogram (lead I, II, III, aVR, aVL and aVF, sensitivity 1, speed 25 mm/s) of a 6-year-old male Pug with recumbency, anemia (Hb. 7.0 g/dL, erythrocytes 3.04 million/mm^3, packed cell volume 19.0%), leukocytosis with neutrophilia (total leucocyte count 39,300/mm^3, neutrophils 86%), thrombocytopenia (26,000/mm^3), and acute renal failure (creatinine 7.0 mg/dL, BUN 56 mg/dL) showing ventricular heart rate as 160 bpm and irregular rhythm (P′ wave is blocked at many places—second degree AV block), small q (0.2 mV) in lead I, normal P (0.2 mV, 0.04 s. duration), normal P-R interval (0.08 s), normal R(0.6–0.75 mV), broad QRS (0.08 s in complex 4, 5 and 9), inverted QRS in aVR, normal S-T segment (0.07 s), negative T (0.15 mV, 0.06–0.08 s), and mean electrical axis in frontal plane as 104° indicating intermittent left bundle branch block (broad QRS) with slight right axis deviation (>100°) and probable right bundle branch block (second-degree AV block) also

16.20 **Snake Bite** (Figs. 16.55, 16.56 and 16.57)

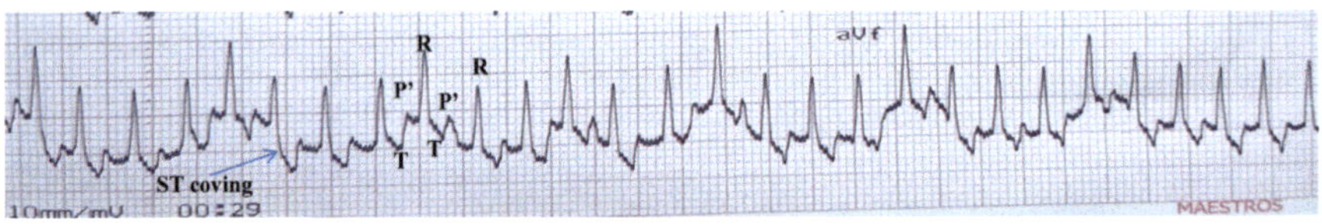

Fig. 16.55 Electrocardiogram (lead II, sensitivity 1, speed 25 mm/s) of an 8-year-old female Pomeranian dog bitten by snake (viper) having three fangs mark on right side of the neck. The tracing, taken within 2.5 h. Post bite, shows regular ventricular heart rate of 260 bpm, narrow QRS(0.3 mV) and nonvisible P wave. All complexes have P′ (premature) waves of varying configuration rather than normal P wave. S-T coving is also visible. Level of cTn-I was increased (3.25 ng/mL).The tracing is suggestive of multifocal continuous atrial tachycardia with myocardial insult (Varshney and Monapara 2019)

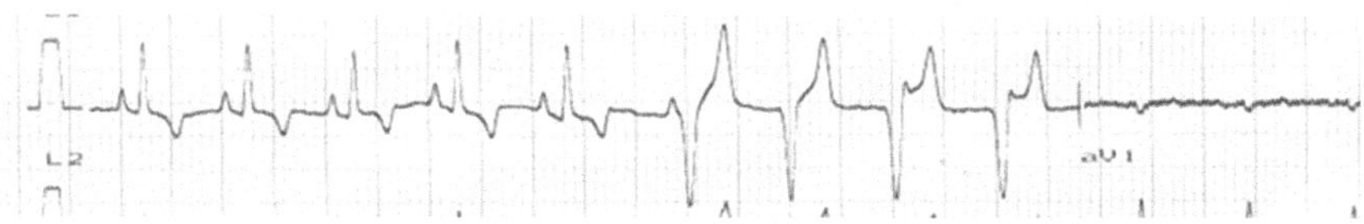

Fig. 16.56 Electrocardiogram (sensitivity 1, speed 25 mm/s) of a Labrador dog bitten by viper taken at the time of referral (within 6 h of bite) showing ventricular tachycardia (Varshney and Monapara 2019)

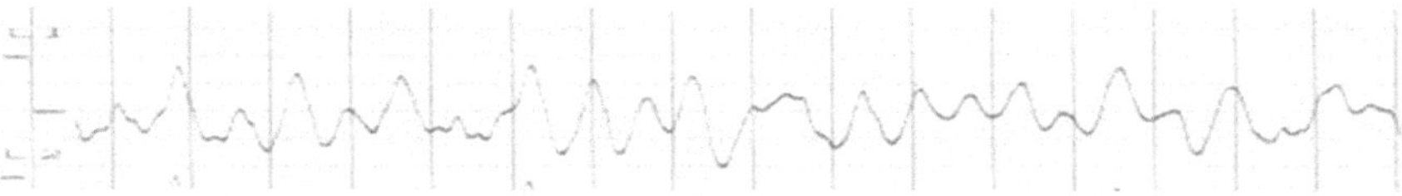

Fig. 16.57 Electrocardiogram (sensitivity 1, speed 25 mm/s) of a German Shepherd dog bitten by viper taken at the time of referral (within 3 h of bite) showing ventricular flutter (Varshney and Monapara 2019)

16.21 Status Epilepticus (Fig. 16.58)

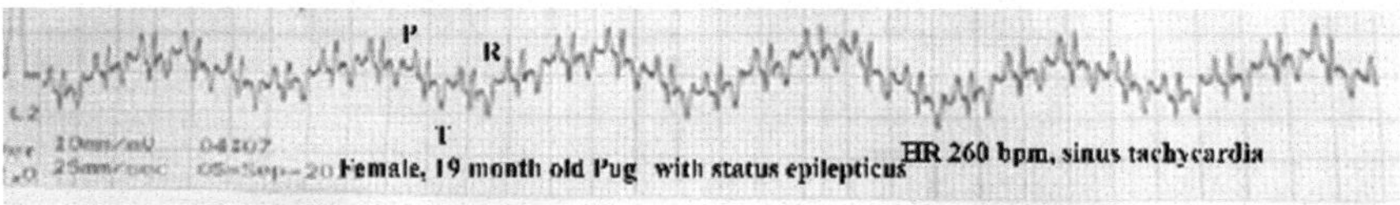

Fig. 16.58 Electrocardiogram (lead, II, sensitivity 1, speed 25 mm/s) of a 19-month-old female Pug with status epilepticus showing sinus tachycardia (heart rate 260 bpm), wandering base line and small complexes

16.22 Syncope (Figs. 16.59, 16.60 and 16.61)

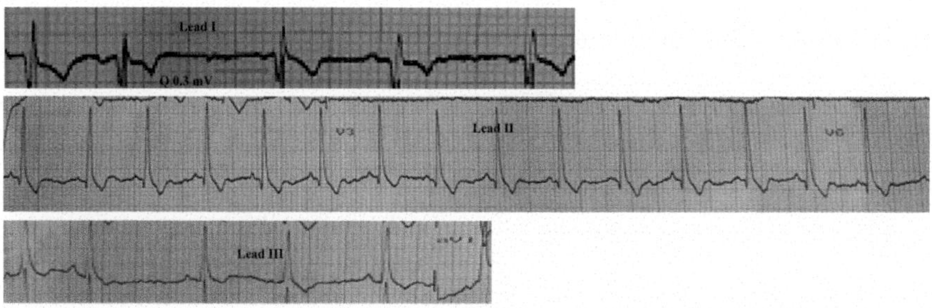

Fig. 16.59 Electrocardiogram (lead I, II, III, sensitivity 1, speed 25 mm/s) of a 10-year-old male Cocker Spaniel with kerato-conjunctivitis sicca (tear production < 5 mm/min both eyes), leukocytosis (36.6×10^3/cubic mm), fever (104.4 °F), and syncope showing sinus rhythm (heart rate 140 bpm), normal R (2.0 mV), broad QRS (0.08 s), S-T coving/ slurring in lead II; 0.3 mV Q in lead I; and MEA in frontal plane 95° suggesting left bundle branch block

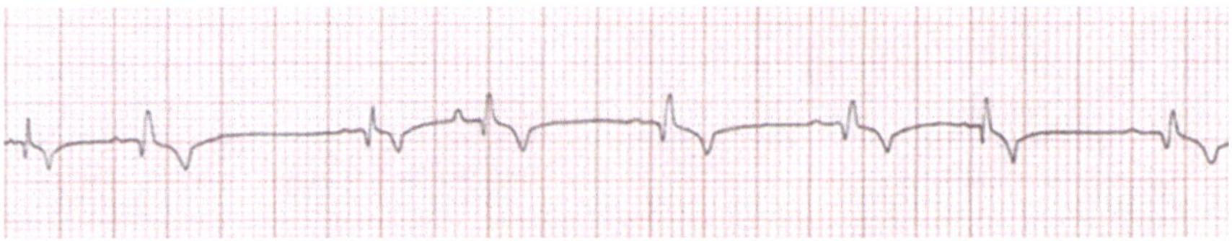

Fig. 16.60 Electrocardiogram (lead II, sensitivity 1, speed 25 mm/s) of a 5-year-old male German Shepherd with syncope showing sinus arrhythmia (HR 120 bpm, R-R interval varying from 0.44 to 0.84 s), small q (0.2 mV), low-voltage complex (R 0.3 mV), broad QRS 0.06 s and wandering pace maker

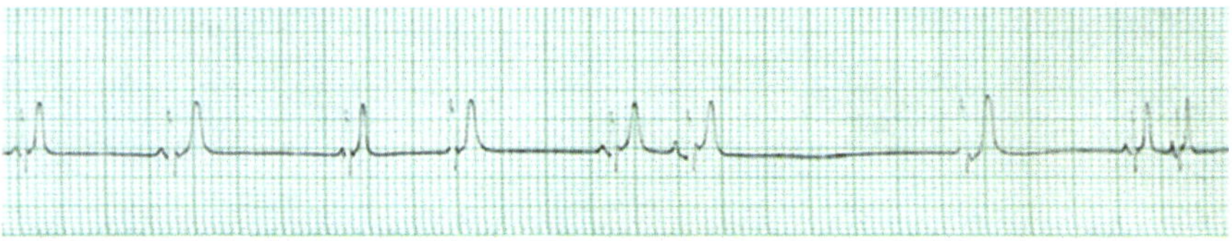

Fig. 16.61 Electrocardiogram (lead II, sensitivity 1, speed 25 mm/s) of a 20-day-old male Pomeranian pup with recumbency and unconsciousness showing heart rate 84 bpm, wandering pace maker (P 0.0–0.2 mV), R wave with descending notch, large T (0.08–0.9 mV, 0.08 s), and sinus arrest (R-R interval 0.48–1.16 s) suggesting myocardial hypoxia

16.23 Toad Poisoning (Fig. 16.62)

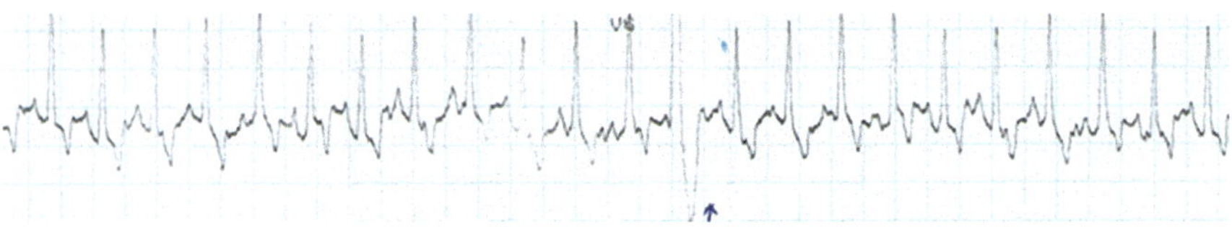

Fig. 16.62 Electrocardiogram (lead II, sensitivity 1, speed 25 mm/s) of a 3-year-old female Pomeranian with toad poisoning showing tachycardia (200 bpm), S-T coving at few places, and ventricular escape complex (Varshney et al. 2011)

16.24 Trauma/Accidental Fall (Figs. 16.63, 16.64, 16.65 and 16.66)

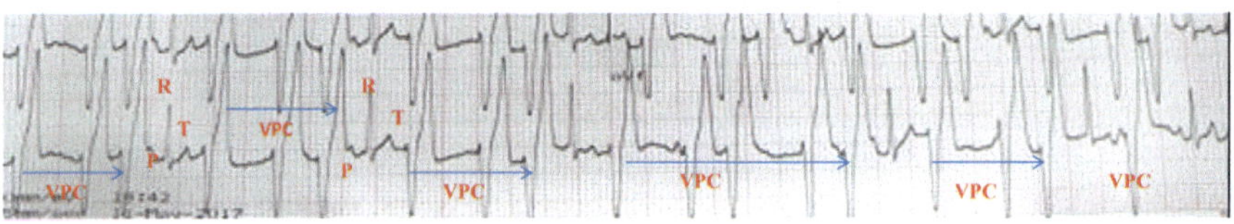

Fig. 16.63 Electrocardiogram (lead II, sensitivity 1, speed 25 mm/s) of a 4-year-old female German Shepherd with accidental chest trauma (radiographic evidence of fluid in chest) showing intermittent ventricular tachycardia. Complex first, second, third VPCs; fourth sinus complex; fifth, sixth, seventh VPCs; eighth sinus complex; ninth, 10th, 11th VPCs; 12th sinus complex; 13th, 14th, 15th, 16th, 17th VPCs; 18th sinus complex; 19th, 20th, 21st VPCs; 22nd sinus complex; 23rd VPC

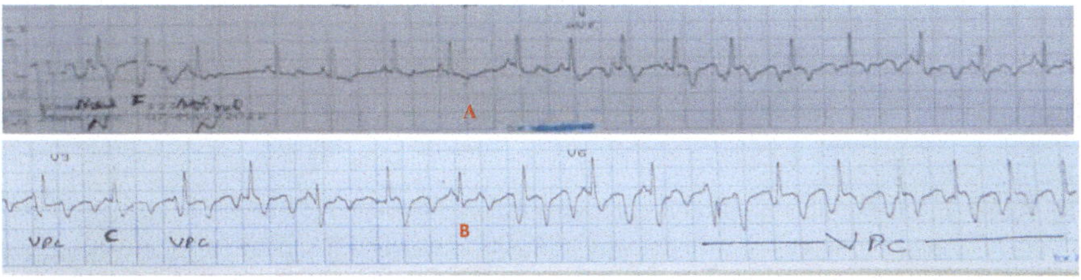

Fig. 16.64 Electrocardiogram (lead II, sensitivity 1, speed 25 mm/s) of a 30-month-old male Cocker Spaniel dog with blunt injuries and dyspnea due to fall from third story showing fusion complex (marked F) in strip A; capture complex (marked C) and a series of ventricular premature complexes (VPC) in strip B suggesting ventricular tachycardia

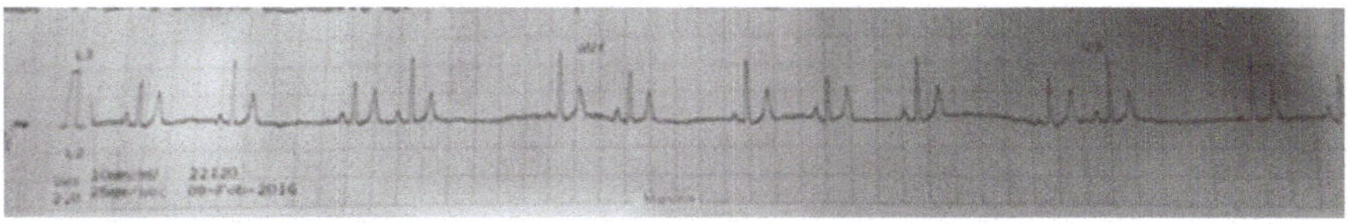

Fig. 16.65 Electrocardiogram (lead II, sensitivity 1, speed 25 mm/s) of an adult Labrador with spinal injury (fracture of first cervical vertebra) showing wandering pace maker, R-alternans (alternate complex), sinoatrial block, and increase in T amplitude in relation to R (Varshney 2016b)

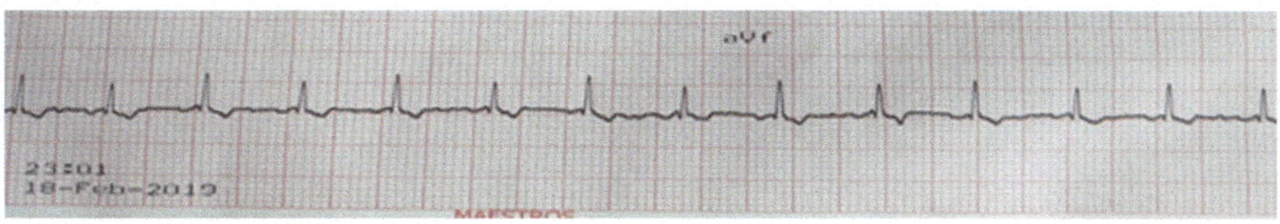

Fig. 16.66 Electrocardiogram (lead II, sensitivity 1, speed 25 mm/s) of a 5-year-old male Pug with chest trauma showing atrial fibrillation

16.25 Tumor/Growth (Figs. 16.67, 16.68, 16.69, 16.70 and 16.71)

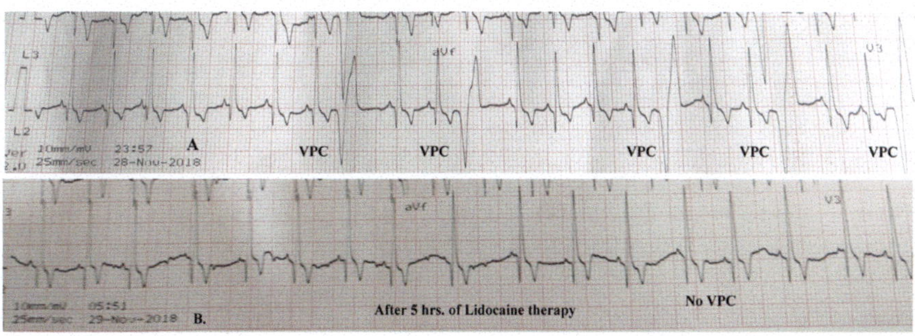

Fig. 16.67 (**a**) Electrocardiogram (lead II, sensitivity 1, speed 25 mm/s) of a male 7.5-year-old Beagle dog with pulmonary tumor showing ventricular premature complexes before treatment. The major deflection of ectopic beat is negative. It suggests that the focus of ectopic beat is in left ventricle. (**b**) The dog (at Fig. 203 A) was treated with lidocaine at 2.0 mg/kg slow IV and then followed by 40 µg/kg/min as constant rate infusion. Electrocardiogram taken 5 h. Post therapy is showing sinus rhythm with no VPC

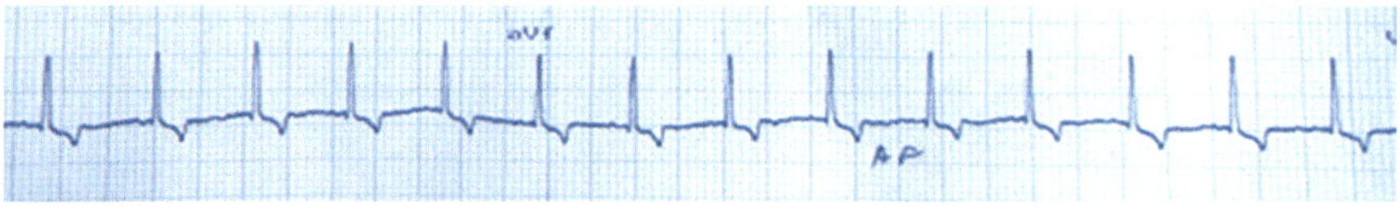

Fig. 16.68 Electrocardiogram (lead II, sensitivity 1, speed 25 mm/s) of a 6-year-old female Lhasa Apso with mammary tumors showing atrial fibrillation. Ventricular rate is 140 per minutes

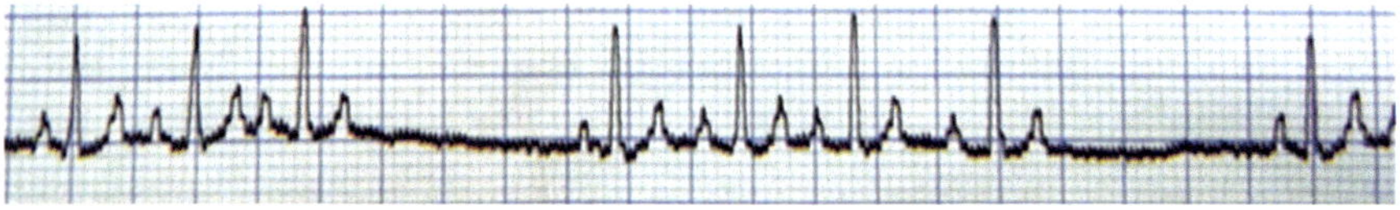

Fig. 16.69 Electrocardiogram (lead II, sensitivity 1, speed 25 mm/s) of an 11-year-old Pomeranian bitch with ulnar tumor having metastasis in lungs showing intermittent sinus arrest

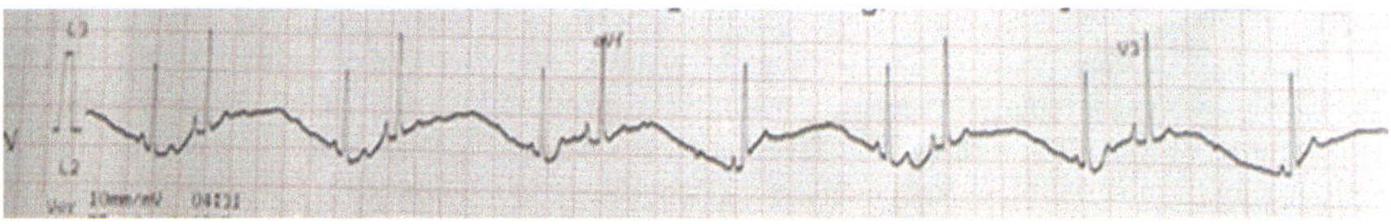

Fig. 16.70 Electrocardiogram (lead II, sensitivity 1, speed 25 mm/s) of a 12-year-old male Cocker Spaniel dog with osteolytic lesions (bone tumor) of left femur distal extremity and left tibia proximal extremity showing almost regular sinus arrest after two normal complexes

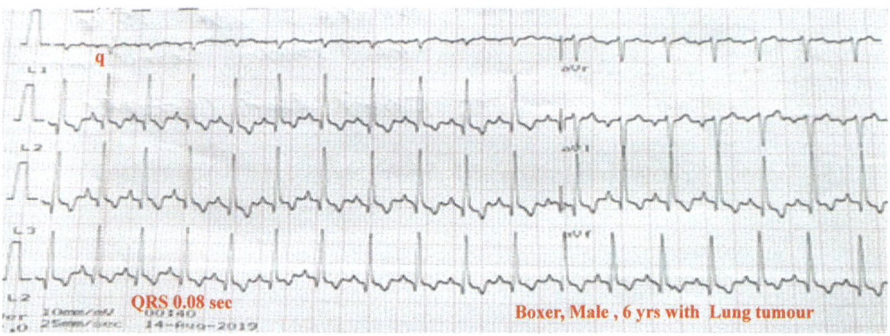

Fig. 16.71 Electrocardiogram (lead I, II, III, aVR, aVL, and aVF, sensitivity 1, speed 25 mm/s) of a 6-year-old male Boxer with severe dyspnea due to lung growth showing sinus tachycardia (180 bpm); small q in lead I, II, III, aVF; P (0.15 mV, 0.04 sec. duration); P-R interval (0.1 s); R(1.3 mV); broad QRS (0.08 s); inverted QRS in aVR; S-T segment (0.04 s); negative T (0.15 mV, 0.08 s) and mean electrical axis in frontal plane +97 suggesting left bundle branch block

16.26 Miscellaneous Electrocardiograms (Figs. 16.72, 16.73, 16.74, 16.75, 16.76, 16.77, 16.78, 16.79, 16.80, 16.81, 16.82, 16.83, 16.84 and16.85)

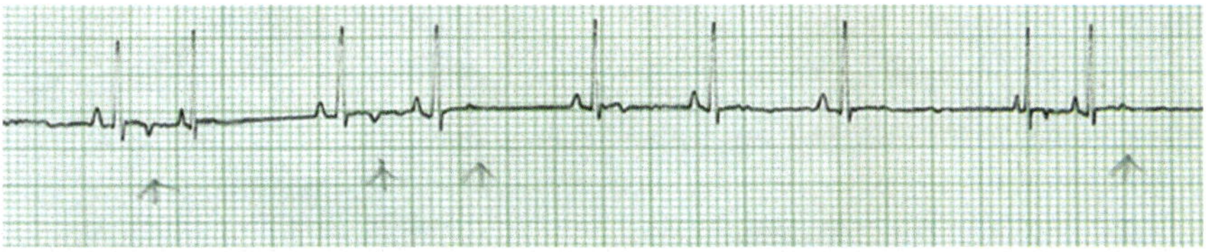

Fig. 16.72 Electrocardiogram (lead II, sensitivity 1, speed 25 mm/s) of a 10-year-old Pomeranian bitch showing sinus arrhythmia (heart rate 120 bpm, R-R interval 0.26–0.76 s) and change in polarity of T wave. T wave is negative in first, third, fifth, and eighth complex and positive in fourth and ninth complex

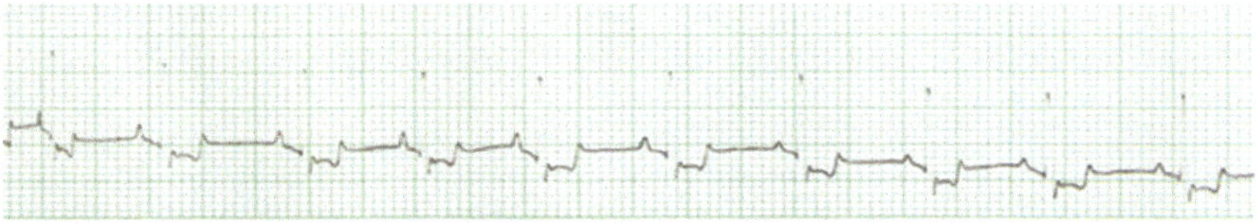

Fig. 16.73 Electrocardiogram (lead II, sensitivity 1, speed 25 mm/s)of a 3-year-old Pomeranian dog showing sinus rhythm (heart rate 90 bpm, R-R interval 0.4 s) and S-T segment depression (0.3 mV)

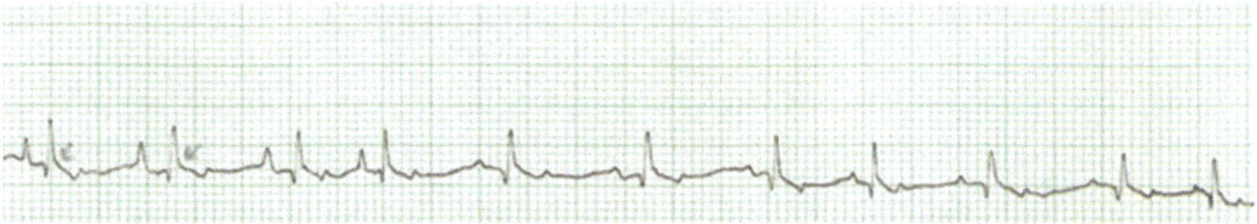

Fig. 16.74 Electrocardiogram (lead II, sensitivity 1, speed 25 mm/s) of a 5-year-old German Shepherd dog showing normal heart rate (120 bpm), wandering pace maker and slurring of S-T segment

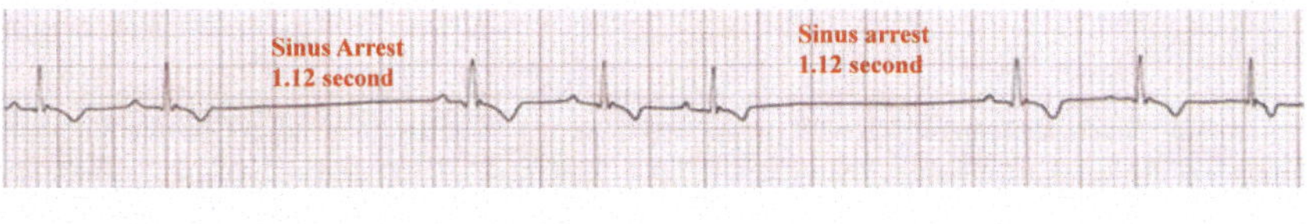

Fig. 16.75 Electrocardiogram (lead II, sensitivity 1, speed 25 mm/s) of a 9-year-old nondescript bitch showing sinus arrest and wandering pace maker

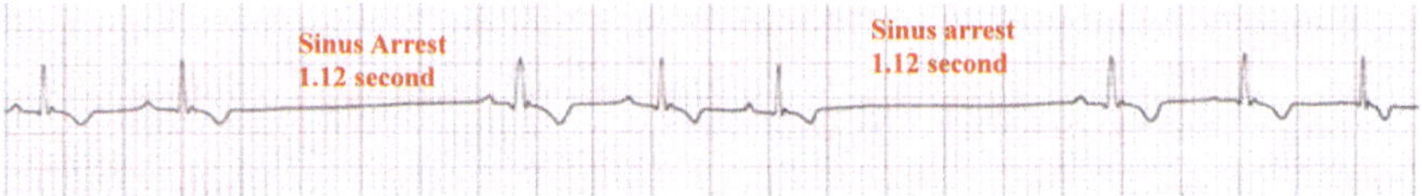

Fig. 16.76 Electrocardiogram (lead II, sensitivity 1, speed 25 mm/s) of a 7-year-old Great Dane showing electrical alternans of R wave, low voltage of R wave (0.2–0.5 mV), peaked and tall T wave and S-T segment depression (0.25 mV) suggesting possible myocardial infarction. S-T segment and T wave changes could be due to myocardial ischemia

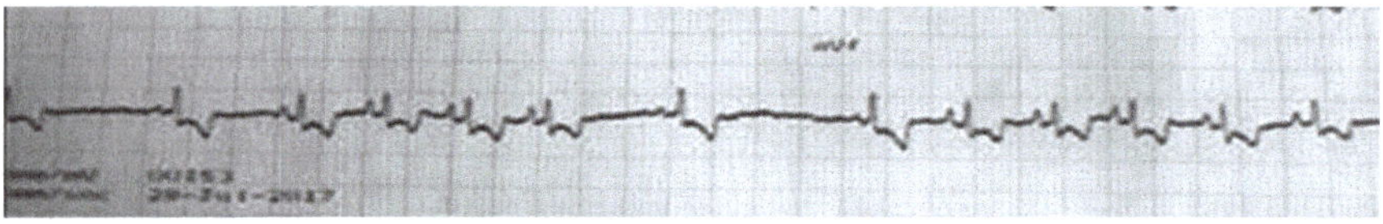

Fig. 16.77 Electrocardiogram (lead II, sensitivity 1, speed 25 mm/s) of a 6-year-old German Shepherd female with the history of severe dyspnea showing low-voltage complexes, sinus arrhythmia (hearty rate 120 bpm, sinus block at 2 places), wandering pace maker and S-T depression (0.1 mV)

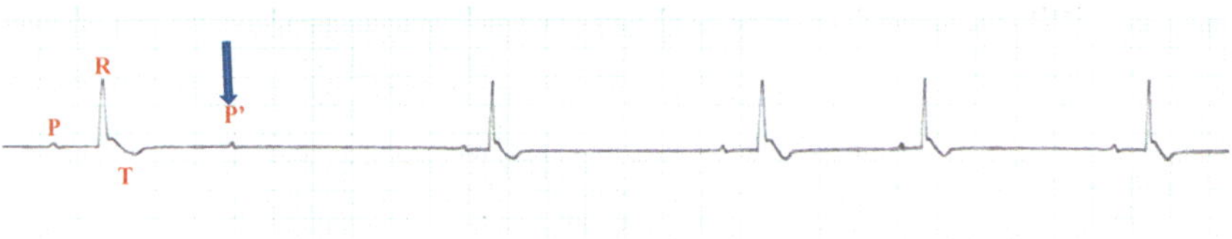

Fig. 16.78 Electrocardiogram (lead II, sensitivity 1, speed 25 mm/s) of a 5-year-old male nondescript dog showing second-degree AV block and bradyarrhythmia (heart rate 48 bpm and variable R-R interval). P′ wave at one point (arrow) is not conducted and P-R interval of the preceding complex is 0.28 s

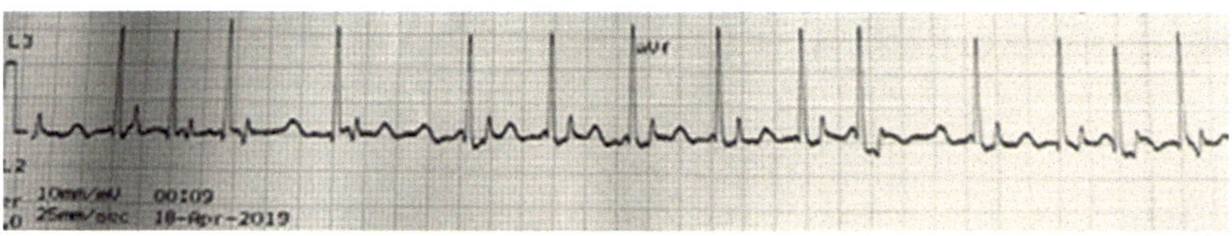

Fig. 16.79 Electrocardiogram (lead II, sensitivity 1, speed 25 mm/s) of an adult German Shepherd bitch showing sinus arrhythmia (heart rate 110 bpm, R-R interval 0.36–0.8 s), broad P (0.08 s), first degree heart block (PR segment 0.28 s), and sudden depression of S-T segment (0.2 mV) at tenth complex from the left

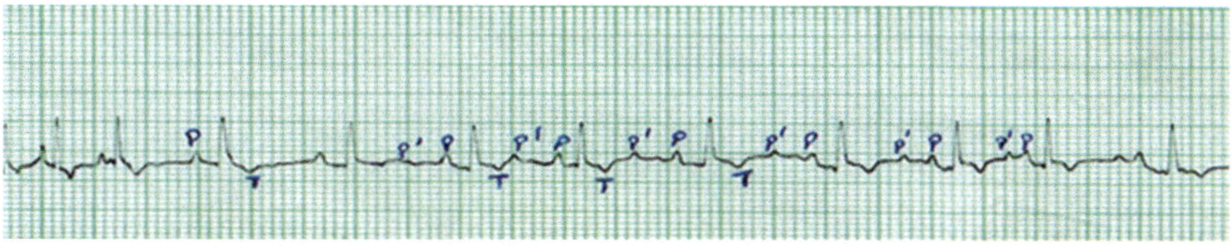

Fig. 16.80 Electrocardiogram (lead II, sensitivity 1, speed 25 mm/s) of a 10-year-old German Shepherd showing second-degree AV block type A (QRS is normal 0.04 s). The site of conduction failure seems to be above the bifurcation of bundle of His (within AV node). Ventricular rate is slower (140 bpm) than the atrial rate (240 bpm) because of blocked P′ waves. P waves are of normal configuration

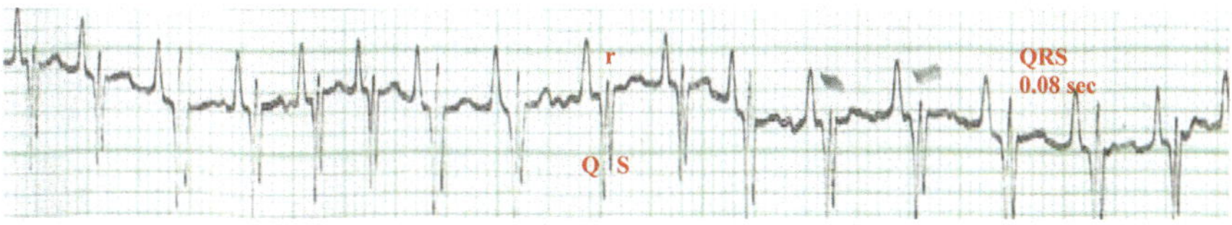

Fig. 16.81 Electrocardiogram (lead II, sensitivity 1, speed 25 mm/s) of a 5-year and 9-month-old German Shepherd dog showing broad QrS pattern (0.08 mV) in lead I, II, III, and MEA 130° (severe right axis deviation) suggesting right bundle branch block

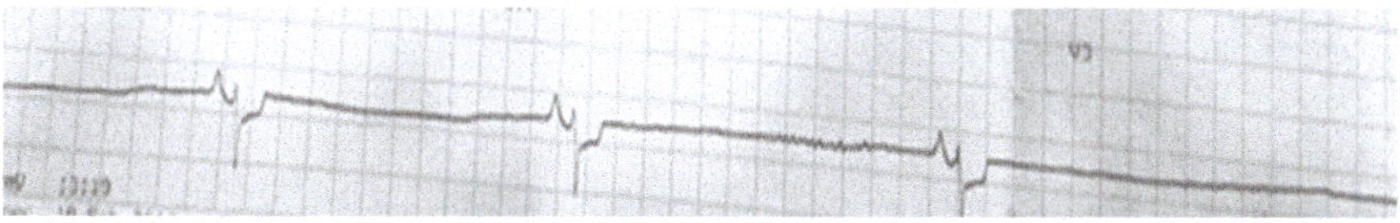

Fig. 16.82 Electrocardiogram (lead II, sensitivity 1, speed 25 mm/s) of a 5-year and 3-month-old Pomeranian bitch with gasping and cyanotic mucus membranes showing bradycardia (heart rate 30 bpm), QRS within normal limit (0.04 s), right axis deviation (−90° calculated from lead I and III), positive QRS in aVR and aVL and large S in lead II, III, and aVF suggesting right bundle branch block with bradycardia

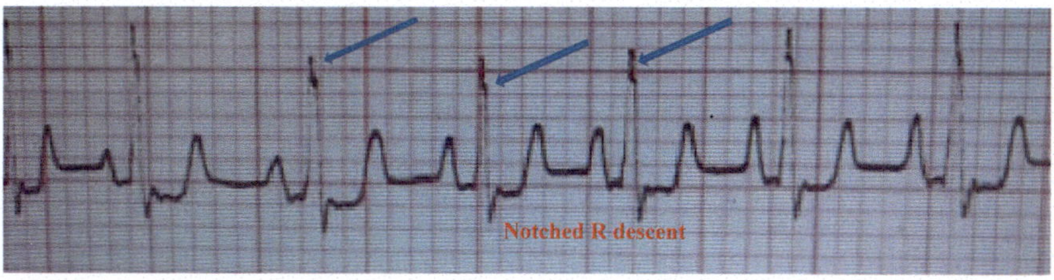

Fig. 16.83 Electrocardiogram (lead II, sensitivity 1, speed 25 mm/s) of a 9-year-old male Mastiff dog with exertion, exercise intolerance, and dyspnea showing notched R descent, broad QRS (0.06 s) suggesting possibility of microscopic intramural myocardial infarction (MIMI)

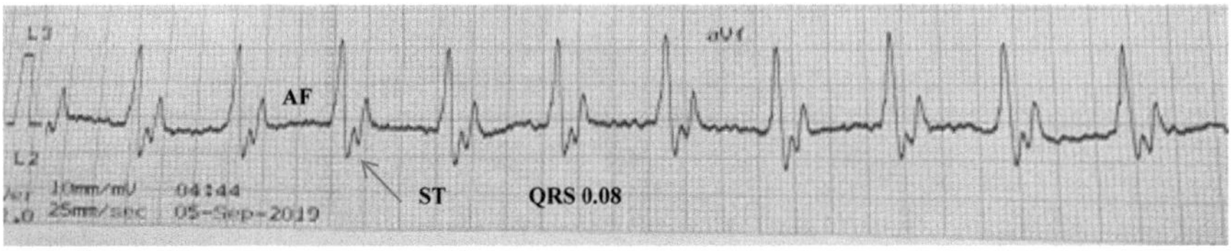

Fig. 16.84 Electrocardiogram (lead, II, sensitivity 1, speed 25 mm/s) of an 8-year-old male Dalmatian with gasping showing ventricular heart rate as 100 bpm, atrial fibrillations (AF), broad QRS 0.08 s, normal R amplitude (1.1 mV) S-T depression T interval as 0.24 s suggesting severe left bundle branch block with atrial fibrillations and ST depression. The dog collapsed within 20 min of taking electrocardiogram. These electrocardiographic changes might be due to severe cardiomyopathy or ischemic cardiomyopathy and led to severe acute heart failure (acute heart attack)

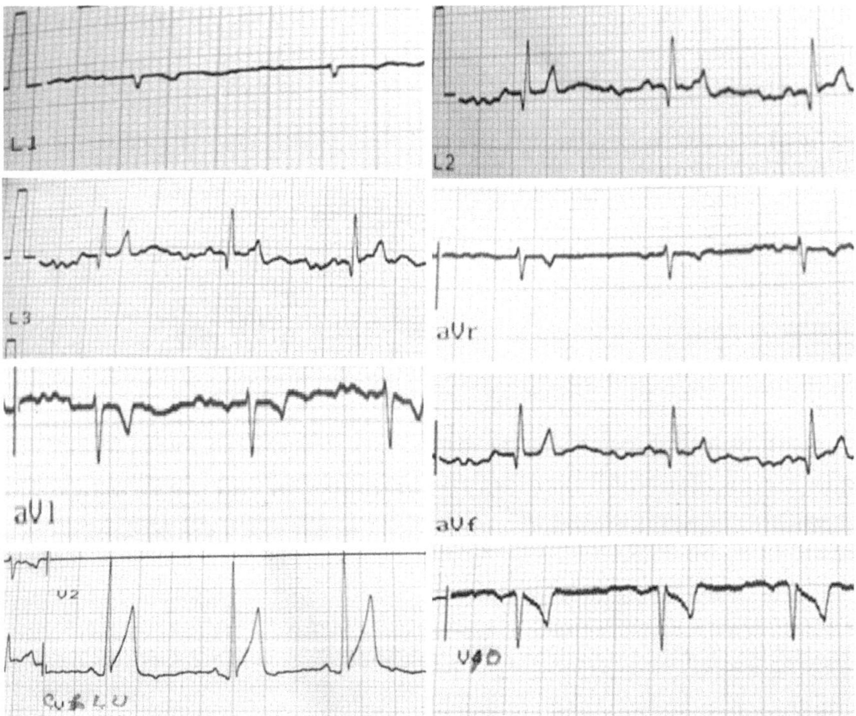

Fig. 16.85 Electrocardiogram (lead I, II, III, aVR, aVL, aVF, CV6LU, V10, sensitivity 1, speed 25 mm/s) of a 15-month-old male German Shepherd with weakness for the last 2.5 month showing normal heart rate 80 bpm, normal P (0.1 mV, 0.03 s, small q (0.2 mV in all limb leads), reduced voltage of R (0.5–0.6 mV) in lead II, III, aVF), <2.5 mV R amplitude in lead CV6LU, broad QRS (0.07 s), slightly elevated S-T segment (0.15 mV), large and broad T wave in comparison of R wave (0.3 mV, 0.12 s) and slightly variable R-R interval suggesting low voltage complex with minor left bundle branch block. X-ray revealed no edema in dorso-caudal hilar region and its VHS was 8.8. There was no ascites. The increased level of cardiac Troponin-I (0.25 ng/mL) suggested cardiac muscle damage (as levels of cTn-I in healthy dogs are <0.07 ng/mL). In the light of all investigations low-voltage complexes with left bundle branch block in this case seems to be due to cardiomyopathy

References

Maiti SK, Dutta A, Varshney JP, Kumar N (2013) Effect of different carbon dioxide pressure gradient in capnoperitoneum for laparoscopic examination in dogs. World J Lap Surg 6:1–10

Varshney JP, Chaudhary PS, Sutaria P (2011) Emergency management of toad poisoning in a dog. Intas Polivet 12:225–226

Varshney JP, Deshmukh VV, Chaudhary PS (2015) Evaluation of myocardial injury in acute canine monocytic ehrlichiosis. Intas Polivet 16:340–344

Varshney JP (2016a) 'J' wave syndrome in dogs: an electrocardiographic study. Indian J Vet Med 36:120–121

Varshney JP (2016b) Spinal shock in a Labrador sustaining an accidental neck trauma. Blue Cross Book 34:112–114

Varshney JP (2018) Right heart failure in an adult Pomeranian suffering from fatal dirofilariasis. Blue Cross Book 37:95–97

Varshney JP, Monapara HD (2019) Evaluation of cardiac toxicity in dogs bitten by viper snake. Vet Pract 20:191–194

Section II

Feline

Electrocardiography in Cats

17

In India, information on cardiac diseases in cats is scanty as number of feline patients in veterinary hospitals is less than that of canines. Therefore, the prevalence rate of cardiac abnormalities in cats in India is unknown. Studies done abroad have revealed that cardiac diseases (hypertrophic cardiomyopathy, mitral valve disease, unclassified cardiomyopathy, thyrotoxic heart disease and congenital heart disease) are commonly diagnosed in cats. Arrhythmias are quite common in cats. It seems that the diseases of heart are not uncommon in feline population in India also. Diagnosis of cardiovascular diseases in cats is not as an easy task because of small size of the body and heart, rapid heart rate, and uncooperative nature. Further abrupt onset of clinical signs makes the situation complex. Some risk factors such as taurine deficiency and hyperthyroidism are associated with feline heart diseases. Breeds such as Maine Coons, Ragdolls, Persians, and Siamese are said to be more predisposed to cardiomyopathies. There are few differences in canine and felines as related to cardiac diseases.

17.1 Differences Between Dogs and Cats with Regard to Cardiac Diseases

17.1.1 Clinical Manifestations

The signs and lesions in cats with cardiac diseases remain asymptomatic until very advanced stage and manifested suddenly in acute form. Whereas in dogs clinical signs of heart ailments are insidious and slowly progressive, cats with congestive heart failure suddenly develop severe dyspnea. In some cases per acute hind limb paralysis is developed due to arterial thromboembolism. Sometimes cats die suddenly due to hypertrophic cardiomyopathy without manifesting any clinical signs. Coughing (nocturnal cough and gagging) and lung crackles or wheezing on chest auscultation, as seen in dogs with left heart failure, is not common in cats despite having severe lung edema. Vasoconstriction in extremities leads to slightly cool

J. P. Varshney, *Electrocardiography in Veterinary Medicine*,
https://doi.org/10.1007/978-981-15-3699-1_17

extremities. But it is not a reliable sign of congestive heart failure in cats. Similarly ascites, seen in dogs with right heart failure, is also not very dominant sign in cats. In cats functional murmurs (due to fever, anemia, volume overload, tranquilizing drugs) are very common. There is hardly any change in pulse quality in feline heart disease. Even capillary refill time and mucus membrane color changes are also not very conspicuous in cats with heart diseases. In ventral septal defects and tricuspid valve defects, murmur localization can be of some help.

17.1.2 Cardiac Murmurs

Areas of location of cardiac murmurs in cats are different than dogs. Murmurs are loud in cat and are located along the left or right sternum or cranial/caudal thorax.

17.1.3 Radiographic Features

Cardiac silhouette in normal young cats is slightly elongated (Fig. 17.1), and in geriatric cats, it is more horizontal in lateral radiographs. Dilation of main pulmonary artery (MPA) in cats does not show a bulge at 1–2° clock position, as seen in dogs, in ventro-dorsal chest radiographs. Cardiomegaly in cats results in elongation and widening of cardiac silhouette in general, and detection of specific chamber enlargement in radiographs is rather difficult. Tracheal elevation is not always seen in cats with cardiomegaly. Pulmonary edema in cardiac diseases in cats may be patchy and ventral (Fig. 17.2), while in dogs it is dorsal in caudal perihilar region. Vertebral heart

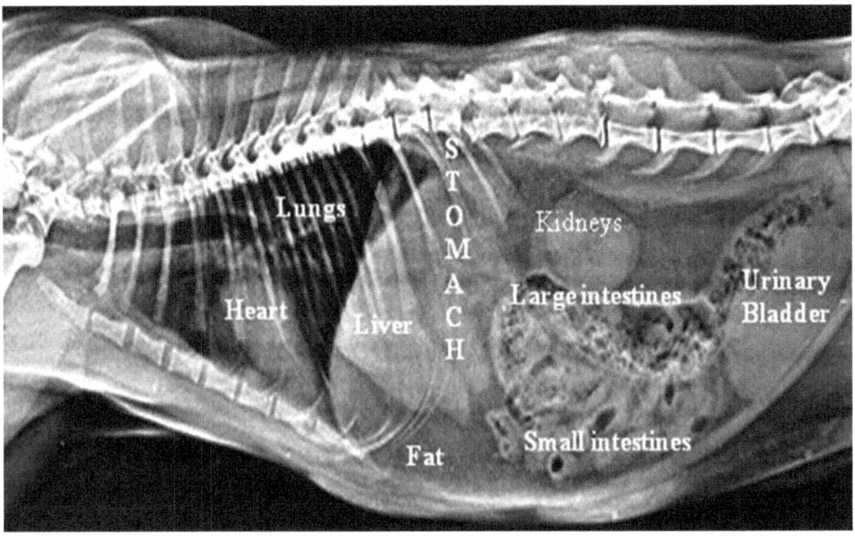

Fig. 17.1 Radiograph of a 15-month-old female cat in right lateral recumbency showing different organs. The heart silhouette is slightly elongated. Cranio-caudal ventricular diameter of cardiac silhouette on this lateral radiograph is approximately 2.25 intercostal spaces

score (VHS) is calculated on the lateral thoracic radiographs to assess cardiac size. Long axis of cardiac silhouette from the carina of the main bronchus to the apex of the heart and short axis at the widest part of the heart are measured. These axes (long and short) are transferred to the vertebrae starting from cranial edge of T4 and count the number of vertebrae fall under each axis. Sum up the number of vertebrae falling under the both axes (Fig. 17.3). Normal vertebral heart score in cats remains within 7.5. Sometimes results of VHS are very illusive as there is no radiographically detectable heart enlargement in hypertrophic cardiomyopathy in cats.

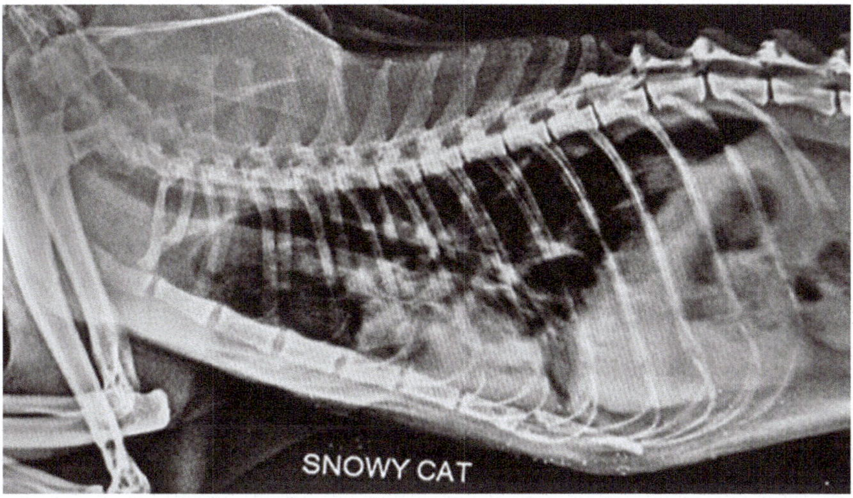

Fig. 17.2 Radiograph of a 5-year-old female cat in right lateral recumbency showing patchy ventral lung edema

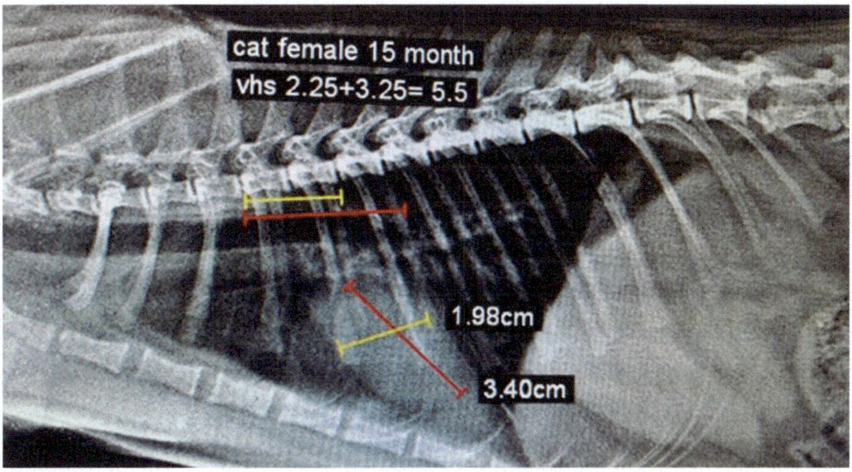

Fig. 17.3 Radiograph of a 15-month-old female cat in right lateral recumbency showing measurement of vertebral heart score (VHS). VHS in this cat is 5.5

17.2 Electrocardiography

Electrocardiography is a noninvasive diagnostic tool routinely employed for monitoring cardiac rhythm, rate, and conduction disturbances in cats also similar to dogs. Electrocardiography in felines is not an effective tool for screening of occult heart diseases. It has been observed that electrocardiograph is not a very sensitive and precise tool for detection of chamber enlargement, though it can detect MEA. Therefore major utility of electrocardiography in cats is limited to the diagnosis of arrhythmias and conduction disturbances. Early detection of arrhythmias of severe nature, electrolyte imbalance, and conduction disturbances facilitates an exact diagnosis of cardiac ailments so that appropriate remedial measures can be undertaken at the earliest to increase the quality of life of cats with heart disease. Holter 24 hour's continuous ambulatory electrocardiography is not very successful in felines.

The generation of electrocardiogram in cats is because of depolarization and repolarization of cardiac muscle fibers as seen in canines. The wave of depolarization begins in sinoatrial node (SA node) and passes through atrium producing "P" wave and reaches to AV node. The delay in conduction at AV node results into P-R segment. Then depolarization of ventricles occurs producing "QRS" complex followed by isoelectric period, i.e., S-T segment. This is followed by repolarization of the heart represented by "T wave. The measurements of amplitude and duration of electrocardiographic complexes and intervals in cats differ from that of dogs.

17.2.1 Positioning of Cats for Electrocardiography

The cat is positioned on a table covered with a foam mattress and nonconductive rubber sheet in right lateral recumbency and is restrained by an attendant putting his right arm over the neck and left over hind quarter. Both limbs (left and right) are kept apart. The forelimbs are kept perpendicular to the long axis of the body. If the left thoracic limb is pulled caudally or back, the form of QRS in lead I appears like that of aVF, whereas if that thoracic limb is pulled cranially, QRS in lead I appears like that recorded in aVR. Some cats are uncooperative in right lateral recumbency. In such cases sternal recumbency may be adopted. In sternal recumbency the size of "P" and "R" is slightly tall than in lateral recumbency.

17.2.2 Placement of the Electrodes

Before putting the electrodes, both electrode and the skin are moistened with electrocardiographic gel, paste, or alcohol. Alcohol works well. Electrodes are attached directly to the skin. In case nothing is available, water can be used to increase contact between electrode and skin in the unfavorable condition. Placement of electrode is shown in Fig. 17.4.

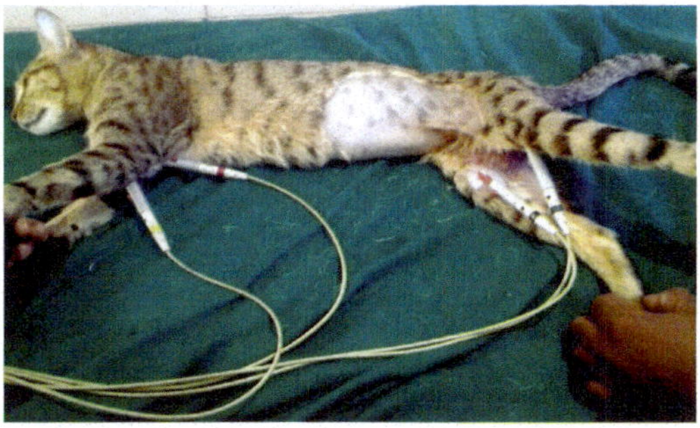

Fig. 17.4 Showing positioning (right lateral recumbency) of the cat and placement of electrodes for electrocardiography

RA	Right forelimb clip or needle electrode is attached proximal to the olecranon on the caudal aspect of the right forelimb.
LA	Left forelimb clip or needle electrode is attached proximal to the olecranon on the caudal aspect of the left forelimb.
RL	Right hind limb clip or needle electrode is attached over patellar ligament on the anterior aspect of the right hind limb.
LL	Left hind limb clip or needle is attached over patellar ligament on the anterior aspect of the left hind leg.

17.2.3 Electrocardiogram and Electrocardiographic Indices

Electrocardiogram of feline is having "P," "QRS," and "T" complexes; P-R interval, S-T segment, Q-T interval, and R-R interval (Fig. 17.5). In lead I "R" is conspicuous, and "P" and "T" are very small (Fig. 17.6). In lead II, III, and aVF (Fig. 17.6),

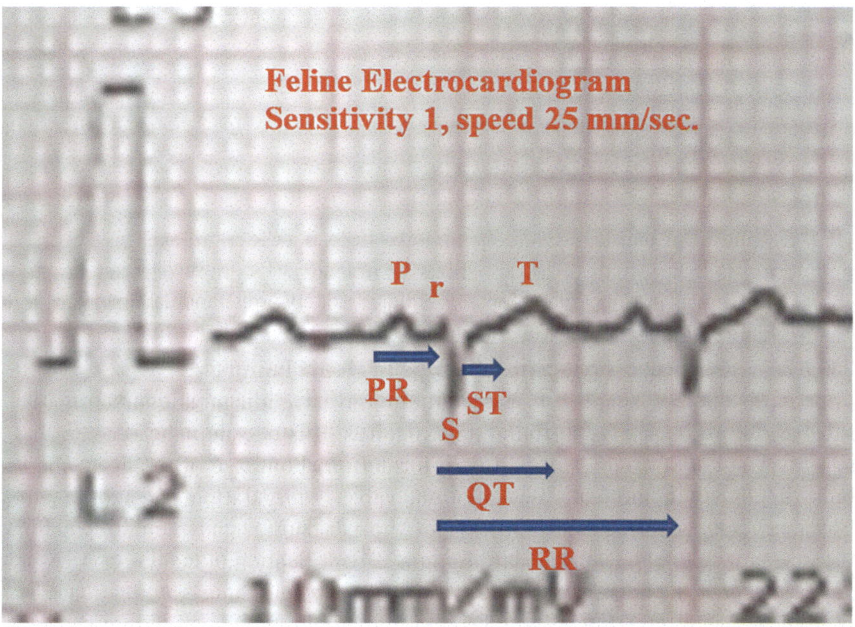

Fig. 17.5 Electrocardiogram (sensitivity 1, speed 25 mm/s) of an adult male cat showing ECG complexes and intervals (lead II)

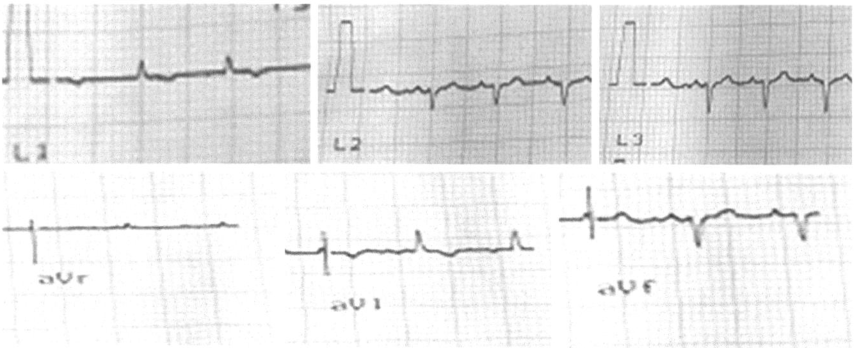

Fig. 17.6 Electrocardiogram of a cat showing ECG complexes in different leads (I, II, III, aVR, aVL, and aVF)

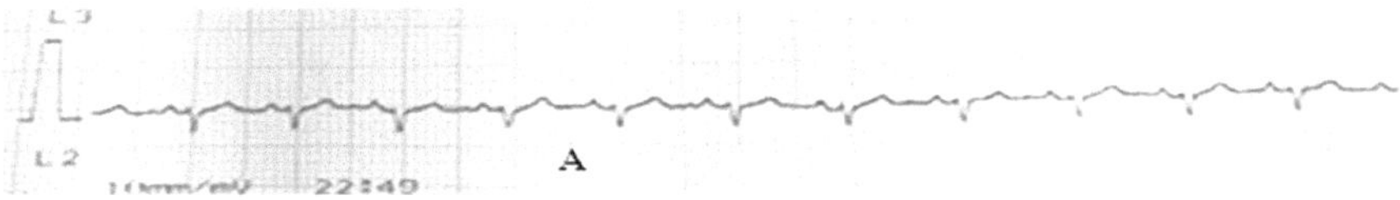

Fig. 17.7 Electrocardiogram (lead II, sensitivity 1, speed 25/s) of a healthy adult male cat showing heart rate as 160 bpm with sinus rhythm; P 0.1 mV, 0.3 s; PR 0.07 s; r 0.1 mV; S 0.2–0.25 mV; QRS 0.03 s; T 0.15 mV, 0.06 s

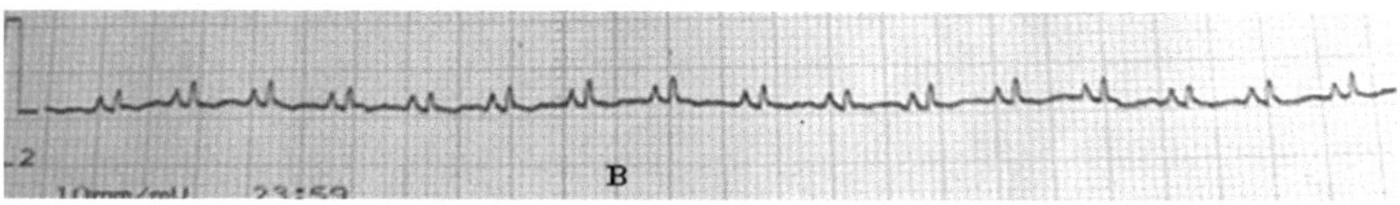

Fig. 17.8 Electrocardiogram (lead II, sensitivity 1, speed 25/s) of another healthy 15-month-old male cat showing heart rate as 200 bpm with sinus rhythm; P 0.1 mV 0.04 s; PR interval 0.08 s; R 0.2–0.3 mV, no Q or S wave; QRS 0.04 s. T wave is not appreciable

"P," "R," and "T" complexes are conspicuous. In lead aVR complexes are reversed in polarity, and very small "s" is recognizable. In lead aVL complexes are recognizable and reversed in polarity as compared to lead II. This pattern is similar to the pattern seen in dogs. Heart rate varies from 140 to 220 beats per minute. Figs. 17.7 and 17.8 show electrocardiograms of healthy cat. Range values of electrocardiographic parameters are given in the table (Table 17.1). "P" amplitude varies from 0.1–0.2 mV with a duration from 0.03 to 0.04 s. P-R interval ranges from 0.05 to 0.07 s. Amplitude of "R" is small than that of dogs and is variable from 0.1 to 0.8 mV. QRS duration is of 0.04 s. S-T segment is short (0.08 s). T wave is sometimes not appreciable and is very, very small. Its amplitude varies from 0.0 to 0.1 mV and duration 0.04 s. Q-T interval is of 0.16 s, and R-R interval is of 0.4 s and varies as per heart rate. Mean electrical axis on frontal plane in normal cats lies between 0 and + 160° (Fig. 17.9).

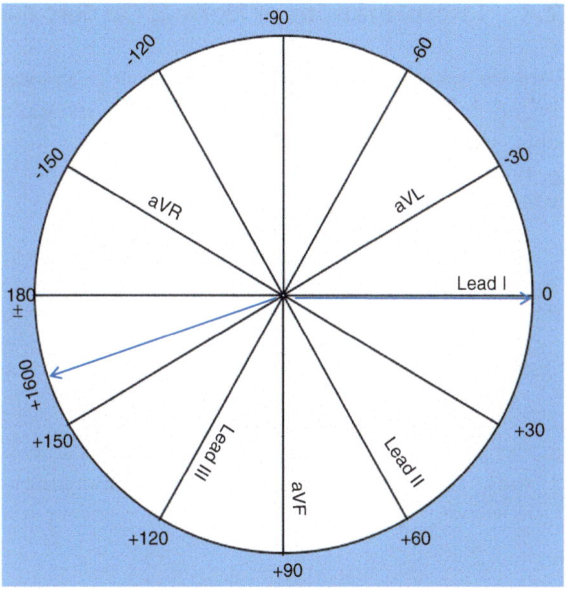

Fig. 17.9 Diagrammatic representation of mean electrical axis on frontal plane in cat

Mean Electrical Axis in Cat

MEA Normal Cat	0 to +160°(area between two blue arrows)
MEA Right Axis Deviation	+160 to - 90°
MEA Left Axis Deviation	0 to - 90°

Table 17.1 Electrocardiographic parameters of healthy cat

Electrocardiographic parameters	Range values
Heart rate	140–220 bpm
P amplitude	0.1–0.2 mV
P duration	0.02–0.03 s
P-R interval	0.05–0.07 s
R amplitude	0.10–0.8 mV
S amplitude	0.0–0.05 mV
QRS duration	0.03–0.04
S-T segment	0.06–0.08 s
T amplitude	Not appreciable −0.2 mV, +ve, −ve, biphasic
T duration	0.0–0.04 s
Q-T interval	0.09–0.16 s

17.3 Electrocardiographic Parameters of Healthy Cat

The range values of different electrocardiographic parameters observed at the hospital are given in the Table 17.1

17.4 Interpretations of Normal Cardiac Wave Forms in Cats

There are many waves, interval, and segment in an electrocardiogram of a cat as is seen in electrocardiograms of other species. Interpretation of these waves, segments, and intervals is given below.

Wave forms segments and intervals	Interpretation
"P" wave	Represents atrial muscle depolarization. Normally the wave is positive in lead II and avF
"P-R" interval	Represents time from the onset of atrial depolarization through conduction over AV node, bundle of His, and Purkinje fibers
"QRS" complex	Represents depolarization of ventricular muscle. Q is first negative deflection, "R" the first positive deflection, and "S" the second negative deflection
"S-T" segment	Represents period of phase 2 of action potential
"T" wave	Represents repolarization of ventricular muscle
"Q-T" interval	Represents total time of ventricular depolarization and repolarization

17.5 Characteristics of Feline Electrocardiogram

Electrocardiogram of different species of animals has certain uniqueness. The main characteristic features of feline electrocardiogram are enumerated below.

1. Complexes are small as compared to dogs.
2. Intervals are short than that of dogs.
3. QRS may be of low voltage in some normal cats.
4. Heart rate is fast as compared to dogs.

17.6 Abnormal Waves, Intervals and Segments with Their Indications

When any wave, segment or interval in an electrocardiogram deviates from the values reported in clinically healthy cats, it is called abnormal wave, segment or interval. Their indications are given below.

Abnormal waves, intervals and segments	Indications
"P" wave	
Broad (more than 0.04 s)	Left atrium enlargement (P mitrale)
Tall (more than 0.2 mV)	Right atrium enlargement (P pulmonale)
Tall and broad (more than0.2 mV and more than 0.04 s)	Biatrial enlargement
Variable amplitude of P wave	Wandering pace maker
Absence of P wave	Atrial stand still or silent atrium

Abnormal waves, intervals and segments	Indications
Ta wave (increased height of descending arm of "P")	Right atrium enlargement
"QRS" complex	
Tall "R" (more than 0.9 mV) in lead II	Left ventricular enlargement (LVE) or left bundle branch block (LBBB)
Wide "QRS" (more than 0.04 s)	LVE or LBBB
Increased amplitude of "S" (more than 0.5 mV in lead I, II, III, aVF)	Right ventricular enlargement (RVE)
	Right bundle branch block (RBBB)
S-T segment/J point	
Elevation (more than 0.10 mV)	Myocardial infarction
	Myocardial hypoxia
	Pericarditis
	Digoxin toxicity
Depression (more than 0.10 mV)	Myocardial ischemia
	Hyperkalemia
	Hypokalemia
Slurring or coving	Left ventricular hypertrophy
P-R interval	P-R interval is inversely proportional to heart rate
Increase (more than 0.09 s)	First-degree heart block
Q-T interval	
Prolong (more than 0.25 s)	Hypocalcemia
	Hypothermia
	Hypokalemia
	CNS disorders
	Ventricular hypertrophy
	Conduction disorders
	Strenuous exercise
	Quinidine toxicity
	Ethylene glycol poisoning
	Secondary to prolonged QRS
Short (<0.15 s)	Hypercalcemia
	Hyperkalemia
	Digitalis toxicity
"T" wave	"T" wave changes may indicate myocardial disease
Sharply pointed and notched	Electrolyte imbalance
Isolated alternans	No specific abnormality if QRS complex is normal
Large T wave	Myocardial hypoxia
	Ventricular enlargement
	Intraventricular conduction abnormalities
	Hyperkalemia
	Metabolic disease
	Respiratory diseases

Type of arrhythmias/ conduction disturbances	Electrocardiographic features
	Some P waves are not conducted
	Normal complexes have consistent P-R interval
	P wave and QRS complexes are of normal configuration
	Second-degree AV block has two types as Mobitz type I and Mobitz type II
(a) Mobitz type I	Progressive prolongation of the P-R interval until a non-conducted "P" wave
(b) Mobitz type II	P-R interval is uniform in preceding the blocked impulse
(3) Third degree	More P waves than QRS complexes
	Varying P-R interval
	No consistent relationship between atrial and ventricular beats
	P waves are of normal configuration
	QRS complexes near normal or bizarre
	P-P and R-R relatively constant
Ashman phenomenon	First reported in humans with atrial fibrillation in 1947 by Gouaux and Ashman
	Characterized by aberrant ventricular conduction due to change in QRS cycle length
	Relatively long QRS cycle is followed by short cycle
	Beat with short cycle has bundle branch block
	Also reported in cat (Nakamura and Zimmerman 2012)
	Characterized by bradycardia, wide QRS (left/right bundle branch block) complex with a longer R-R interval alternate with normal sinus complex with a short R-R interval, second-degree AV block (Fig. 17.10)

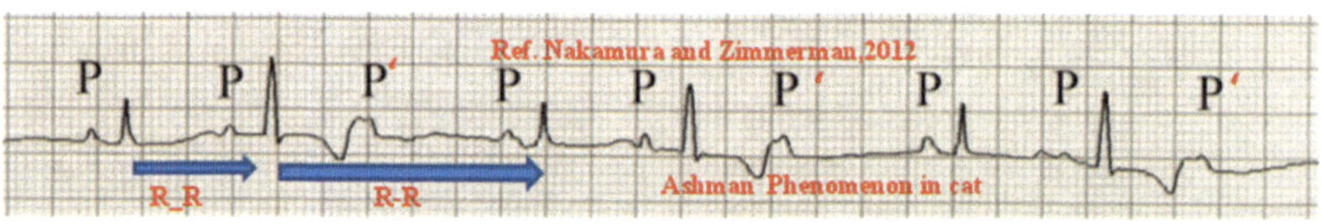

Fig. 17.10 Electrocardiogram of a cat with bradycardia (80 bpm) showing normal sinus complex (1,3,5); alternate with LBBB (2,4,6), short and long R-R interval alternately and second-degree AV block ("P") following LBBB suggesting Ashman phenomenon. (Reference Nakamura,R.K. and Zimmerman, S. 2012. Veterinary Medicine Today. ECG of the month. J Am Vet Med Assoc 241:433)

17.8 Echocardiography in Felines

Definite diagnoses of feline heart diseases can be made by echocardiography. But highly skilled persons are needed. Nowadays much advancement has been done in echocardiography machine. M-mode and color Doppler facility are available to know different dimensions and blood flow pattern. Because of small heart size and high heart rate, high-frequency transducer (7–10 MHz) is generally preferred. To discuss echocardiography in detail is beyond the scope of this book as it is mainly devoted to electrocardiography.

17.9 Cardiac Biomarkers

Cardiac troponin-I (cTn-I) and atrial natriuretic peptide (ANP) and its prohormones NT-pro ANP and NT-pro BNP have been investigated as biomarkers for cardiac muscle damage in humans and dogs. Use of these biomarkers in feline medicine is not extensive and is restricted to screening of subclinical heart disease and to differentiate acute dyspnea of pulmonary and cardiac origin. The level of cTn-I is very low ($\leq$0.04 ng/mL) in clinically healthy cats (Sleeper et al., 2001). With the damage of cardiac muscle in hypertrophic cardiomyopathy, level of cTn-I is significantly increased. Another important cardiac biomarker NT-pro BNP has also been investigated in cats. NT-pro BNP level is low (<50 pmol/L) in healthy cats (Anjos et al. 2015) but increases significantly in cats with heart diseases.

17.10 Abnormal Electrocardiograms in Cats (Figs. 17.11, 17.12, 17.13, 17.14, 17.15, 17.16, 17.17, 17.18, 17.19 and 17.20)

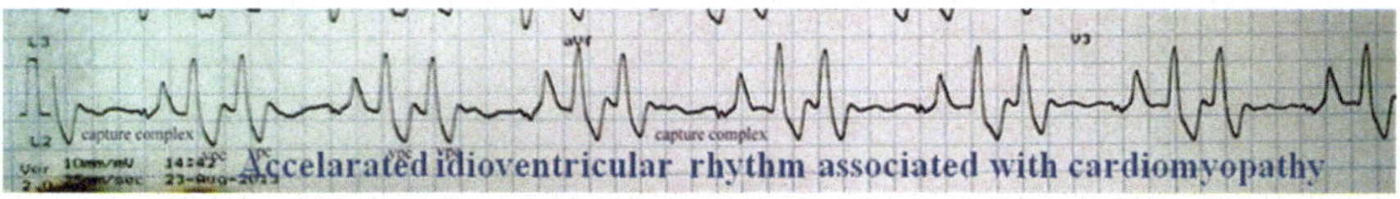

Fig. 17.11 Electrocardiogram of a 5-year-old cat showing VPC at the rate of 90 bpm suggesting accelerated idioventricular rhythm (Varshney 2016)

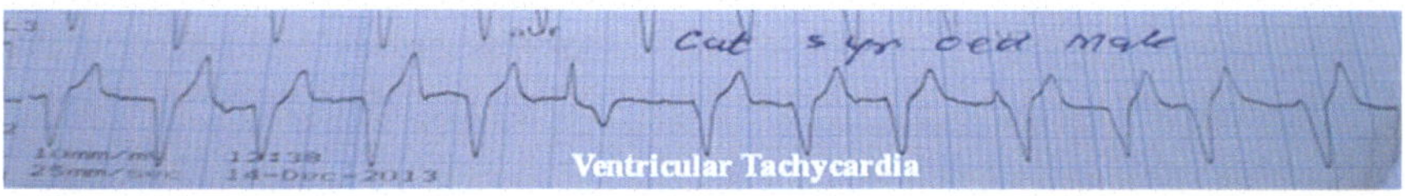

Fig. 17.12 Electrocardiogram of a 7-year-old cat showing ventricular tachycardia (ventricular rate 144 bpm). The sixth complex from the beginning of the strip is a capture complex

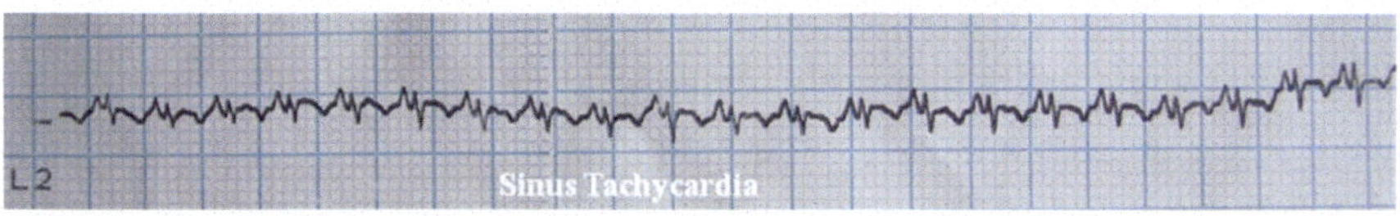

Fig. 17.13 Electrocardiogram of a 3-year-old female Persian cat showing heart rate of 280 bpm suggesting sinus tachycardia. P, QRS, and T complexes are of normal configuration

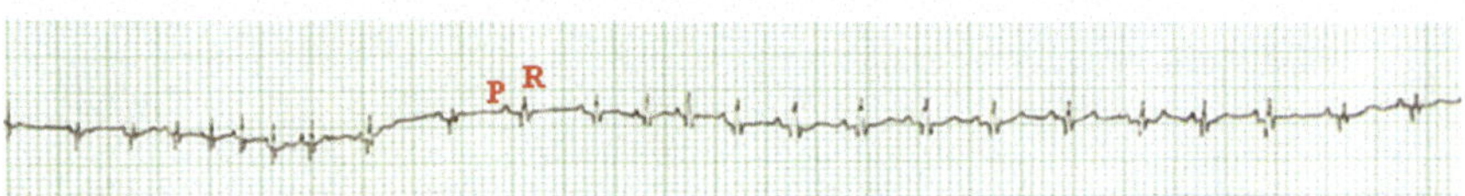

Fig. 17.14 Electrocardiogram (lead II, sensitivity 1, speed 25 mm/s) of a 5-year-old male cat with stud tail syndrome and fever (104. 2 °F) showing heart rate of 300 bpm suggesting sinus tachycardia. Small q (0.15 mV) and small s (0.1 mV) are also visible

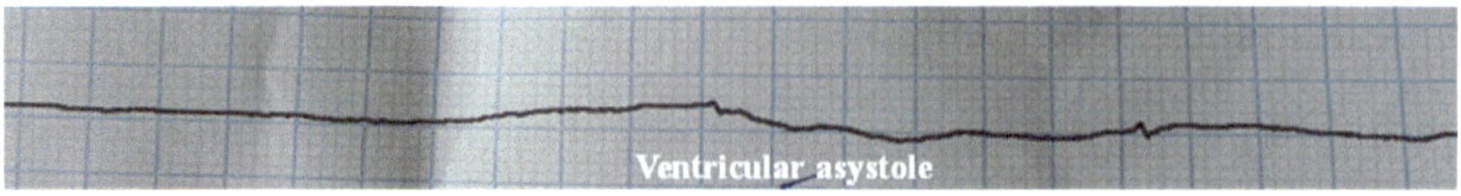

Fig. 17.15 Electrocardiogram (lead II, sensitivity 1, speed 25 mm/s) of a 9-year-old female cat showing ventricular asystole. P is of normal configuration but not conducted. There is no ventricular rhythm (QRS)

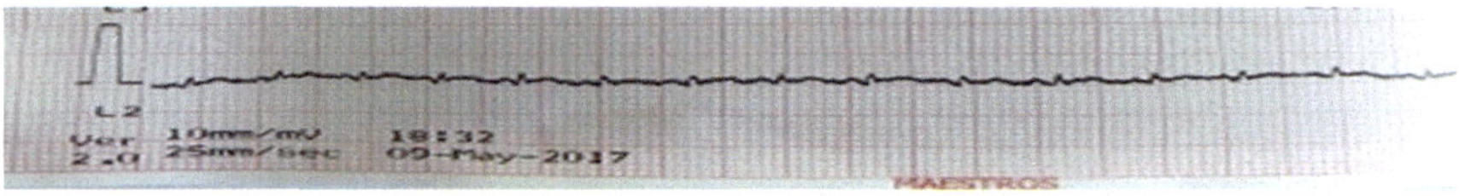

Fig. 17.16 Electrocardiogram (lead II, sensitivity 1, speed 25 mm/s) of a cat with hyperthyroidism (serum T_4 5.4 µg/dL, T_3 1.12 ng/mL) showing increased heart rate (H.R. 220 bpm) with regular R-R interval suggesting sinus tachycardia (Varshney 2018)

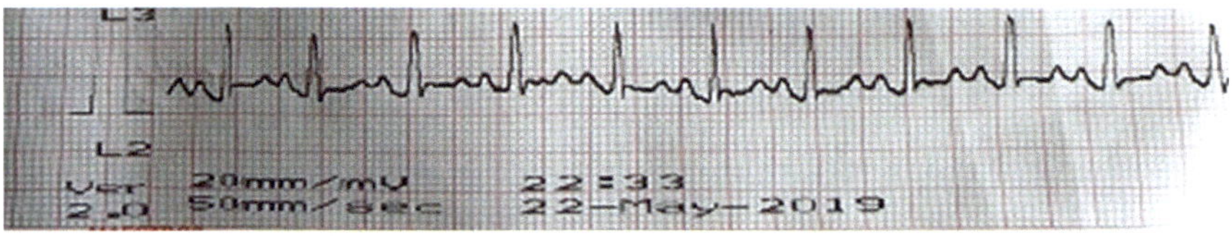

Fig. 17.17 Electrocardiogram (sensitivity 2, speed 50 mm/s) of a 2-month-old female kitten (weighing 750 g) with acute renal failure grade 4 (serum creatinine 5.3 mg/dL, BUN 48.7 mg/dL, ALT 56 IU/L, SAP 85 U/L, T bilirubin 0.9 mg/dL) caused by high doses of meloxicam (2.5 mg PO for 3 days) showing sinus tachycardia (heart rate 300 bpm, sinus rhythm, P 0.1 mV, 0.03 s; PR interval 0.07 s; R 0.5 mV, QRS 0.03 s; ST segment 0.06 s; Q-T 0.12 s; and R-R interval 0.2 s)

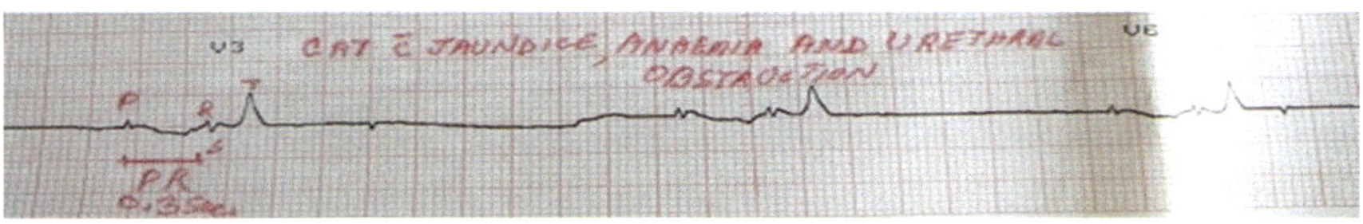

Fig. 17.18 Electrocardiogram (sensitivity 1, speed 25 mm/s) of a 1.5-year-old male cat (1.6 kg) with urethral obstruction (uroliths and crystalluria), jaundice (total bilirubin 2.0 mg%, direct bilirubin 0.6 mg%), anemia (hemoglobin 4.0 g/dL, PCV 12.4%), renal failure (creatinine 8.4 mg/dL, BUN 40.0 mg/dL), and hyperkalemia (7.2 mEq/L) showing bradycardia (heart rate 36 per minute), first-degree AV block (prolonged P-R interval 0.3 s), and tall and pointed T wave (0.5 mV)

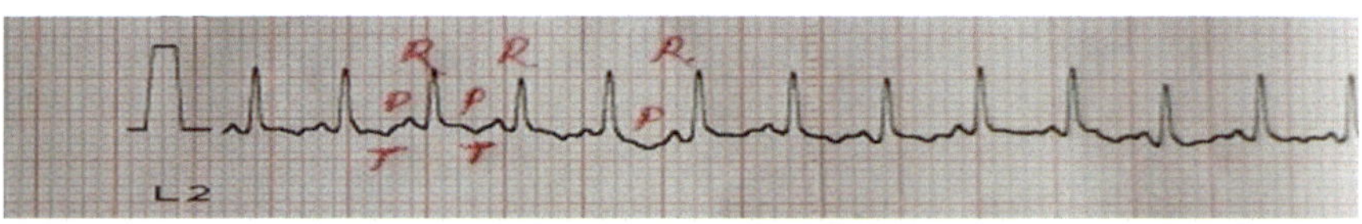

Fig. 17.19 Electrocardiogram (sensitivity 1, speed 25 mm/s) of a 1-month-old male kitten with anemia (hemoglobin 5.5 g/dL, erythrocyte 1.4 million/mm^3, packed cell volume 16%) and severe dyspnea showing sinus tachycardia (heart rate 260 bpm sinus rhythm), P 0.1 mV, 0.04 s; constant PR interval (0.06 s); and R 0.7 mV, 0.03 s

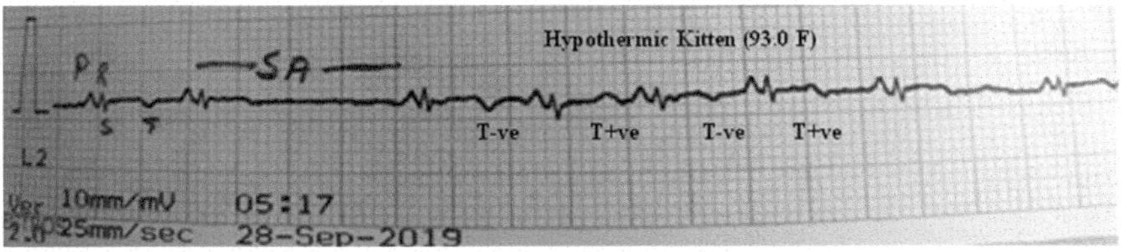

Fig. 17.20 Electrocardiogram (sensitivity 1, speed 25 mm/s) of a 40-day-old female kitten with hypothermia (93.0 °F), recumbency, and severe dehydration (skin tenting 5 s) showing bradycardia (heart rate 120 bpm), varying R-R interval (0.34–0.84 s, sinus arrest between second and third complexes as R-R interval is 0.84 s; sinus block between seventh and eighth complexes as R-R interval is 0.64 s), normal P (0.1–0.15 mV, 0.04 s), normal P-R interval (0.07 s), small r (0.1 mV), small s (0.1 mV), normal QRS (0.04 s), no elevation/depression of S-T segment (0.12–0.14 s duration), and changing polarity of T wave (1,3,5 complex negative; 4 and 6 positive). Sinus block/sinus arrest and bradycardia are not common in cats. Presence of radycardia, sinus arrest, and polarity change of T wave in cats indicates severe underlying disease of the heart

17.11 Clinical Conditions Associated With Arrhythmias and Conduction Disturbances

Arrhythmias and conduction disturbances are quite common in feline practice. Clinical conditions with which they are associated are given below.

Arrhythmias and conduction disturbances	Associated clinical conditions
Normal sinus rhythm	Normal rhythm of the heart varies with respiration, excitement, and vagal stimulation
Sinus tachycardia	Physical exercise
	Pain
	During examination procedure
	Fever
	Shock
	Anemia
	Infection
	Hypoxia
	Congestive heart failure
	Renal failure
	Hyperthyroidism
Sinus bradycardia	Drugs (atropine, epinephrine, ketamine, propranolol, digoxin, lidocaine)
	Renal failure
	Cardiomyopathy
Sinus arrhythmia	May be associated with respiration
Wandering pace maker	Sinus arrhythmia
	Normal in most cases
Atrial premature complexes	Hypertrophic cardiomyopathy (atrial enlargement)
	Hyperthyroidism

Arrhythmias and conduction disturbances	Associated clinical conditions
	Normal in aged cats
	Atrial disease
Atrial tachycardia	Atrial enlargement (hypertrophic cardiomyopathy)
Atrial flutter	Atrial disease
Atrial fibrillation	Atrial enlargement
	Hypertrophic cardiomyopathy
	Restrictive cardiomyopathy
Ventricular premature complexes	Congestive heart failure
	Myocardial infarction
	Cardiac neoplasia
	Hypoxia
	Anemia
	Uremia
	Pyometra
	Drugs (digitalis, diazepam, phenytoin)
Ventricular tachycardia	Feline cardiomyopathy
	Acute myocardial infarction
Ventricular fibrillation	Shock
	Anoxia
	Myocardial infarction or trauma
	Acid-base imbalance
	Hyperkalemia
Ventricular asystole	Ventricular fibrillation and complete AV block
	Severe acidosis
	Severe hyperkalemia
Sinus arrest	Stimulation of vagus nerve
	Atrial dilatation
First-degree AV block	May occur in healthy cats
	Digitalis toxicity
	Reaction to propranolol
	Hyper- or hypokalemia
Second-degree AV block	Increased vagal tone
	Cardiomyopathy
	Neoplasm
Third-degree AV block	Cardiomyopathy
Electrical alternans	Pericardial effusion
	Supraventricular tachycardia
	Cardiomyopathy

17.12 Anti-arrhythmic Drugs in Feline Arrhythmias

Atrial tachyarrhythmia, ventricular tachyarrhythmia, and ventricular arrhythmias are commonly seen in cats. Their management requires anti-arrhythmic drugs. Common drugs used in arrhythmia management in cats are given in the Table 17.2.

Table 17.2 Drugs for feline arrhythmias: at a glance

Arrhythmias	Drugs	Doses	Brand names available in India
Atrial tachyarrhythmias			
	Digoxin	0.0025–0.004 mg/kg PO BID	Cardin, Cardioxin, Dixin, Lanoxin, Sangoxin (tablets of 0.25 mg)
	Diltiazem	7.5 mg/cat PO TID	Angizem, Cardem, Dil Time, Dil Gard, Iski (30 and 60 mg tab)
Ventricular tachyarrhythmias			
	Atenolol	6.25–12.5/cat PO OD	Alonol, Altol, Aten, Betacard, Catenol, Lonol, Tecard (25 and 50 mg tab)
	Sotalol	2.0 mg/kg PO BID	Sotagard, Solet, Sotalar (40 and 80 mg tab), Sotalar inj. (10 mg/mL)
Ventricular arrhythmias			
	Lidocaine	0.25–0.5 mg/kg slow IV. Can be repeated two times more if need be	Gasicard, Xylocard (5 mL or 50 mL vials)
	Propranolol	0.02 mg/kg slow IV. Repeat if need be	Betabloc (10 mg tab), Betapro (10, 20, 40 mg tab), Ciplar (10, 40 mg tab.), Poplol (20, 40 mg tab)
	Esmolol	200–500 μg/kg slow IV. Followed by 25–100 μg/kg/min as CRI	Esocard (10 mL vial 100 mg/mL), Miniblock (10 mL vial 100 mg/mL)

17.13 Cardiomyopathies in Felines

- Cardiomyopathies are common heart disease in cats.
- Types of cardiomyopathies, viz., hypertrophic, dilated, restrictive, unclassified, arrhythmogenic right ventricular, excess moderator band, or endocardial fibroelastosis.
- Hypertrophic cardiomyopathy is more common.

17.13.1 Hypertrophic Cardiomyopathy

- Local or diffuse concentric hypertrophy of ventricular muscle.
- Etiology largely obscure.
- Genetic mutations, as observed in humans, are speculated.
- Prevalence studies are lacking in India.
- Clinical picture highly variable.
- Murmurs audible but nonspecific.
- Atrial and ventricular arrhythmias may be seen.
- Signs of left heart failure in some cats.
- Coughing is not a classical sign of left heart failure in cats. In majority of cases, it may be associated with feline asthma.
- Pulmonary edema or pleural effusions or both may be seen.
- Tachydyspnea is the most marked clinical manifestation with severe congestive heart failure.

- Subnormal body temperature along with signs of congestive heart failure is fatal.
- Radiography and electrocardiography are insensitive tools for the diagnosis of hypertrophic cardiomyopathy in cats.
- Echocardiography revealing increased left ventricular wall thickness at standard submitral location is of diagnostic value in cases of hypertrophic cardiomyopathy in cats.
- Therapy of HCM in cats is controversial.
- No drugs seem to delay the progression or reversal of hypertrophy in cats.
- Congestive heart failure can be managed by avoiding undue stress, placing the affected cat in an oxygen cage, doing pleurocentesis to relieve excessive fluid, and resorting to diuretic therapy. Furosemide is a diuretic of choice and can be given @ 2–4 mg/kg body weight intramuscularly at 2–4 h interval till respiration is comfortable and there is no dyspnea. After relief the dose of the furosemide should be tapered off to avoid dehydration and depletion of potassium. Fluid transfusion should be avoided, and ad lib drinking water should be provided.

17.13.2 Feline Dilated Cardiomyopathy

- Uncommon in cats.
- Increased left ventricular dimension on echocardiographic examination.
- Melamine contaminated pet foods for long periods may cause taurine associated dilated cardiomyopathy.
- Clinical picture is similar to HCM.
- Most of the cats remain subclinical for most of the time.
- Treatment of subclinical dilated cardiomyopathy is controversial.
- Manage congestive heart failure.

17.13.3 Arrhythmogenic Right Ventricular Cardiomyopathy (ARVC)

- Characterized by fibro-fatty infiltration of right ventricular wall (Fox et al. 2000).
- Right ventricle contractile dysfunction.
- Severe tricuspid regurgitation.
- Right heart congestive failure.
- Effusions are reduced by abdominocentesis and thoracocentesis.

17.13.4 Excess Moderator Band Cardiomyopathy

- A rare disease of cats.
- Etiology and pathophysiology are unknown.
- Characterized by a network of fibroid bands in left ventricular chamber.
- Endocardial thickening in some cases.
- Treatment is directed against congestive heart failure.

17.14 Cardiogenic Arterial Thromboembolism (ATE) in Cats

- An uncommon complication of cardiomyopathy in cats (Smith et al. 2003).
- It is occlusive vasculopathy of systemic arteries.
- Distal aorta is most common site of occlusion.
- Occlusion can be partial or complete.
- Secondary thrombosis in visceral arteries.
- Thromboemboli in forelimb arteries is seen occasionally.
- Clinical signs depend on the occlusion of arteries.
- Acute paresis/paralysis with pain is hallmark.
- Loss of pulse in affected limb.
- Nails/foot pad may become cyanotic.
- Increase in creatine kinase and aspartate aminotransferase levels.
- Congestive heart failure is not a common feature.
- Cats with infarction have poor prognosis.
- Definite treatment protocol is yet to be developed.
- Therapy with thrombolytic drug (streptokinase, urokinase, tissue plasminogen activator) or intravascular thrombectomy; fluid (to promote collateral circulation); and unfractionated heparin (to prevent thrombosis during acute phase) has been suggested. But recovery rate is not very encouraging.

17.15 Glucocorticoid-Associated Congestive Heart Failure

- Casual association of glucocorticoid administration and development of congestion heart failure in cats has been suggested (Smith et al. 2002, 2004). But it remains to be confirmed.
- No other concrete proof is available about association of glucocorticoid administration and development of congestive heart failure in cats.

References

Anjos DS, Cintra CA, Roch AJR, Junior DP (2015) Cardiac biomarkers—an ally in the prognosis of heart disorders in small animals. Rev Investig Med Vet 14:38–45

Fox PR, Maron BJ, Basso C (2000) Spontaneously occurring arrhythmogenic right ventricular cardiomyopathy in the domestic cat: a new animal model similar to the human disease. Circulation 102:1863

Nakamura RK, Zimmerman S (2012) Veterinary medicine today. ECG of the month. J Am Vet Med Assoc 241:433

Sleeper MM, Cliffoer CA, Laster LL (2001) Cardiac troponin I in the normal dog and cat. J Vet Intern Med 15:501–503

Smith S, Tobias A, Fine D (2002) Corticosteroid-associated congestive heart failure in 29 cats (abstract). J Vet Intern Med 16:371

Smith S, Tobias A, Jacob K (2003) Arterial thromboembolism in cats: acute crisis in 127 cases (1992–2001) and long-term management with low-dose aspirin in 24 cases. J Vet Intern Med 17:73

Smith S, Tobias A, Fine D (2004) Corticosteroid-associated congestive heart failure in 12 cats. Intern J Appl Res Vet Med 2:159
Varshney JP (2016) Ventricular arrhythmias associated with cardiomyopathy in cats. Indian J Vet Med 36:66–67
Varshney JP (2018) Hyperthyroidism- a cause of excessive meowing in a young female cat. Blue Cross Book 37:123–125

Further Reading

Connolly DJ, Cannata J, Boswood A, Archer J, Groves EA, Neiger R (2003) Cardiac troponin I in cats with hypertrophic cardiomyopathy. J Feline Med Surg 5:209–216
Falconer L, Atwell R (2003) Feline aortic thromboembolism. Aust Vet Pract 33(20):2003
Fox PR (1999) Feline cardiomyopathies. In: Fox PR, Sisson DD, Moise NS (eds) Textbook of canine and feline cardiology, 2nd edn. Saunders, Philadelphia, PA
Fox PR, Maron BJ, Basso C (2000) Spontaneously occurring arrhythmogenic right ventricular cardiomyopathy in the domestic cat: A new animal model similar to the human disease. Circulation 102:1863
Harpster NK (1992) Feline arrhythmias: diagnosis and management. In: Kirk RW, Bonagura JD (eds) Current veterinary therapy XI small animal practice. Saunders, Philadelphia, PA
Jackson CB (2001) Feline cardiomyopathy, Hanley and Belfus Inc. In: Lappin MR (ed) Feline internal medicine secrets, Philadelphia, PA
Koyama H, Matsumoto H, Fukushima RU (2010) Local intraarterial administration of urokinase in the treatment of a feline distal aortic thromboembolism. J Vet Med Sci 72:1209
Meurs KM, Fox PR, Norgard M (2007a) A prospective genetic evaluation of familial dilated cardiomyopathy in the Doberman Pinscher. J Vet Intern Med 21:1016–1020
Meurs KM, Norgard MM, Ederer MM (2007b) A substitution mutation in the myosin binding protein C gene in ragdoll hypertrophic cardiomyopathy. Genomics 90:261
Moore KE, Morris N, Dhupa N (2000) Retrospective study of streptokinase administration in 46 cats with arterial thromboembolism. J Vet Emerg Crit Care 10:245
Pion PD (1988) Feline aortic thromboemboli: a thrombolysis followed by aspirin therapy and rethrombosis. Vet Clin North Am Small Anim Pract 18(262):1988
Reimer SB, Kittleson MD, Kyles AE (2006) Use of rheolytic thrombectomy in the treatment of feline distal aortic thromboembolism. J Vet Intern Med 20:290
Rishniw M (2012) Cardio-vascular diseases. In: Little SE (ed) The cat. Clinical medicine and management. Elsevier, Saunders, St. Louis, MO

Section III

Ruminants

Electrocardiography in Ruminants 18

18.1 Electrocardiography

Though electrocardiography is routinely employed for monitoring cardiac rhythm, rate, conduction disturbances, and size of the chambers of the heart in man, dogs, and cats, its use in ruminants is restricted to monitoring of heart rate and rhythm only. Detection of cardiomyopathy from electrocardiogram is not feasible in ruminants (cows, buffaloes, sheep, goat) and other hoofed mammals (pigs and horses) due to deep penetration of Purkinje fibers from endocardium to epicardium and cross-fiber bridges and depolarization and repolarization occurring over multiple minor fronts at a time. Nevertheless, identification of arrhythmias and conduction disturbances is of immense clinical value in ruminants also. Early detection of arrhythmias of severe nature and electrolyte imbalance might enable large animal practitioners to take appropriate remedial measures and to assess an accurate prognosis of ailing animals.

The generation of electrocardiogram in ruminants is due to depolarization and repolarization of cardiac muscle fibers similar to canines and felines. The wave of depolarization begins in sinoatrial node (SA node) and passes through atrium producing "P" wave and reaches to AV node. The delay in conduction at AV node results into P-R segment. Then depolarization of ventricles occurs producing "QRS" complex followed by isoelectric period, i.e., S-T segment. This is followed by repolarization of the heart represented by "T" wave. The values of amplitude and duration of complexes and intervals in ruminants vary from that of dogs and cats.

Positioning of animals: The animals (cow, buffalo, goat, sheep, mithun) are kept standing on a rubber mat, and the sites for applying alligator clips are prepared by shaving and cleaning with alcohol. Gel is applied liberally on the areas of attachment of electrodes.

Placement of electrodes and recording of ECG: Traditional lead system in large animals was based on Einthoven's triangle using standard bipolar limb leads and augmented unipolar limb leads as in case of humans, dogs, and cats. In this lead system, ECG wave forms are affected by the variations in limb position, and

J. P. Varshney, *Electrocardiography in Veterinary Medicine*,
https://doi.org/10.1007/978-981-15-3699-1_18

complexes are small. Other lead system, i.e., base-apex lead system, where such variations of limb positioning have less impact, is more appropriate in large animals. In base-apex lead system, the right arm electrode is attached to the neck in the left jugular furrow two third down in the neck, left arm electrode is placed over the apex of the heart just behind the left elbow, and earth electrode (right hind leg electrode) is attached at wither (Figs. 18.1, 18.2, 18.3, 18.4, and 18.5). Electrocardiogram is recorded in a calm and quite surroundings using electrocardiographic machine at a paper speed of 25 mm per second and amplitude of 1 mV per 10 mm. The recording is done in lead I. In hexaxial lead system, forearm electrodes are attached at anterior aspect of respective olecranon on medial side of the elbow joint, and hind limb electrodes are placed on medial aspect of respective stifle joint (Fig. 18.6) similar to canines.

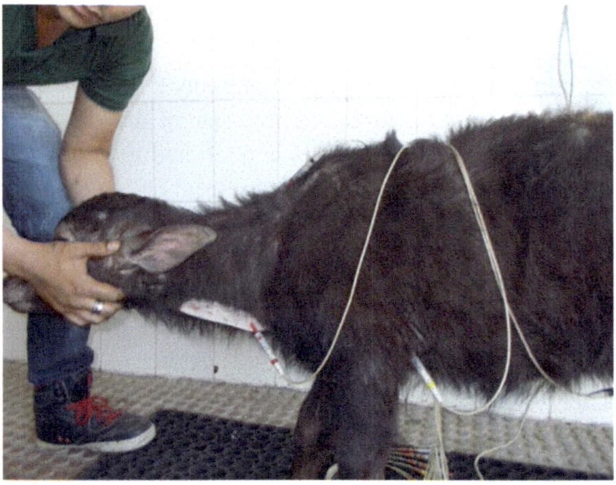

Fig. 18.1 Placement of electrodes in buffalo calf (base-apex system)

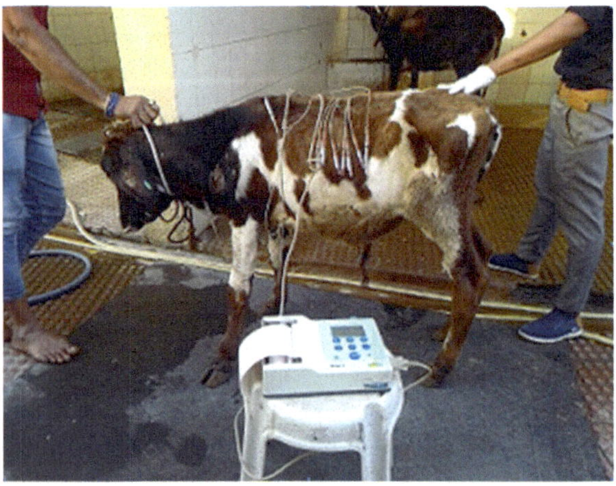

Fig. 18.2 Placement of electrodes in cow calf (base-apex system)

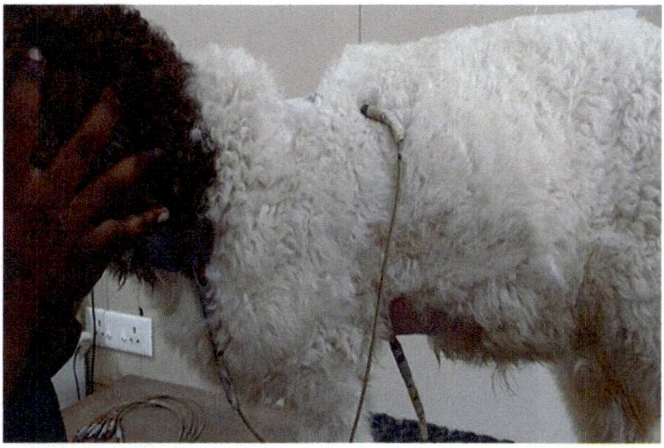

Fig. 18.3 Placement of electrodes in sheep (base-apex system)

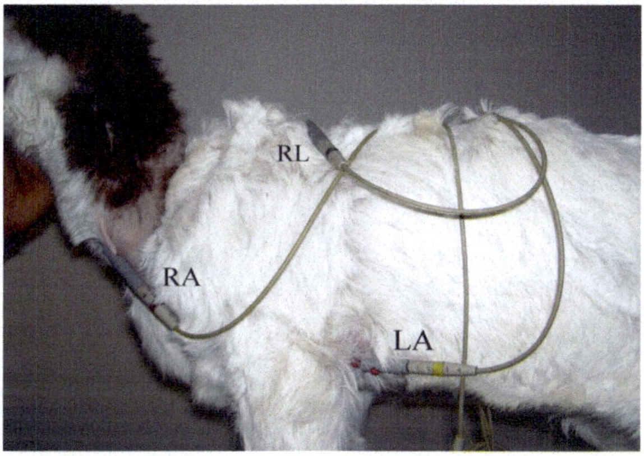

Fig. 18.4 Placement of electrodes in goat (base-apex system)

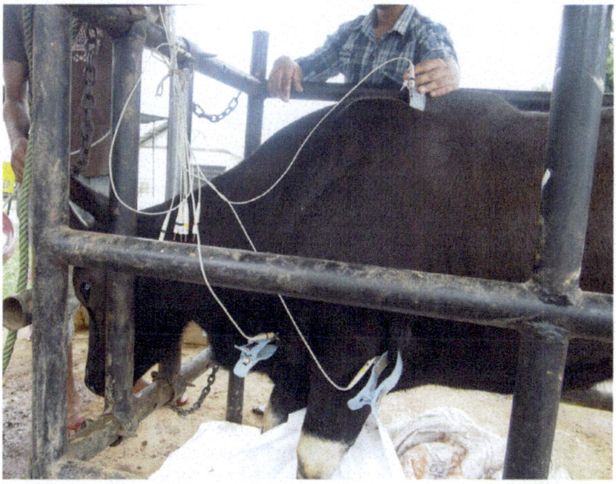

Fig. 18.5 Placement of electrodes in mithun (base-apex system) (Courtesy Dr. Akhalesh Kumar)

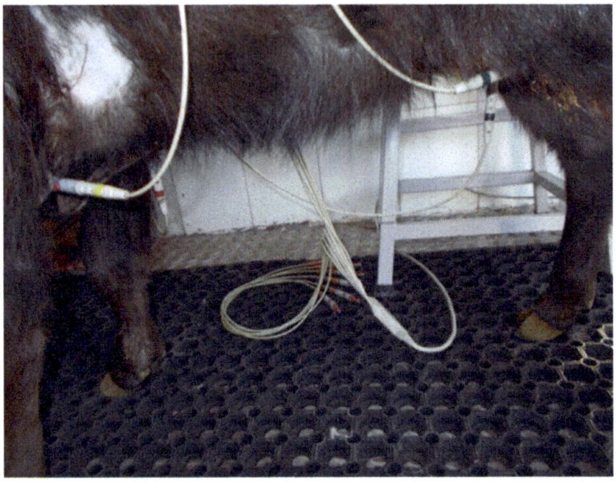

Fig. 18.6 Placement of electrodes in buffalo calf (hexaxial system)

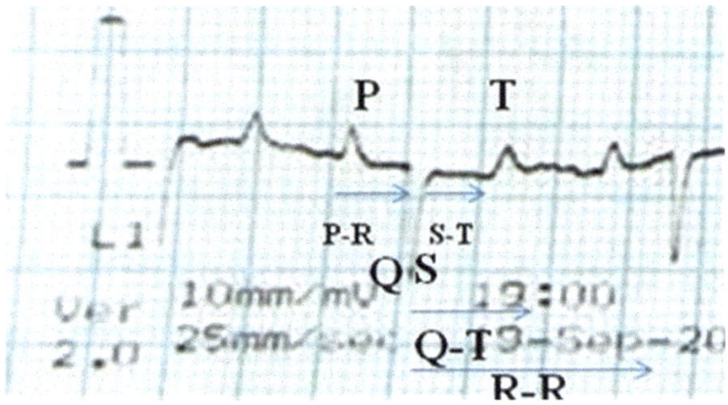

Fig. 18.7 Electrocardiogram (base-apex lead I, speed 25 mm/s, and sensitivity of 1.0) of a 1-year-old calf showing complexes, duration, and intervals. QRS complex is QS type and predominantly negative

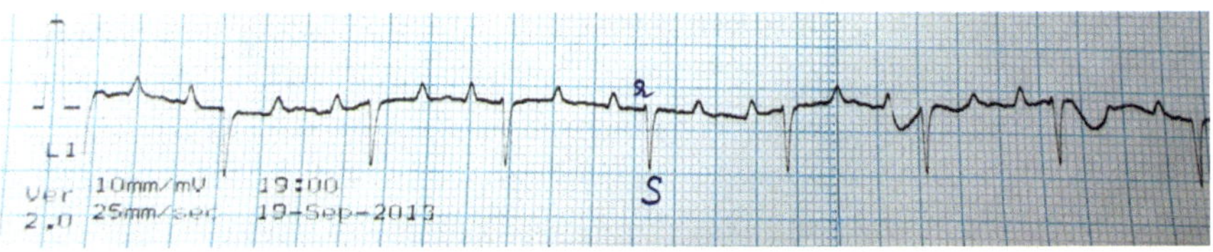

Fig. 18.8 Electrocardiogram (base-apex lead I, speed 25 mm/s, and sensitivity of 1.0) of Surti buffalo calf showing "rS"-type QRS pattern

18.1.1 Electrocardiogram in Buffalo and Cow Calves

Electrocardiogram of ruminants is having "P," "QRS," and "T" complexes, P-R interval, S-T interval, Q-T interval, and R-R intervals (Fig. 18.7) as also seen in other animals. But "QRS" complex in ruminants is predominantly negative in the form of "rS" (Figs. 18.8 and 18.13) or "QS" (Figs. 18.9 and 18.14), and the "T" wave is either positive (Fig. 18.9), negative (Fig. 18.10), or biphasic. "P" waves may be either round, peaked, or slightly isoelectric in configuration.

Complexes are significantly ($P < 0.05$) large in lead I of base-apex (Fig. 18.11) than those of lead II of hexaxial lead (Fig. 18.12) system.

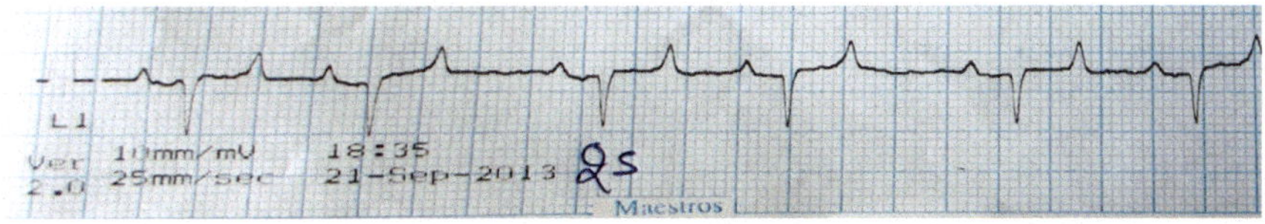

Fig. 18.9 Electrocardiogram (base-apex lead I, speed 25 mm/s, and sensitivity of 1.0) of Surti buffalo calf showing "QS"-type QRS pattern and positive T wave

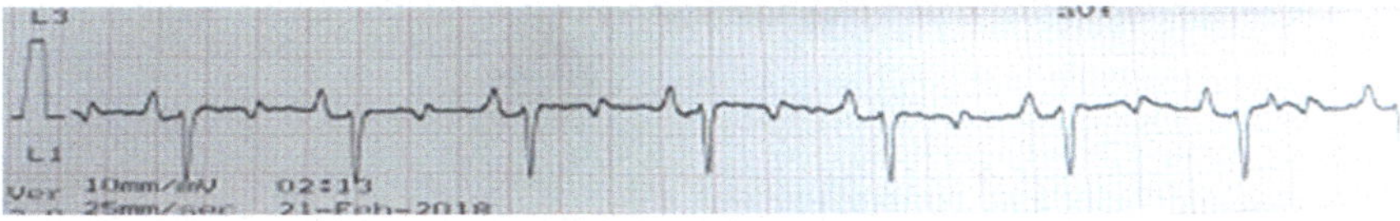

Fig. 18.10 Electrocardiogram (base-apex lead I, speed 25 mm/s, and sensitivity of 1.0) of a Holstein crossbred calf showing negative "T" wave

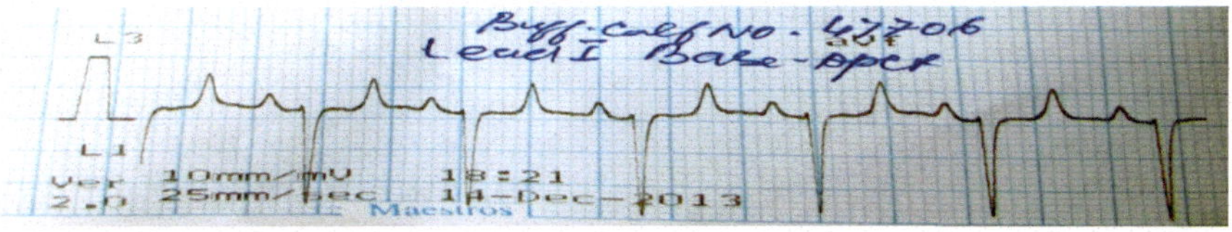

Fig. 18.11 Electrocardiogram (base-apex lead I, speed 25 mm/s, and sensitivity of 1.0) of Surti buffalo calf. Note the size of "P," "QRS," and "T" waves

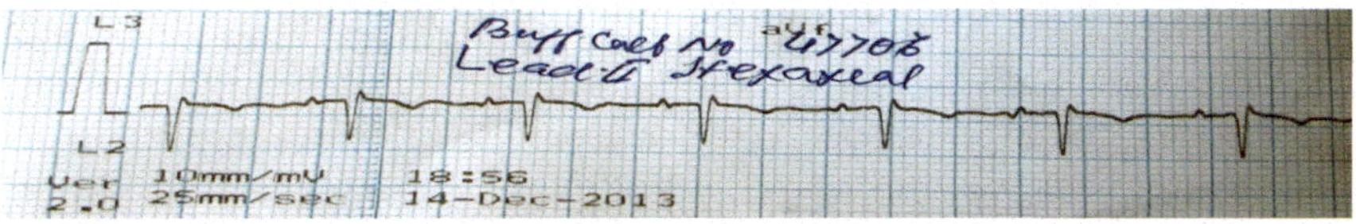

Fig. 18.12 Electrocardiogram (hex axial lead system, lead II, speed 25 mm/s, and sensitivity of 1.0) of the same Surti buffalo calf (Fig. 252) employing hexaxial lead system (lead II). Note the size of the complexes. The size of "P," "QRS," and "T" is smaller as compared to that of base-apex lead system (Fig. 18.11)

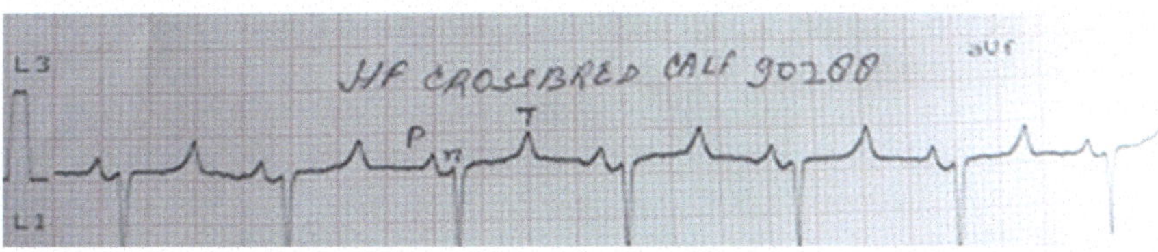

Fig. 18.13 Electrocardiogram (base-apex lead I, speed 25 mm/s, and sensitivity of 1.0) of Holstein crossbred calf showing "rS"-type QRS pattern and + ve "T" wave

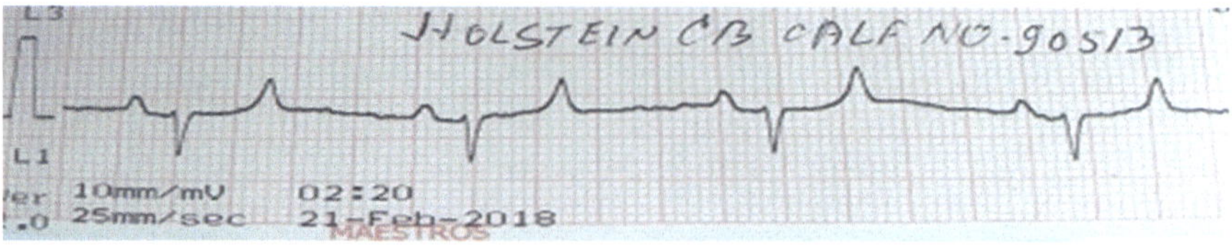

Fig. 18.14 Electrocardiogram (base-apex lead I, speed 25 mm/s, and sensitivity of 1.0) of Holstein crossbred calf showing "QS" type QRS pattern

Table 18.1 Electrocardiographic indices in Surti buffalo calves (Base-apex lead system)

ECG indices	Range	Mean ± S.E.	Median
Heart rate (bpm)	90–130	90.39 ± 3.44	90.0
P amplitude (mV)	0.1–0.35	0.195 ± 0.010	00.20
P duration (s)	0.04–0.07	0.0513 ± 0.0014	00.05
P-R segment (s)	0.011–0.16	0.11 ± 0.0558	00.12
P-R interval (s)	0.061–0.24	0.166 ± 0.056	00.165
QRS amplitude (mV)	0.45–1.750	0.877 ± 0.062	00.70
QRS duration (s)	0.04–00.08	0.051 ± 0.0017	00.05
ST segment (s)	0.08–0.36	0.167 ± 0.034	00.165
T amplitude (mV)	0.10–1.00	0.278 ± 0.043	00.020
T duration (s)	0.04–0.10	0. 0.065 ± 0.027	00.06
Q-T interval	0.20–0.50	0.284 ± 0.087	00.28
R-R interval (s)	0.40–1.36	0.671 ± 0.041	00.64

18.2 Electrocardiographic Indices in Buffaloes

Values of electrocardiographic indices in healthy Surti buffalo calves are given in Table 18.1. Mean values of heart rate, "P" amplitude, "P" durations, "R" segment, "P-R" interval, "QRS" amplitude, "QRS" duration, "S-T segments" amplitude, "T" duration, "Q-T" interval, and "R-R" interval in buffaloes have been reported as 90.39 ± 3.44 bpm, 0.195 ± 0.010 mV, 0.0513 ± 0.0014 s, 0.11 ± 0.0558 s, 0.166 ± 0.056 s, 0.877 ± 0.062 mV, 0.051 ± 0.0017 s, 0.167 ± 0.034 s, 0.278 ± 0.043 mV, 0.065 ± 0.027 s, 0.278 ± 0.043 s, and 0.278 ± 0.043 s, respectively (Varshney et al. 2013).

18.3 Electrocardiographic Indices in Holstein Crossbreds

Values of electrocardiographic indices in healthy Holstein crossbred calves are given in Table 18.2. Mean values for heart rate, "P" amplitude, "P" durations, "R" segment, "P-R" interval, "QRS" amplitude, "QRS" duration, "S-T segments" amplitude, "T" duration, "Q-T" interval, and "R-R" interval have been reported as 85.2 ± 4.23 bpm, 0.234 ± 0.031 mV, 0.05 ± 0.0024 s, 0.125 ± 0.025 s, 0.884 ± 0.06 mV, 0.054 ± 0.0001 s, 0.182 ± 0.038 s, 0.318 ± 0.088 mV, 0.069 ± 0.0037 s, and 0.308 ± 0.0017 s, respectively (Varshney 2018c). Electrocardiographic complexes in buffaloes and cows are almost of similar pattern.

Table 18.2 Electrocardiographic indices in Holstein crossbred calves (Base-apex lead system)

ECG indices	Range	Mean ± S.E.	Median
Heart rate (BPM)	60.0–140.0	85.2 ± 4.23	80.0
P amplitude (mV)	0.10–0.30	0.234 ± 0.031	00.2
P duration (s)	0.04–0.08	0.05 ± 0.002	0.02
P-R interval (s)	0.08–0.18	0.125 ± 0.025	0.12
S amplitude (mV)	0.20–1.70	0.884 ± 0.06	0.90
QRS duration (s)	0.04–0.08	0.054 ± 0.0001	0.06
S-T segment (s)	0.12–0.28	0.182 ± 0.038	0.20
T amplitude (mV)	0.1–0.6	0.318 ± 0.088	0.30
T duration (s)	0.04–0.12	0.069 ± 0.0037	0.08
Q-T interval (s)	0.24–0.39	0.308 ± 0.0017	0.32

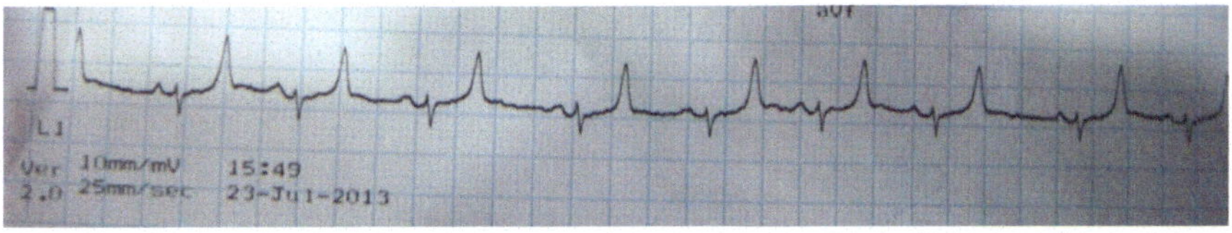

Fig. 18.15 Electrocardiogram (base-apex lead I, speed 25 mm/s, and sensitivity of 1.0) of an adult male goat showing sinus rhythm with heart rate as 90 bpm. The complexes ("P," "QRS," "T") and intervals (P-R, S-T, "Q-T," and R-R or S-S) are similar to other ruminants (cows and buffaloes as shown in above electrocardiograms)

18.4 Electrocardiographic Indices in Goats

In goats placement of electrodes (Fig. 18.4) and recording of electrocardiogram are similar to other ruminants. The complexes ("P," "QRS," "T") and intervals (P-R, S-T, "Q-T," and R-R or S-S) in goats (Fig. 18.15) are similar to other ruminants (cows and buffaloes as shown in above electrocardiograms). Nevertheless, values for complexes and intervals vary slightly. Range of electrocardiographic indices of healthy goats is given in Table 18.3.

18.5 Electrocardiographic Indices In Mithuns

Placement of electrodes (Fig. 18.5) and recording of electrocardiogram in mithuns are similar to other ruminants. Generally base-apex lead system is more appropriate in mithuns also as complexes are of larger size. The complexes ("P," "QRS," "T") and intervals (P-R, S-T, "Q-T," and R-R or S-S) in mithuns (Fig. 18.16) are of similar pattern as seen in other ruminants. Nevertheless, values for complexes and intervals differ slightly. The values of electrocardiographic waves, duration, and different intervals of healthy mithuns are given in Table 18.4.

Table 18.3 Electrocardiographic indices in healthy goats (Base-apex lead system)

Electrocardiographic parameters	Range values
Hear rate	60–130 bpm
P amplitude	0.1–0.2 mV
P duration	0.04–0.06 s
P-R interval	0.06–0.10 s
R amplitude	0.05–0.1 mV
Q/S amplitude	0.40–1.4 mV
QRS duration	0.06–0.10 s
S-T segment	0.16–0.40 s
T amplitude	0.10–0.70 mV
T duration	0.07–0.16 s
Q-T interval	0.22–0.56 s
R-R interval	0.10–0.46 s

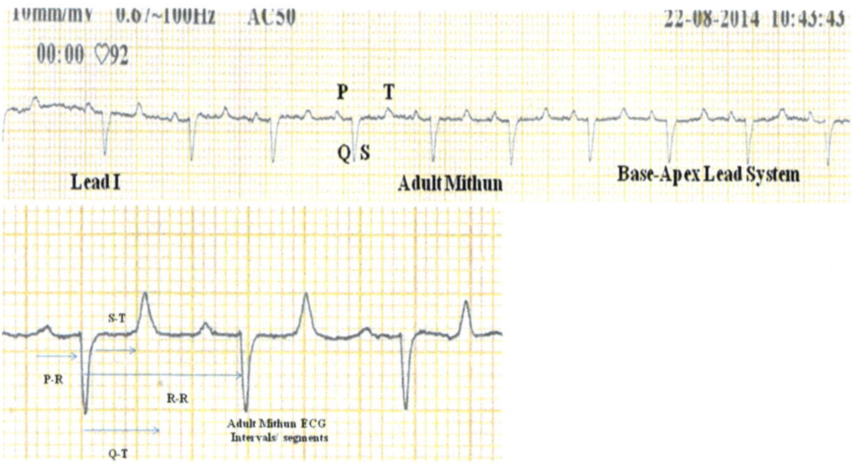

Fig. 18.16 Electrocardiogram (lead I, sensitivity 1, speed 25 mm/s) of an adult mithun showing sinus rhythm with heart rate as 75 bpm; P wave 0.15 mV, 0.04 mV; P-R interval 0.16 s; Q 0.7 mV; S 0.7 mV; QRS 0.06 s; S-T 0.14 s; T 0.3.5 mV, 0.07 s; Q-T interval 0.28 s; and R-R interval 0.58 s. (Courtesy Dr. Akhalesh Kumar, Scientist Mithun Centre, Dimapur, Nagaland)

18.6 Electrocardiographic Indices in Sheep

In sheep placement of electrodes (Fig. 18.3) and recording of electrocardiogram are similar to other ruminants. Generally base-apex lead system is followed in sheep as complexes are of larger size. The complexes ("P," "QRS," "T") and intervals (P-R, S-T, "Q-T," and R-R or S-S) in sheep (Fig. 18.17) are of similar pattern as seen in goats. Range of electrocardiographic indices of healthy sheep is given in Table 18.5.

Table 18.4 Electrocardiographic indices in healthy mithuns (Base-apex lead system) (Courtesy Dr. Akhlesh Kumar, Scientist, Mithun Centre, Dimapur, Nagaland)

Electrocardiographic indices	Range
Heart rate	80–150 bpm
"P" wave amplitude	0.05–0.35 mV
"P" wave duration	0.03–0.06 s
P-R segment	0.05–0.12 s
P-R interval	0.09–0.16 s
"Q" wave	0.00–0.40 mV
"R" wave amplitude	0.00–0.20 mV
"S" wave	0.00–0.80 mV
"QRS" duration	0.03–0.08 s
"S-T" segment	0.12–0.24 s
"Q-T" interval	0.15–0.30 s
"T" wave amplitude	0.05–0.40 mV
"T" wave duration	0.02–0.08 s
"R-R" interval	0.39–0.75 s

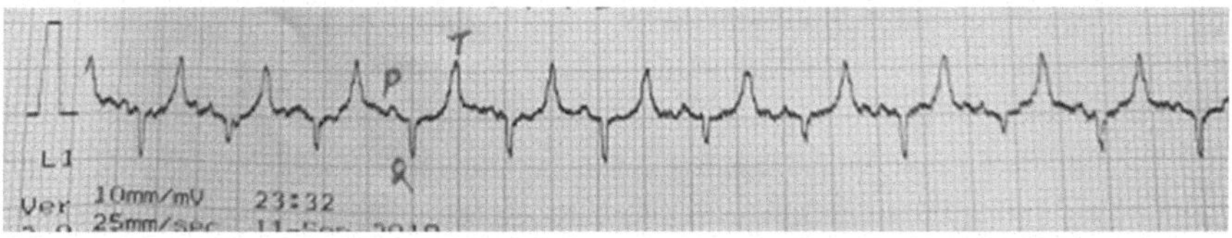

Fig. 18.17 Electrocardiogram (lead I, sensitivity 1, speed 25 mm/s) of a 3-month-old lamb showing sinus rhythm with heart rate as 140 bpm; P wave 0.10 mV, 0.04 mV; P-R interval 0.09 s; Q wave 0.4 mV; QRS 0.03 s; S-T 0.12 s; T 0.4 mV, 0.07 s; Q-T interval 0.24 s; and R-R interval 0.42 s

Table 18.5 Electrocardiographic parameters in healthy adult sheep (Base-apex lead system)

Electrocardiographic parameters	Range values
Hear rate	75–115 bpm
P amplitude	0.1–0.2 mV
P duration	0.03–0.06 s
P-R interval	0.08–0.16 s
R amplitude	0.00–0.1 mV
Q/S amplitude	0.40–0.85 mV
QRS duration	0.04–0.08 s
S-T segment	0.12–0.14 s
T amplitude	0.20–0.40 mV
T duration	0.08–0.12 s
Q-T interval	0.24–0.28 s
R-R interval	0.52–0.8 s

18.7 Characteristics of Ruminants' Electrocardiogram

1. Heartbeat is slow as compared to canines and felines.
2. "QRS" is predominantly negative, rS or QS type.
3. R wave is very small (r) and sometimes no r wave.
4. Heart chamber enlargement pattern cannot be ascertained by electrocardiography.

18.8 Arrhythmias in Ruminants (Figs. 18.18, 18.19, 18.20, 18.21, 18.22, 18.23, 18.24, 18.25, 18.26, 18.27, 18.28, 18.29, 18.30, 18.31, 18.32, 18.33, 18.34, 18.35, and 18.36)

Variations in heart rate, rhythm, and both are not uncommon in ruminants and have been observed in health and diseases. These variations include bradycardia (slow heart rate), tachycardia (increased heart rate), conduction disturbances, and irregularity of both rate and rhythm. Though in most cases arrhythmias can be detected on clinical examination, their categorization is rather difficult purely on the basis of physical examination. This can only be ascertained by electrocardiographic examination. In fact occurrence of arrhythmias and conduction disturbances in ruminants is more common than expected. Despite common occurrence of cardiac arrhythmias, their clinical significance is not always clear. Sinus tachycardia, sinus bradycardia, sinus arrest, atrial fibrillation, ventricular premature complexes, ventricular tachycardia, accelerated idioventricular rhythm, ventricular asystole, low-voltage complexes, right bundle branch block, and second-degree heart block are common arrhythmias in ruminants (Sarita Devi et al. 2014, 2015; Varshney et al. 2015; Varshney 2019).

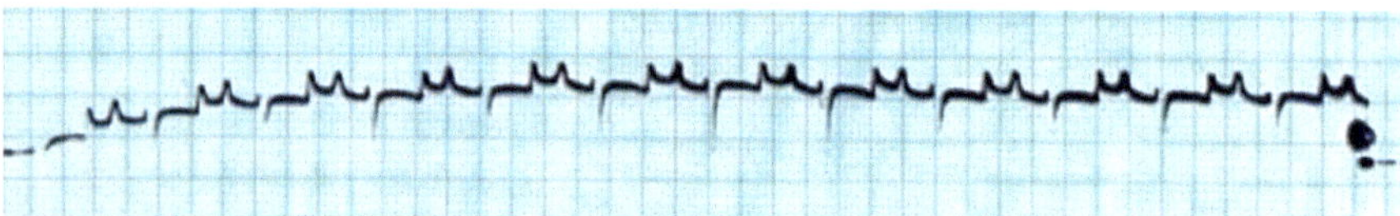

Fig. 18.18 Electrocardiogram (lead I, sensitivity 1, speed 25 mm/s) of an adult buffalo with traumatic pericarditis showing low-voltage complex, alternans, and heart rate as 100 bpm suggesting pericardial effusions (Sarita Devi et al. 2014)

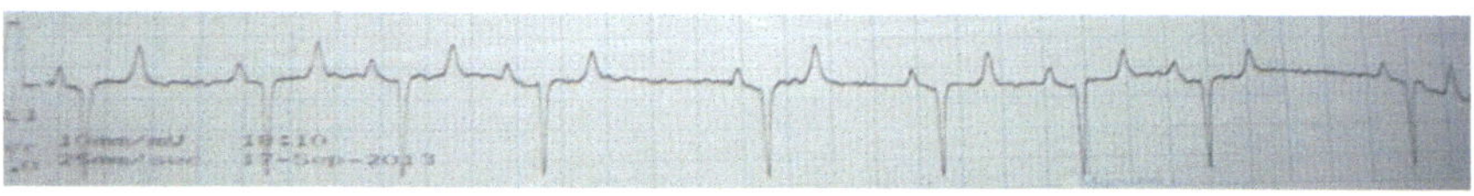

Fig. 18.19 Electrocardiogram (lead I, sensitivity 1, speed 25 mm/s) of an adult buffalo with traumatic pericarditis showing heart rate as 80 bpm and varying QRS interval (0.56–1.0 s) suggesting sinus arrhythmia (Sarita Devi et al. 2014)

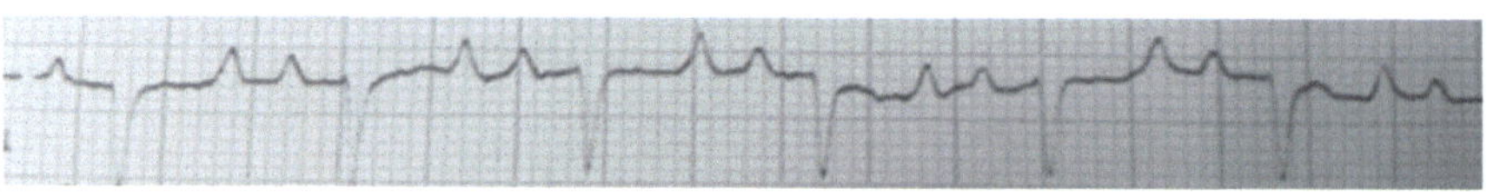

Fig. 18.20 Electrocardiogram (lead I, sensitivity 1, speed 25 mm/s) of an adult buffalo with traumatic pericarditis showing increased heart rate (120 bpm) suggesting sinus tachycardia (Sarita Devi et al. 2014)

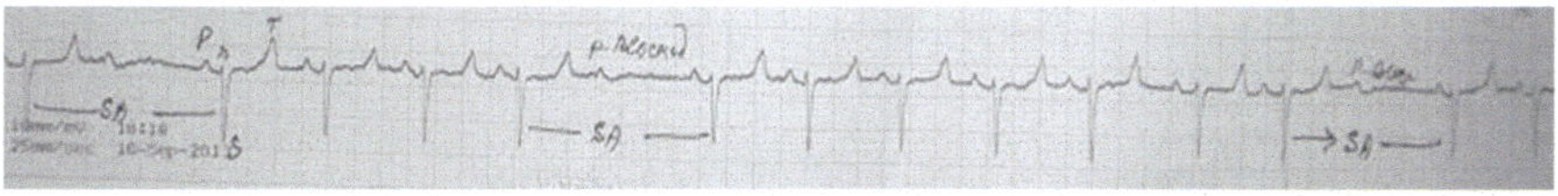

Fig. 18.21 Electrocardiogram (lead I, sensitivity 1, speed 25 mm/s) of an adult buffalo with traumatic pericarditis showing heart rate as 72 bpm. At three places R-R (S-S) interval is 1.44 s against regular interval of 0.6 sec suggesting sinus arrest (Sarita Devi et al. 2014)

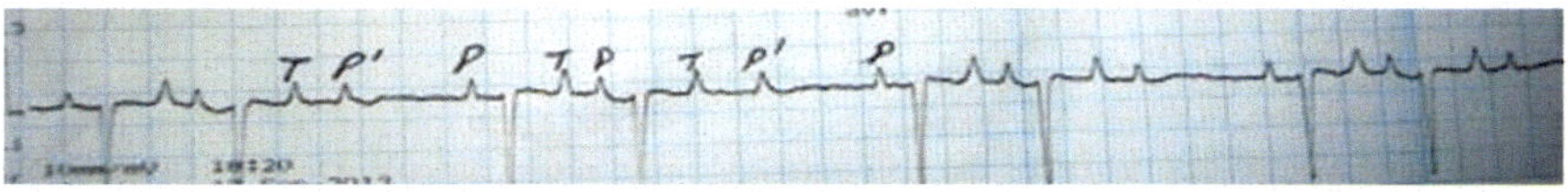

Fig. 18.22 Electrocardiogram (lead I, sensitivity 1, speed 25 mm/s) of an adult buffalo with traumatic pericarditis showing ventricular rate as 68 bpm. Alternate beat has "P" wave that is blocked making the atrial rate as 94 bpm. P-R interval is 0.16 s (almost constant for conducted P wave). QRS interval is alternately 0.6 and 1.32 s. ECG findings are suggestive of second- degree Mobitz type II. AV block (Sarita Devi et al. 2014)

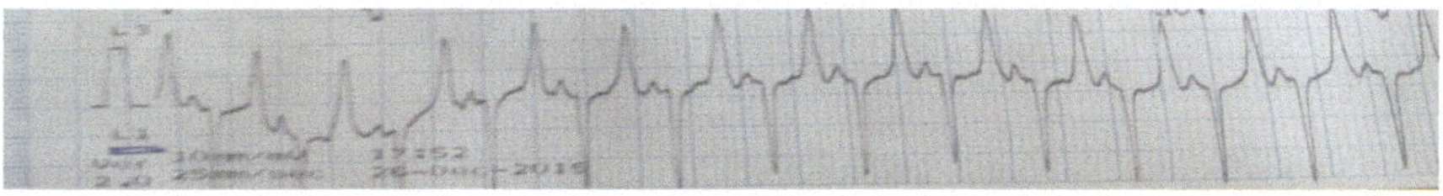

Fig. 18.23 Electrocardiogram (lead I, sensitivity 1, speed 25 mm/s) of a calf with pneumonia showing tachycardia (heart rate 160 bpm)

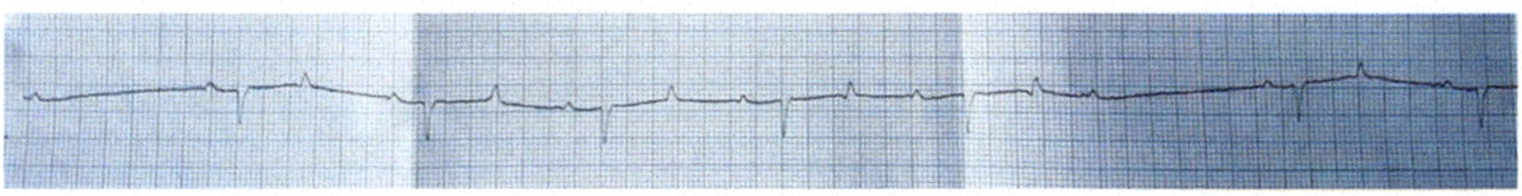

Fig. 18.24 Electrocardiogram (lead I, sensitivity 1, speed 25 mm/s) of a crossbred calf showing sinus bradycardia (heart rate 40 bpm). P of normal configuration is followed by QRS

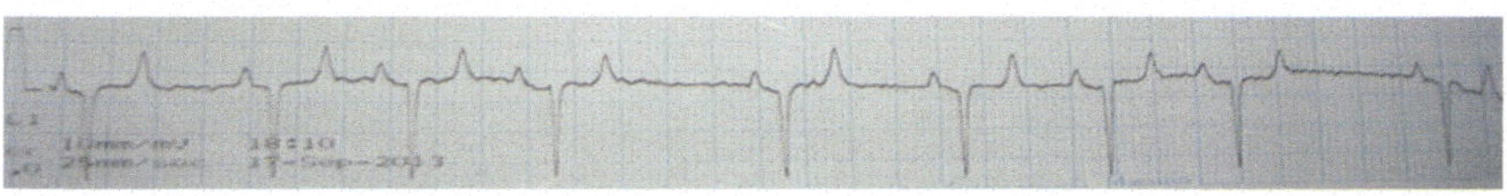

Fig. 18.25 Electrocardiogram (lead I, sensitivity 1, speed 25 mm/s) of a crossbred cattle showing variable QS interval/R-R interval (0.56–1.0 s) suggesting sinus block

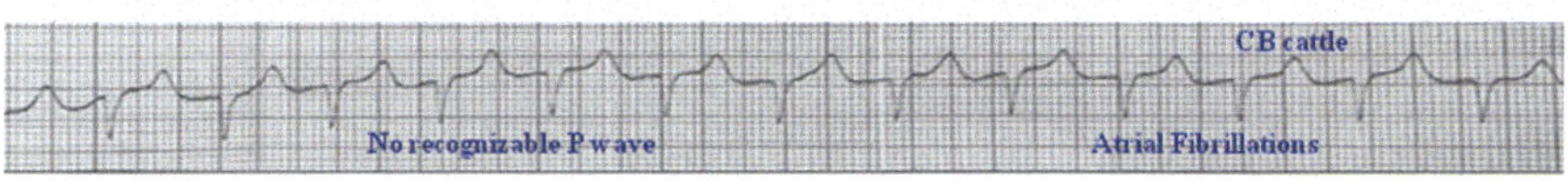

Fig. 18.26 Electrocardiogram (lead I, sensitivity 1, speed 25 mm/s) of a crossbred cattle showing atrial fibrillations (there is no recognizable "P" wave)

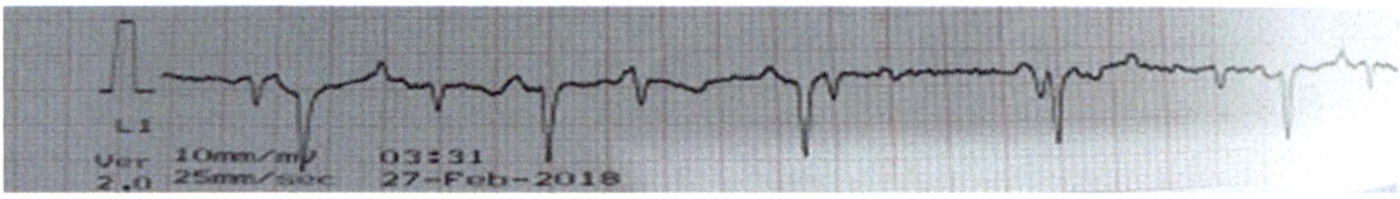

Fig. 18.27 Electrocardiogram (lead I, sensitivity 1, speed 25 mm/s) of a crossbred calf with FMD lesions showing ventricular premature complexes

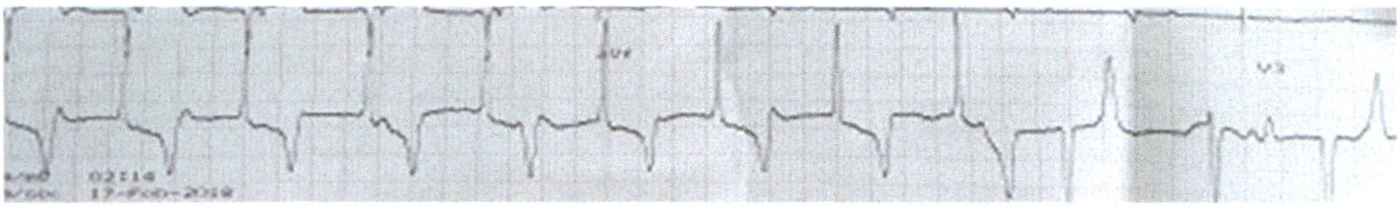

Fig. 18.28 Electrocardiogram (lead I, sensitivity 1, speed 25 mm/s) of a crossbred calf with foot and mouth disease lesions showing ventricular tachycardia

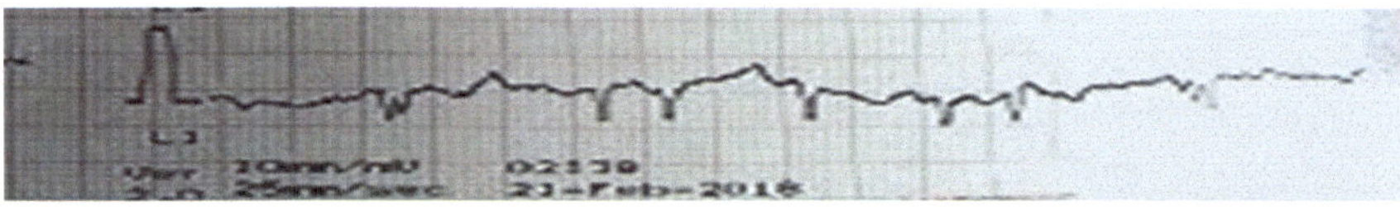

Fig. 18.29 Electrocardiogram (lead I, sensitivity 1, speed 25 mm/s) of a crossbred calf with foot and mouth disease lesions showing RBBB

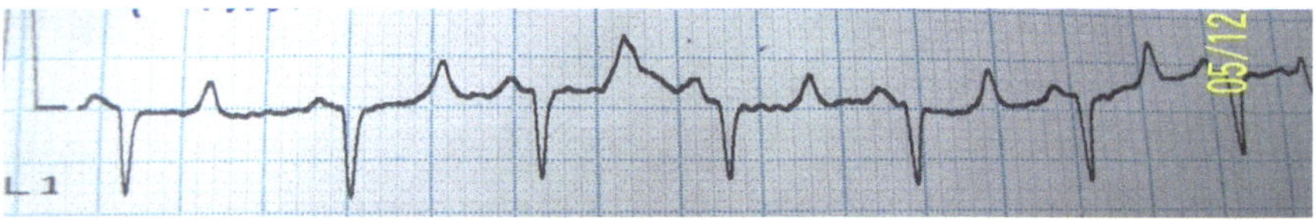

Fig. 18.30 Electrocardiogram (lead I, sensitivity 1, speed 25 mm/s) of an adult male goat with rumen acidosis (rumen pH 4.5) showing HR 140 bpm, P 0.15 mV, 0.04 s, P-R interval 0.08 s, P-R segment 0.03 s, QRS –ve 0.8 mV 0.04 s, T + ve 0.3–0.5 mV, 0.08 s, suggesting sinus tachycardia (Varshney et al. 2015)

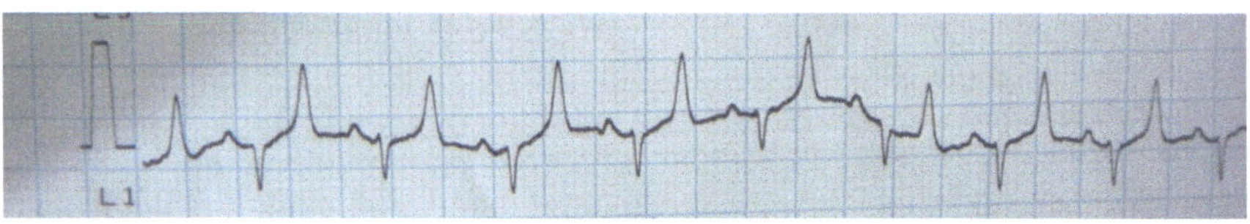

Fig. 18.31 Electrocardiogram (lead I, sensitivity 1, speed 25 mm/s) of an adult goat with rumen acidosis (pH 4.5) showing an increased amplitude (0.7 mV) of T wave (Varshney et al. 2015)

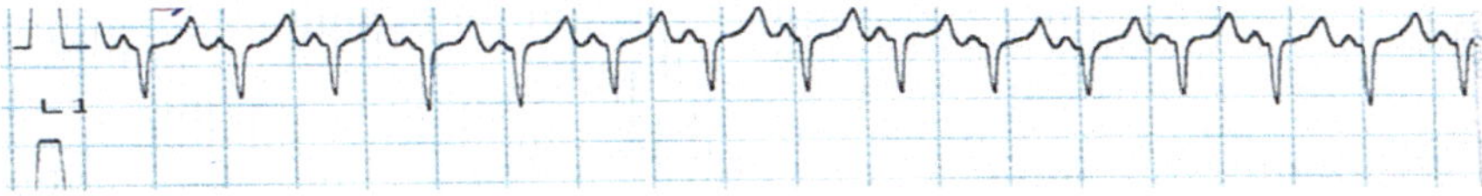

Fig. 18.32 Electrocardiogram (lead I, sensitivity 1, speed 25 mm/s) of a 10-month-old goat with pneumonia showing tachycardia (heart rate 240 bpm) and sinus rhythm suggesting sinus tachycardia (Sarita Devi et al. 2015)

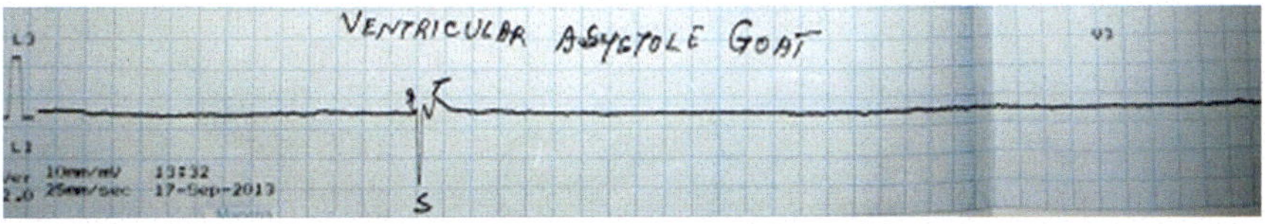

Fig. 18.33 Electrocardiogram (lead I, sensitivity 1, speed 25 mm/s) of an adult goat in terminal stage showing ventricular asystole

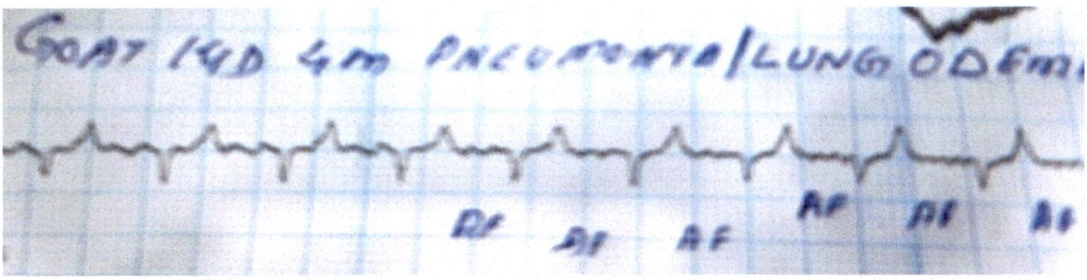

Fig. 18.34 Electrocardiogram (lead I, sensitivity 1, speed 25 mm/s) of a 4-month-old goat kid with lung edema/pneumonia showing atrial fibrillation

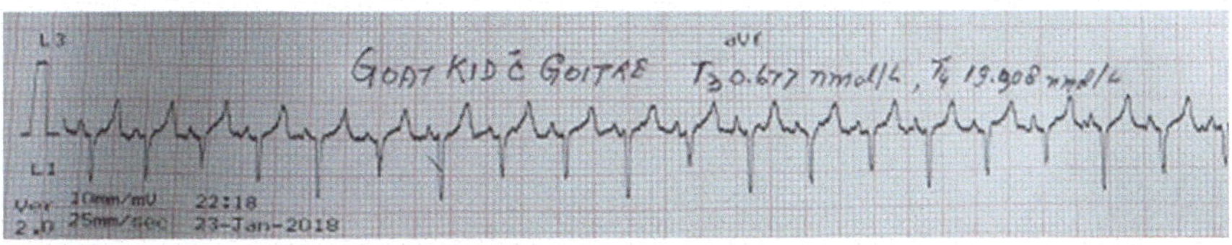

Fig. 18.35 Electrocardiogram (lead I, sensitivity 1, speed 25 mm/s) of a goiter-affected kid showing sinus rhythm with heart rate as 180 bpm, variable amplitude of "S" wave (0.4–0.8 mV), and predominant positive "T" wave (0.4 mV, 0.12 s)

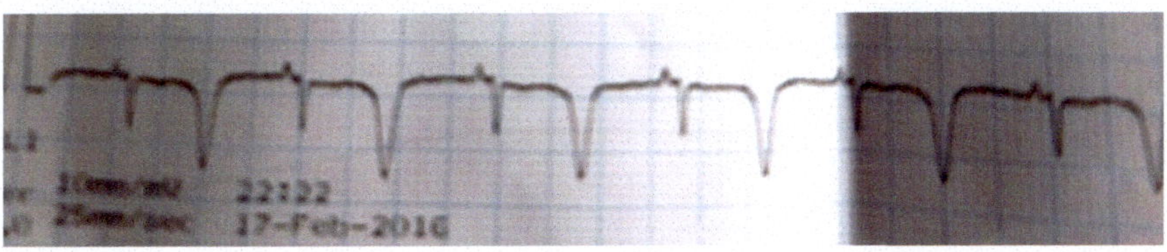

Fig. 18.36 Electrocardiogram (lead I, sensitivity 1, speed 25 mm/s) of a 15-day-old hypothermic (temperature 92 °F) and hypoglycemic kid (blood glucose 40 mg/dL) showing sinus bradycardia (heart rate 80 bpm sinus rhythm), "QS" (0.5 mV, 0.04 s), and predominant negative "T" wave (0.8 mV, 0.12 s) suggesting myocardial hypoxia

18.9 Electrocardiographic Abnormalities and Their Association with Diseases/Conditions

Arrhythmias and conduction disturbances are commonly seen in various clinical diseases/conditions in ruminants. Electrocardiographic abnormalities and their association with diseases/conditions are given below.

Electrocardiographic abnormalities	Associated diseases/conditions
Sinus bradycardia	Mass in cranium
	Hypothermia
	Hypoglycemia
	Pituitary abscess
	Xylazine injection
	Vagus indigestion
	Diaphragmatic hernia
	Anorexia
	CNS disorder
	Increased parasympathetic stimulation
Sinus tachycardia	Hyperthermia/fever
	Pain syndrome
	Exercise
	Respiratory distress
	Pneumonia
	Rumen acidosis
	Adrenergic drugs
	Hypotension
	Nervousness
	Stress
	Excitement
	Hypoxia
	Anemia
Sinus arrhythmias	During an early stage of treatment for hypocalcemia
	In young ones associated with respiration
	Traumatic reticulopericarditis
SA block	Increased vagal tone
Sinus arrest	Normal incidental finding
AV bock second degree	Electrolyte imbalances
	Myocardial disease
	Incidental finding
AV block third degree	Acid-base imbalance
	Electrolyte imbalance
	Hypoxia
	Hypercapnia
	Anesthesia

Electrocardiographic abnormalities	Associated diseases/conditions
Atrial premature beats	Gastrointestinal disease (cattle)
	Acid-base imbalance
	Electrolyte imbalance
Atrial fibrillation	Myocardial disease (cattle)
	Endocarditis (cattle)
	Abdominal pain
	Acute enteritis
	Left-sided abomasal displacement
	Uterine torsion
	Change in vagal tone
	Metabolic disease
	Interstitial pneumonia (goat)
	Gastrointestinal problems
Ventricular arrhythmias (premature beat, ventricular tachycardia, accelerated idioventricular rhythm, ventricular asystole)	Cardiac abnormality
	VPCs in normal cattle
	Focal myocarditis
Low-voltage complexes	Traumatic pericarditis
Right bundle branch block	Incidental finding

References

Devi S, Varshney JP, Undirwade SC, Jadav KM (2014) Cardiac Troponin-I and Electrocardiographic findings in buffaloes with traumatic reticuloperitonitis. Ruminants Sci 3:123–126

Devi S, Varshney JP, Undirwade SC, Jadav KM (2015) A Clinical study on electrocardiographic changes in goats suffering from pneumonia. Indian J. Small Ruminants 21:138–140

Varshney JP, Parmar SJ, Deshmukh VV, Chaudhary PS (2013) Electrocardiographic studies in Surti buffalo calves. Ruminant Sci 2:167–170

Varshney JP, Chaudhary PS, Deshmukh VV (2015) Therapeutic management of rumen acidosis and its electrocardiographic evaluation both pre and post treatment. Intas Polivet 16:107–109

Varshney JP (2018) Electrocardiographic studies in Holstein crossbred calves. Intas Polivet 19:16–18

Varshney JP (2019) Cardiac arrhythmias and conduction disturbances in crossbred cattle. Intas Polivet 20:44–46

Section IV

Other Animals

Electrocardiography in Other Animals 19

19.1 Chelonians

Without evaluation of the heart, detailed clinical examination is not complete in any individual animal. In turtles/tortoises (chelonian), hard shell (carapace and plastron) does not allow low amplitude heart sounds to be heard through auscultation. Hence, cardiac assessment is generally omitted in chelonians' clinical practice. Assessing heartbeat in chelonians with respiratory or cardiovascular arrest is of vital importance. Heartbeat can be determined and monitored using ultrasonic Doppler flow probe, electrocardiogram, and/or ultrasound. Electrocardiography is an invaluable technique that can be employed for monitoring cardiac function in chelonians too. As compared to canines and felines, electrocardiography in chelonians is underdeveloped, and information is scanty. It is sparingly used in clinical practice because of limited understanding regarding its interpretation. Small number of observations in most of the studies restrict usefulness of electrocardiogram in the clinical examination of chelonians. Though electrocardiography is routinely employed for evaluation of heart in canines and felines, its application in chelonians and reptiles is yet to find routine place as diagnostic technique probably because of the scarcity of electrocardiographic reference values for healthy chelonian species.

19.1.1 Positioning of Turtles/Tortoises for Electrocardiogram

The turtles/tortoises are subjected to electrocardiography by placing ventral side up. For obtaining a standard bipolar lead ECG, needle electrodes/crocodile clips are attached in a modified Einthoven lead system. Yellow and red electrodes are attached to the proximal aspects of left and right forelimbs, respectively; green and black electrodes are attached to the left and right hind limbs (Fig. 19.1), respectively. Electrocardiographic machine is standardized at the sensitivity of 1 or 2 (2 cm = 1 mV or 1 cm = 1 mV) and at a speed of 25 mm per second, and electrocardiogram is recorded in lead I, II, III, aVR, aVL, and aVF leads at room temperature

J. P. Varshney, *Electrocardiography in Veterinary Medicine*,
https://doi.org/10.1007/978-981-15-3699-1_19

19.1.4 Abnormal Electrocardiograms (Figs. 19.4, 19.5, 19.6, 19.7, 19.8 and 19.9)

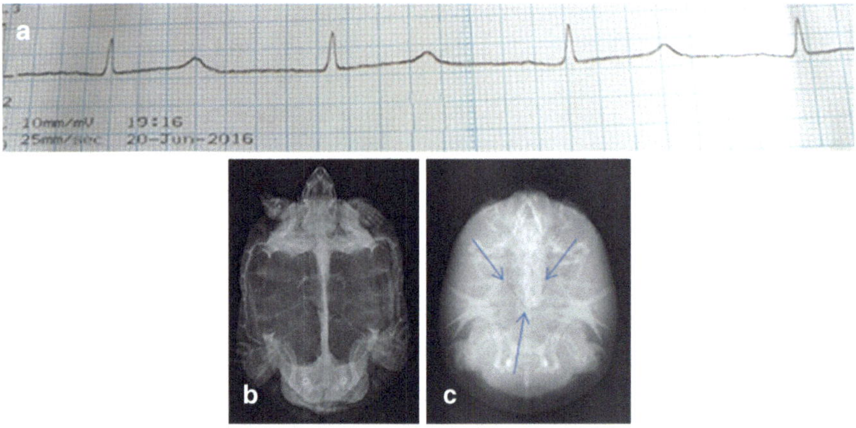

Fig. 19.4 (**a**) Electrocardiogram (lead II, sensitivity 1, speed 25 mm/s, at room temperature of 30 °C) of an adult male conscious pet turtle weighing 900 g, showing sinus rhythm with heart rate as of 40 bpm; very low amplitude (0.05 mV) but broad (60 ms) "P"; P-R interval of 240 ms; "QRS" (0.7 mV, 80 ms); prolonged S-T interval (440 ms); +ve, small and broad "T" (0.2 mV. 120 ms); prolonged Q-T interval (640 ms); and R-R interval as 1480 ms suggesting cardiomyopathy. (**b**) Dorso-ventral radiograph of a healthy turtle showing lungs and no radio-opacity of the heart as heart is poorly observed structure situated in the cranial part of coelomic cavity in healthy turtles/tortoises. (**c**) Dorso-ventral radiograph of the turtle showing enlarged shadow of the heart and fluid in whole of the coelomic cavity. The visible shadow (marked with arrows) is of heart due to heart enlargement or cardiac effusion (Varshney and Monapara 2019). Heart enlargement in this turtle is also reflected in electrocardiogram in Fig. 19.4a

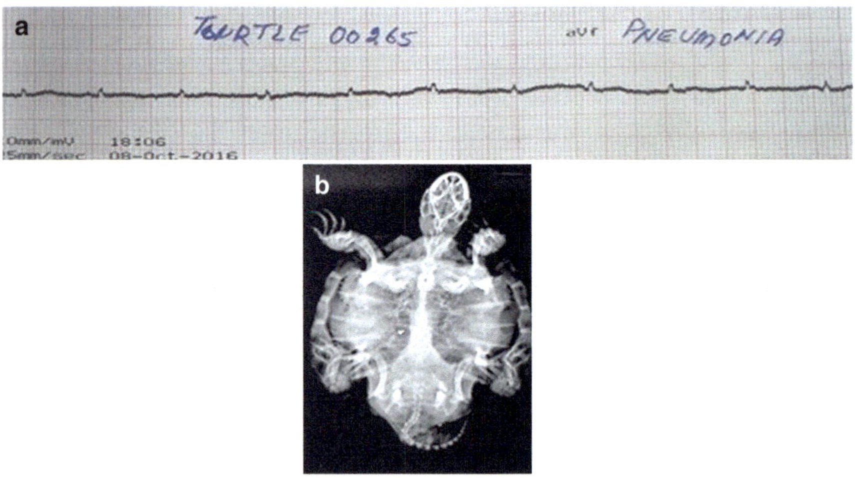

Fig. 19.5 (**a**) Electrocardiogram (lead II, sensitivity 1, speed 25 mm/s, at room temperature of 30 °C) of a 5-year-old turtle with pneumonia showing sinus tachycardia (heart rate 120 bpm, R-R interval 0.56 s), R wave 0.15 mV, QRS 0.04 s, "P" and "T" waves are not appreciable. (**b**) Dorso-ventral radiograph of the same turtle showing increased radio-opacity of both lungs suggesting lung congestion/pneumonia

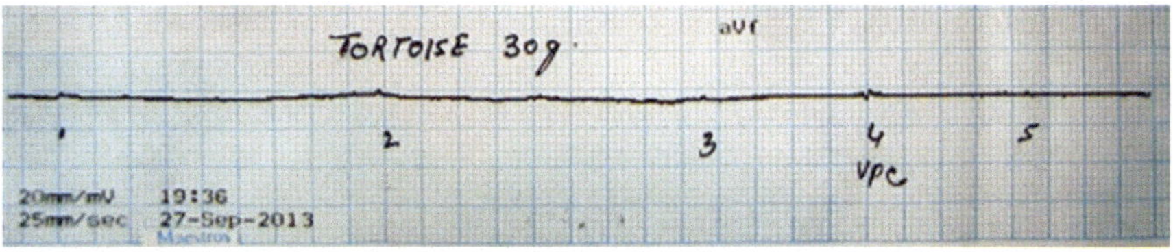

Fig. 19.6 Electrocardiogram (lead II, sensitivity 2, speed 25 mm/s, at room temperature of 30 °C) of a 30 g tortoise showing ventricular premature complex (4th). Other complexes 1, 2, 3, and 5 are normal

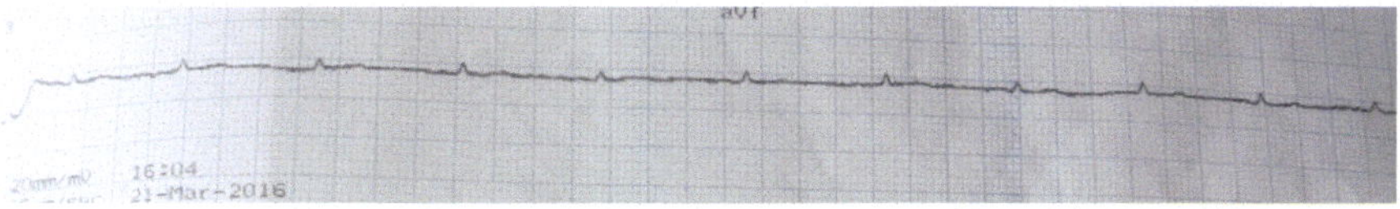

Fig. 19.7 Electrocardiogram (lead II, sensitivity 2, speed 25 mm/s, at room temperature of 30 °C) of 1.6 kg star tortoise showing sinus tachycardia (heart rate 90 bpm with almost constant R-R interval). "P" and "T" complexes are very very small

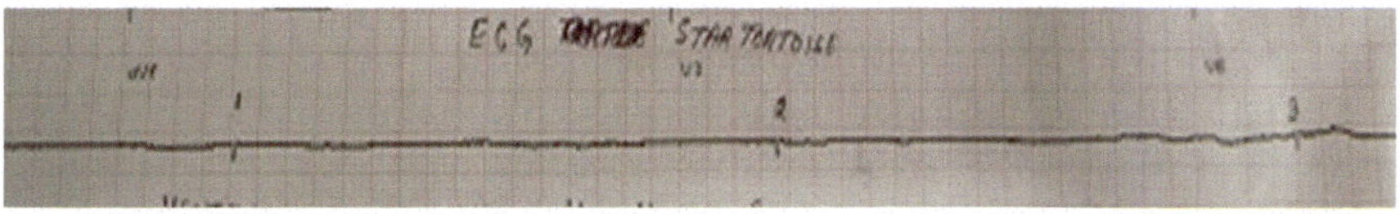

Fig. 19.8 Electrocardiogram (lead II, sensitivity 2, speed 25 mm/s, at room temperature of 30 °C) of a 200 g star tortoise with almost no activity showing ventricular escape rhythm

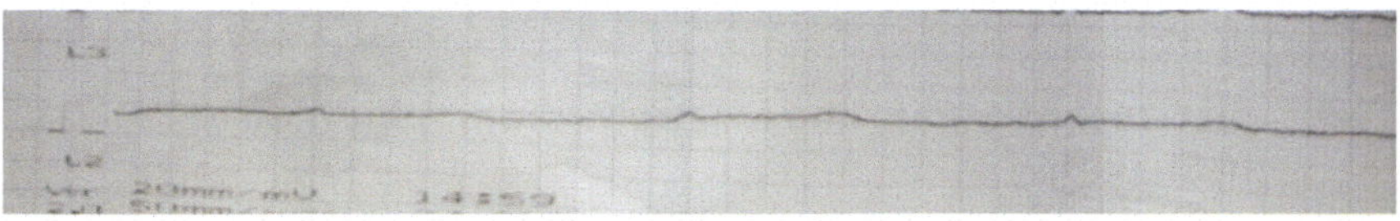

Fig. 19.9 Electrocardiogram (lead II, sensitivity 2, speed 25 mm/s at room temperature of 30 °C) of a female 4-year-old turtle with anasarca showing sinus rhythm with a heart rate of 60 bpm; imperceptible "P" wave; broad (0.06 s) and small (0.05–0.1 mV) "R" wave; imperceptible "T" wave; and almost regular R-R interval of 1.0 s suggesting sinus tachycardia (Varshney 2016)

19.2 Snakes

Reptile cardiology is in infancy. The heart of snakes and chelonians is three-chambered organ. Some workers consider sinus venosus as a separate chamber. Ventricular septum or ridge is incomplete. Blood flow pattern is little different in reptiles owing to three-chambered organ. Reptile heart needs to be examined during routine clinical examination as many signs of cardiac disease are nonspecific. Auscultation may help in evaluating heart to some extent. By putting a wet cloth over heart region, friction is reduced, and auscultation is facilitated. Detection and differentiation of arrhythmias and heart enlargement in snakes can be diagnosed by electrocardiography. It is very sparingly used in the clinical examination of reptiles as knowledge about reptile electrocardiogram is very limited. Since heart rate of reptiles is related to environmental temperature, it is mandatory to maintain reptile within its preferred optimum temperature zone during electrocardiographic examination.

19.2.1 Signs of Heart Disease

Dullness.
Peripheral edema.
Ascites.
Cyanosis.
Anorexia.
Weight loss.
Sudden death.

19.2.2 Placement of Electrodes for Electrocardiogram

The heartbeat is located visually in dorsal recumbency and right arm (red) and left arm (yellow) electrodes (needle) are placed two heart length cranial to heart on right and left side, respectively; right limb (black) and left limb (green) electrodes are placed two heart length caudal to the heart on right and left side, respectively (Fig. 19.10). Electrocardiographic machine is standardized at 10 mm = 1 mV and at a speed of 25 mm per second, and electrocardiogram is recorded in lead I, II, III, aVR, aVL, and aVF leads at 28–30 °C room temperature.

19.2.3 Electrocardiogram and Electrocardiographic Indices in Snakes

Electrocardiogram of snakes (Fig. 19.11) is comprised of three main complexes ("P," "QRS," and "T") as seen in canines, felines, humans, and ruminants. In snakes,

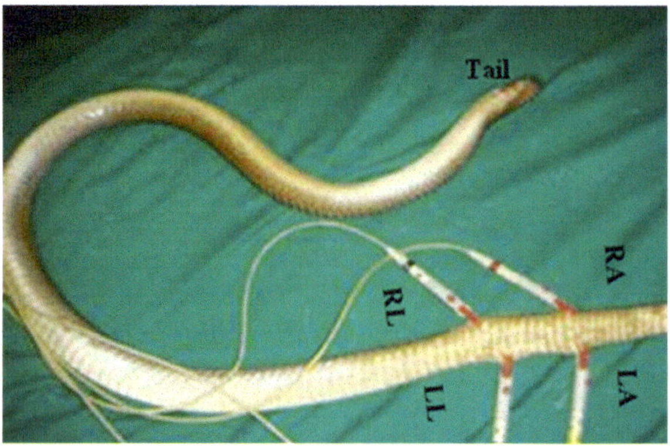

Fig. 19.10 Placement of electrode in a snake. Right arm (RA red), left arm (LA yellow) electrodes are placed two heart length cranial to heart on right and left side, respectively; right limb (RL black) and left limb (LL green) electrodes are placed two heart length caudal to the heart on right and left side, respectively

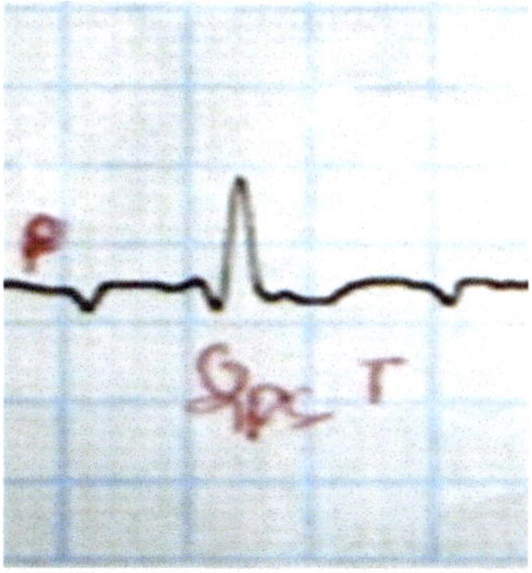

Fig. 19.11 Electrocardiogram (lead II, sensitivity 1, speed 25 mm/s at room temperature of 30 °C) of a rattle snake showing "P," "QRS," and "T" complexes

depolarization begins with SV, and the "P" wave is created by atrial contraction. It is followed by ventricular contraction resulting into "R" wave. Repolarization of ventricle is represented by "T" wave. Low amplitudes of the complexes are the characteristic of snake electrocardiogram as is the case with chelonians.

19.2.4 Factors Influencing Electrocardiogram of the Snakes

1. Heart rate is dependent on body temperature.
2. Body temperature is dependent on environmental temperature.
3. Intervals such as "P-R" and "Q-T" are dependent on heart rate.

19.2.5 Uniqueness of Reptile Electrocardiogram

1. Low heart rate.
2. Very low electrical amplitude of the waves.
3. Many times P and T waves imperceptible.
4. P may be negative.
5. No Q and S waves.

19.2.6 Cardiovascular Diseases in Snakes

1. Primary heart diseases in snakes are not common.
2. Mostly heart ailment is secondary to systemic diseases.
3. Congenital defects of atrioventricular valves and stenosis of vessels may cause primary heart disease.
4. Endocarditis and congestive heart failure have been reported in python.
5. Cardiomyopathy has been reported in snakes.
6. Increased repolarization time on calcium-deficient diet leading to secondary nutritional hyperparathyroidism has also been reported.

19.2.7 Abnormal Electrocardiogram in a Cobra Snake (Fig. 19.12)

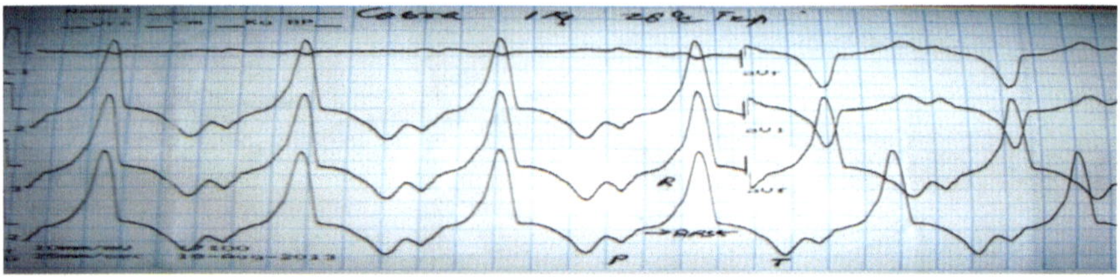

Fig. 19.12 Electrocardiogram (lead II, sensitivity 1, speed 25 mm/s at room temperature of 30 °C) of the cobra snake (weighing 1.0 kg) showing negative "P" wave, increased amplitude of the "R" wave (3.0 mV), broad "QRS" (280–320 ms), and negative "T" wave suggesting heart enlargement (Varshney and Chaudhary 2014)

19.3 Avian Species

Avian cardiology is a developing area in veterinary medicine. Despite a great magnitude of cardiac diseases in avian species, its diagnosis poses a great challenge to veterinarians. Cardiovascular diseases often remain undiagnosed in avian species until the heart decompensate because of nonspecific clinical signs and limited diagnostic techniques. For evaluation of the heart in birds, auscultation is unrewarding owing to fast heart rate. Though electrocardiography is routinely employed for evaluation of heart in canines and felines, its application in avian patients is yet to find routine place as diagnostic technique probably because of the scarcity of electrocardiographic reference values for healthy avian species. Though pigeons are high-endurance birds having an elevated cardiopulmonary capacity, they are also prone to cardiac problems. For detecting cardiac arrhythmias and heart enlargement in pigeons, electrocardiography is an important noninvasive cost-effective potential tool. It is all the more important that reference values for each species/breeds of birds are worked out because of the differences in the electrocardiographic parameters between different breeds.

19.3.1 Placement of Electrodes for Electrocardiogram in Pigeons

The pigeons are manually restrained in dorsal recumbency without any anesthetic or tranquilizer. Leads are attached on the right wing (RA), left wing (LA) and left leg (LL), and right leg(RL) at gastrocnemius muscle as shown in Fig. 19.13. Feathers are clipped on the proximal part of the rachis of the feathers, and gel is applied liberally between skin and clips. The standard bipolar limb leads (lead I, II, III) and

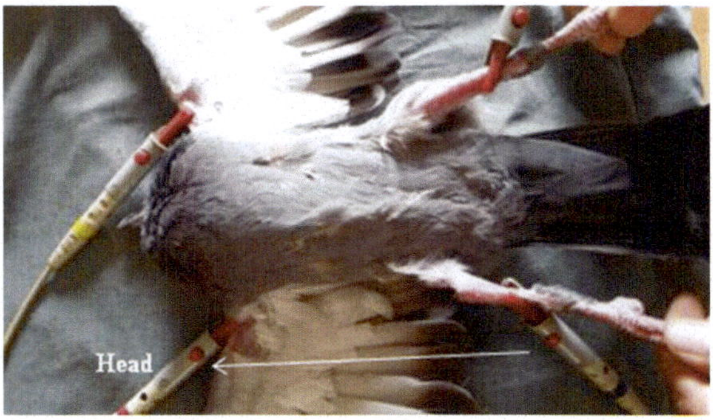

Fig. 19.13 Showing attachment of electrodes in pigeon. Electrodes are attached on the right wing (RA), left wing (LA) and left leg (LL), and right leg (RL) at gastrocnemius muscle

Fig. 19.14 Electrocardiogram of clinically healthy pigeon (lead II, sensitivity 2, speed 50 mm/s) showing "P," "QRS," and "T" complexes

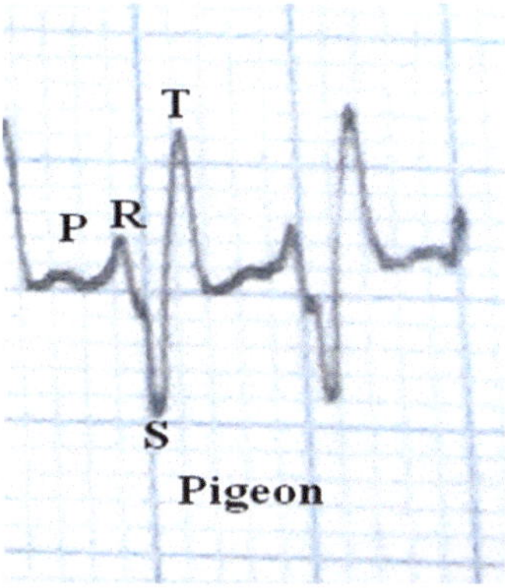

augmented leads (aVR, aVL and aVF) are recorded in a quiet place, after comfortably settling of the pigeons for 5–10 min, on a multichannel electrocardiographic machine with a paper speed of 25 or 50 mm/s and calibration of 10 or 20 mm equal to 1 mV. A frequency filter may be used to avoid muscular tremor artifacts. ECG recordings are analyzed for amplitude and duration of "P," "QRS," and "T" waves; P-R, Q-T, and R-R intervals in lead II.

19.3.2 Electrocardiogram and Electrocardiographic Indices in Pigeons

Electrocardiogram of the pigeon is comprised of "P," "QRS," and "T" complexes (Figs. 19.14 and 19.15) as seen in other animals. "P" wave is small, peaked, or slightly isoelectric, positive, and monophasic. PR segment is either depressed or isoelectric owing to fast heart rate. "Q" wave is absent. "R" wave is very small (r),

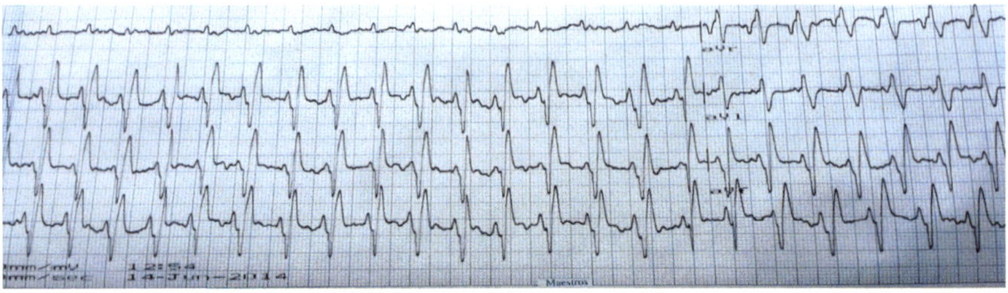

Fig. 19.15 Electrocardiogram of clinically healthy pigeon in limb leads (lead I, II, III) and augmented leads (aVR, aVL, and aVF) on a multichannel electrocardiographic machine with a paper speed of 50 mm/s and calibration of 20 mm equal to 1 mV

Table 19.2 Electrocardiographic indices in conscious healthy domestic pigeons (lead II)

ECG indices (*n* = 180)	Range	Mean ± S.E.	Median
Heart rate (bpm)	200–400	315.40 ± 26.72	300.0
"P" amplitude (mV)	0.05–0.35	0.156 ± 0.0056	0.15
"P" duration (s)	0.02–0.06	0.032 ± 0.0006	0.03
P-R interval (s)	0.02–0.06	0.036 ± 0.0005	0.04
"S" amplitude (mV)	0.10–0.80	0.445 ± 0.0027	0.40
"S" duration (s)	0.02–0.08	0.033 ± 0.0005	0.03
S-T segment (s)	0.00–0.03	0.0083 ± 0.0011	0
Q-T interval (s)	0.02–0.10	0.0417 ± 0.0005	0.04
"T" amplitudes (mV)	0.10–1.30	0.370 ± 0.0139	0.35
"T" duration (s)	0.02–0.12	0.051 ± 0.0006	0.05
R-R interval (s)	0.11–0.42	0.206 ± 0.0041	0.20

but "S" (negative deflection) is quite appreciable. In most of the pigeons the "QRS" is of rS pattern in lead II. The polarity of "QRS" is negative in lead II, III, and aVF and positive in avR and aVL (Fig. 19.15). The "QRS" complex in lead I is very small or isoelectric. "ST" segment is slurring. "T" wave is generally positive in lead I, II, III, and aVF and negative in lead aVR and aVL leads.

Electrocardiographic indices of healthy pigeons are given in Table 19.2 (Varshney 2017a). The heart rate in pigeons varies from 200 to 400 bpm with a mean of 315.40 ± 26.72 and a median of 300. "P" waves amplitude ranges from 0.05 to 0.35 mV with a mean of 0.156 ± 0.0056 and a median of 0.15 mV. Its duration varies from 0.02 to 0.06 s with a mean duration of 0.032 ± 0.0006 s (median 0.03 s). The mean "P-R" interval is 0.036 ± 0.0005 s (range 0.02–0.06 s; median 0.04 s). "QRS" complex is predominantly negative in the form of rS. The "S" amplitude ranges from 0.10 to 0.80 mV with a mean of 0.445 ± 0.0027 mV and a median of 0.40 mV. Its duration varies from 0.02 to 0.08 s with a mean of 0.033 ± 0.0005 s and a median of 0.03 s. The "T" wave is generally positive with an amplitude of 0.370 ± 0.0139 mV (range 0.10–1.3 mV and median as 0.35 mV) and duration of 0.051 ± 0.0006 s (range of 0.02–0.12 s and median of 0.05 s). The "QT" interval varies from 0.02 to 0.12 s (0.0417 ± 0.0005 s and median of 0.04 s). "R-R" interval varies from 0.0.30 to 0.15 s (mean 0.194 ± 0.0041 s; median 0.20 s) depending upon heart rate. High heart rate has shorter R-R interval.

19.3.3 Uniqueness of Pigeon Electrocardiogram

1. High heart rate.
2. Low electrical amplitude of "P" waves.
3. P-R segment very small depression of isoelectric line.
4. "QRS" predominantly negative "rS" type
5. S-T segment negligible or slurring.

19.3.4 **Parrot** (Fig. 19.16)

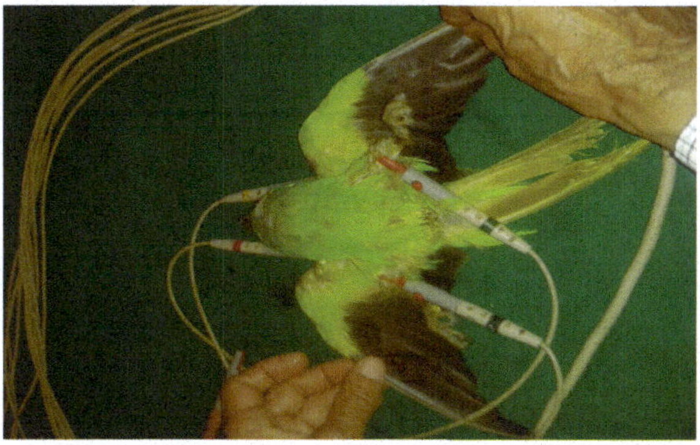

Fig. 19.16 Placement of electrodes in a parrot for recording electrocardiogram

19.3.5 **Arrhythmias and Other Electrocardiographic Changes in Pigeons** (Varshney 2017a) (Figs. 19.17, 19.18, 19.19, 19.20, 19.21, 19.22, 19.23, 19.24, 19.25, 19.26, 19.27 and 19.28)

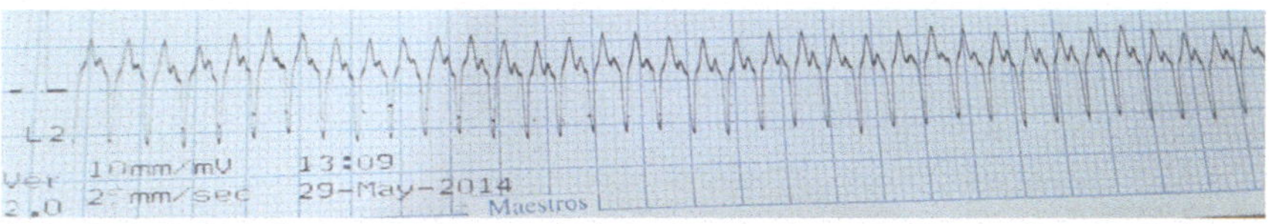

Fig. 19.17 Electrocardiogram (lead II, sensitivity 1, speed 25 mm/s) of a pigeon showing heart rate as 440 bpm suggesting tachycardia

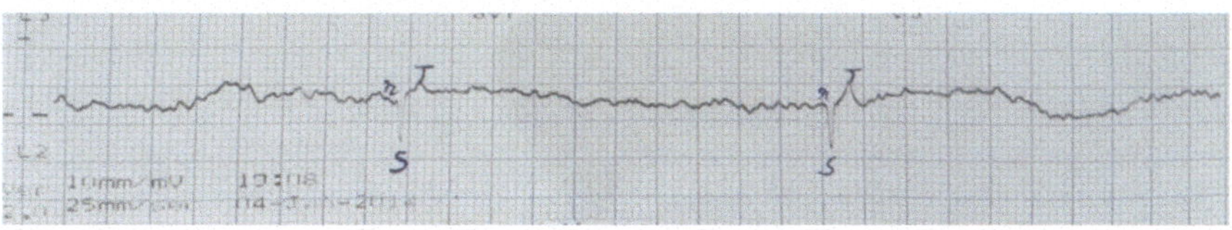

Fig. 19.18 Electrocardiogram (lead II, sensitivity 1, speed 25 mm/s) of a pigeon

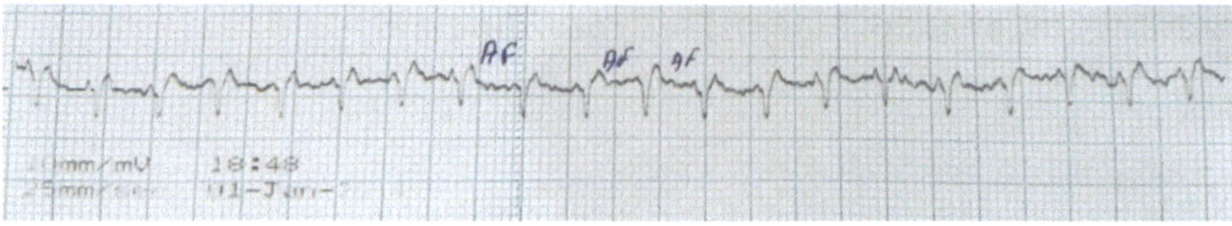

Fig. 19.19 Electrocardiogram (lead II, sensitivity 1, speed 25 mm/s) of a pigeon showing rapid atrial rhythm in chaotic manner and "P" wave is replaced by fine "f" wave oscillations suggesting atrial fibrillation

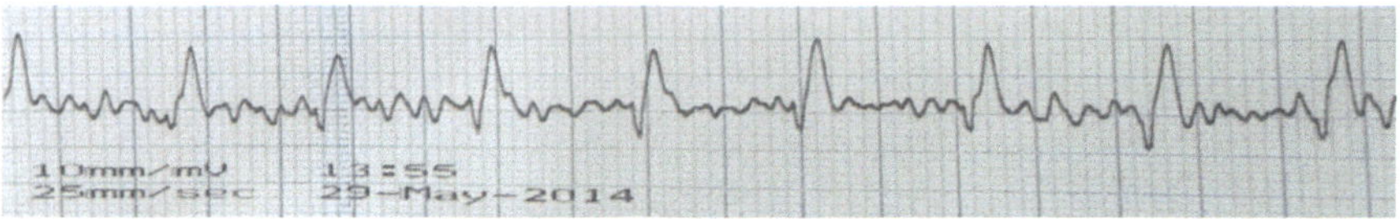

Fig. 19.20 Electrocardiogram (lead II, sensitivity 1, speed 25 mm/s) of a pigeon showing rapid atrial rhythm in chaotic manner and "P" wave is replaced by saw tooth wave suggesting atrial flutters

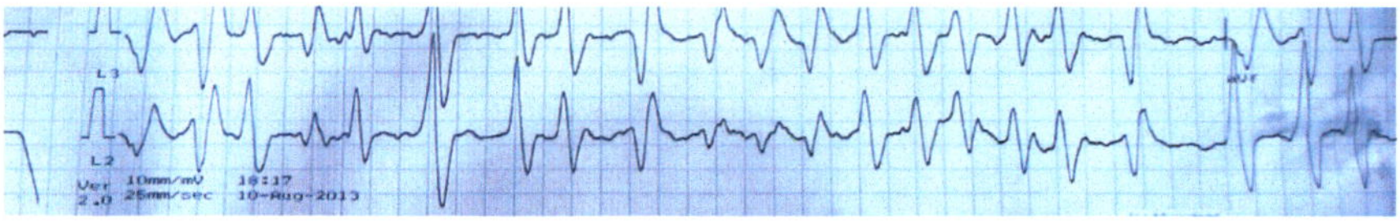

Fig. 19.21 Electrocardiogram (lead II, sensitivity 1, speed 25 mm/s) of a pigeon with heart failure showing multiform (varying configuration of ventricular complexes) ventricular premature complexes at an average of 140/min. The fourth, 10th, 11th, and 12th complexes from the beginning of the strip are a capture complex. The normal sinus impulse has reached the AV junction capturing the ventricles for one complex

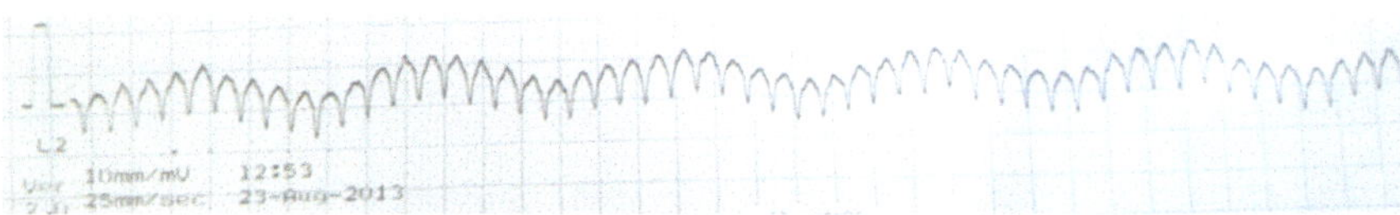

Fig. 19.22 Electrocardiogram (lead II, sensitivity 1, speed 25 mm/s) of a pigeon showing ventricular tachycardia

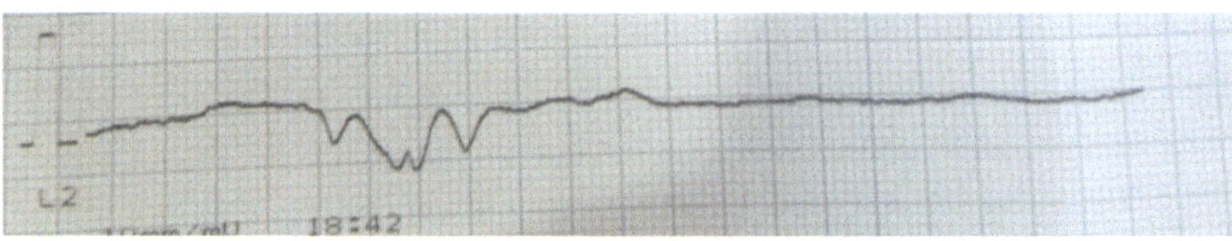

Fig. 19.23 Electrocardiogram (lead II, sensitivity 1, speed 25 mm/s) of a pigeon showing ventricular flutter

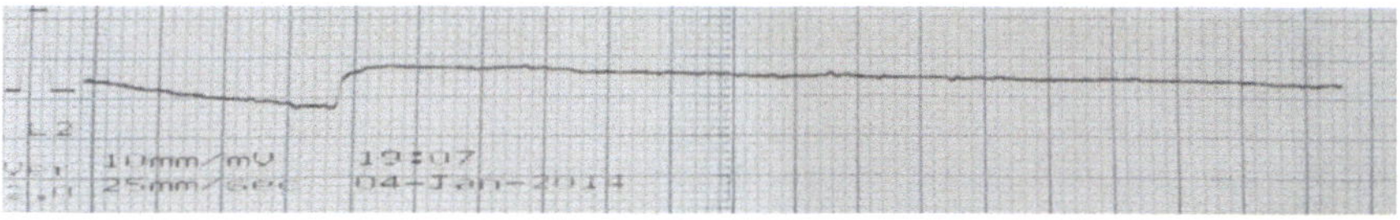

Fig. 19.24 Electrocardiogram (lead II, sensitivity 1, speed 25 mm/s) of a pigeon showing ventricular asystole with severe A-V block

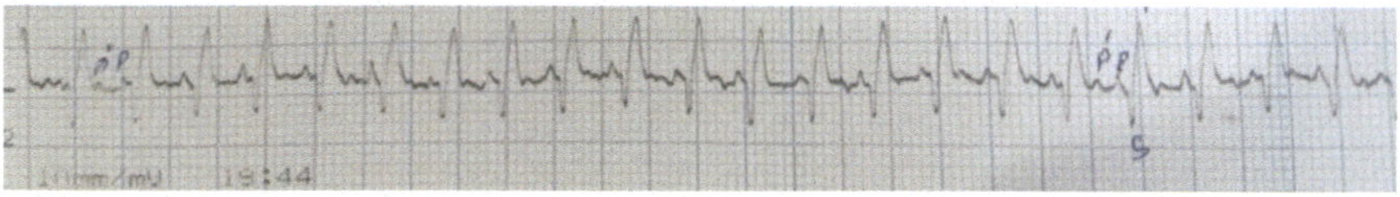

Fig. 19.25 Electrocardiogram of (lead II, sensitivity 1, speed 25 mm/s) a pigeon showing second-degree AV block. "P" wave is without QRS and T complexes

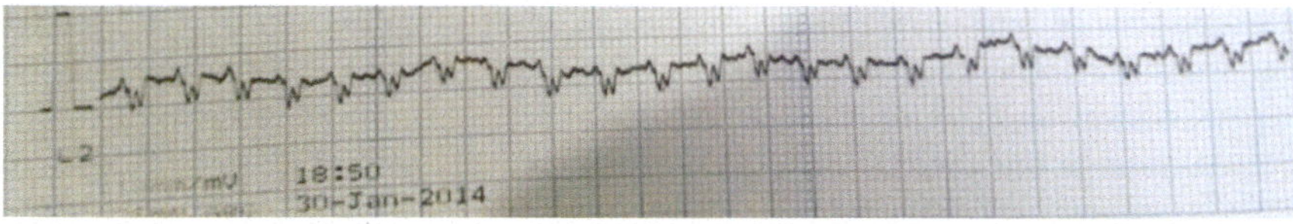

Fig. 19.26 Electrocardiogram (lead II, sensitivity 1, speed 25 mm/s) of a pigeon showing right bundle branch block (RBBB). "QRS" duration is large and wide

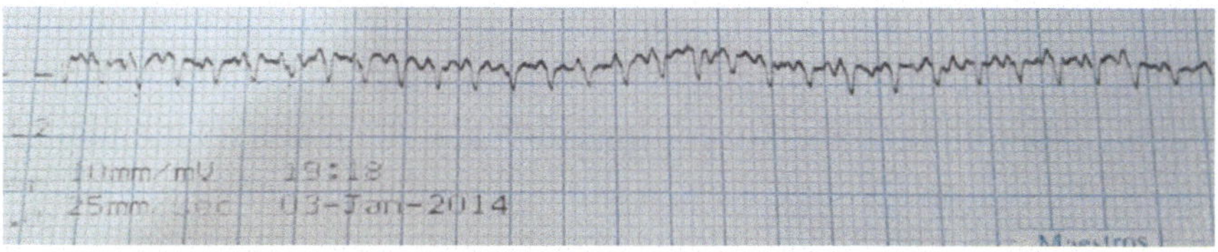

Fig. 19.27 Electrocardiogram (lead II, sensitivity 1, speed 25 mm/s) of a pigeon showing low voltage "QRS." "QRS" complex is small

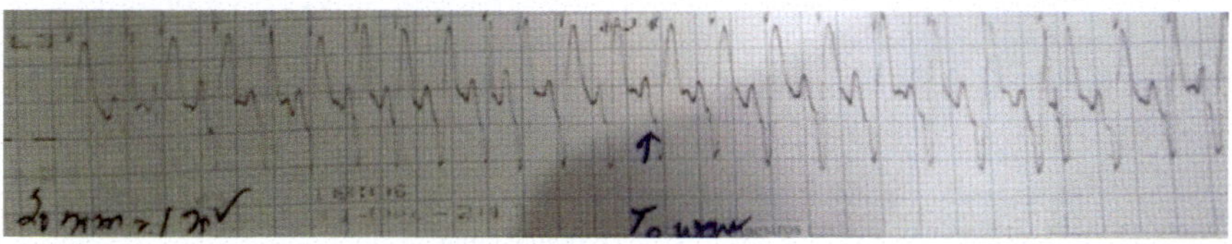

Fig. 19.28 Electrocardiogram (lead II, sensitivity 2, speed 25 mm/s)of a pigeon showing "Ta" wave. Right descending arm of "P" wave is falling below baseline and is longer than the left ascending arm

19.3.6 Abnormal Electrocardiogram in a Parrot (Fig. 19.29)

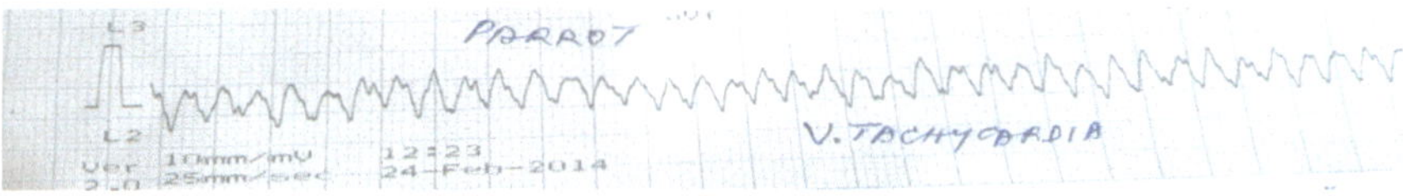

Fig. 19.29 Electrocardiogram (lead II, sensitivity 1, speed 25 mm/s) of a parrot with syncope showing aberrant wave forms in run, wide "QRS" unrelated to "P" suggesting ventricular tachycardia at an average of 300 VPC per minute (Varshney 2014)

19.4 Equines

The electrocardiogram in dogs, chelonians, reptiles, and avian is being used to assess the status of myocardium, its hypertrophy, arrhythmias, and conduction disturbances. Whereas in equines, the use of electrocardiogram is restricted to the detection and characterization of arrhythmias and conduction disturbances because of different natures of ventricular depolarization. Ventricular depolarization of large myocardial muscle mass in equines is not recognized by surface electrocardiogram owing to deep penetration of Purkinje fibers in ventricular muscles and depolarization and repolarization occurring over multiple minor fronts at a point of time. Because of these characteristics, base-apex system is used in equines confining to detection of arrhythmias and conduction disturbances as in ruminants.

19.4.1 Placement of Electrodes for Electrocardiogram in Equines

In equines base-apex lead system is quite appropriate as is in ruminants for detecting arrhythmias and conduction disturbances. The equines are kept in standing position on a rubber mat, and the sites for applying alligator clips are prepared by shaving and cleaning with alcohol. Gel is applied liberally on the areas of attachment of electrodes. Electrocardiogram is recorded in calm and quite surroundings on an electrocardiographic machine at a paper speed of 25 mm per second and amplitude of 1 mV per 10 mm. In base-apex lead system, the right arm electrode is attached to neck in the right jugular furrow two-third down in the neck, left arm electrode is placed over the apex of the heart just behind left elbow, and earth electrode (right hind leg electrode) is attached at wither (Figs. 19.30 and 19.31) or in the right side in front of scapula. ECG recording is done in lead I and is analyzed for amplitudes and duration of "P," "QRS," and "T" waves; P-R interval, ST segment, and Q-T interval; and heart rate.

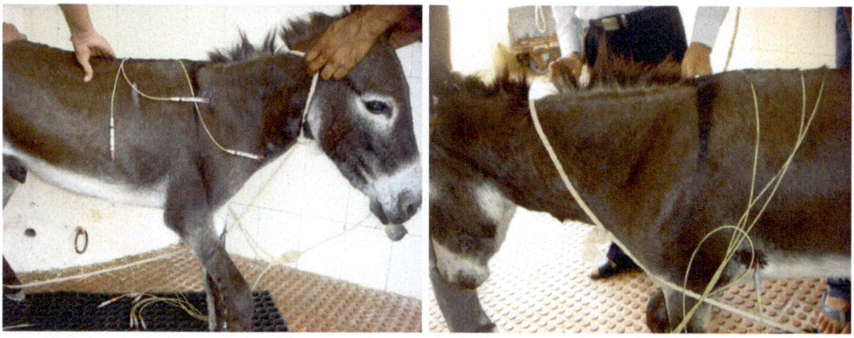

Fig. 19.30 Placement of electrodes in donkey in base-apex lead system. Right arm electrode (red) is placed in right jugular furrow two-third (2/3) down in the neck, left arm electrode is placed over the apex of the heart just behind left elbow, and earth electrode (right hind leg electrode) is attached at wither or in the right side in front of scapula

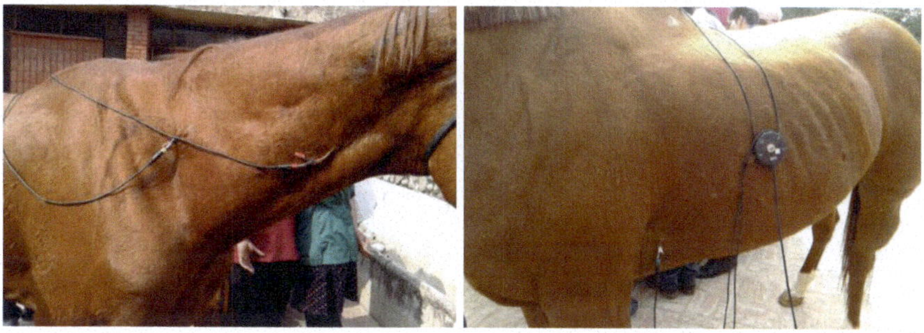

Fig. 19.31 Placement of electrodes in horse in base-apex lead system. Right arm electrode (red) is placed in right jugular furrow two-third (2/3) down in the neck, left arm electrode is placed over the apex of the heart just behind left elbow, and earth electrode (right hind leg electrode) is attached in the right side in front of scapula
(Photos of horse showing electrode placement are with the Courtesy of Dr. Neetu Saini Department of Medicine, Guru Angad Dev Veterinary and Animal Science University, Ludhiana, Punjab)

19.4.2 Electrocardiogram and Electrocardiographic Indices in Equines

Electrocardiogram of equines is similar to ruminants in shape, and its QRS is predominantly negative. "P" wave is either simple monophasic or biphid in configuration (Fig. 19.32). The QRS complex is predominantly negative in the form of rS or QS (Fig. 19.32, donkeys; Fig. 19.33, horse). The T wave is either biphasic (Fig. 19.32) or monophasic. In donkeys heart rate varies from 32 to 58 bpm with a mean of 40.16 ± 3.72, and a median of 38.0. "P" wave is of 0.15–0.20 mV amplitude (mean 0.175 ± 0.0110 mV, median of 0.175 mV), and its duration varies from 0.07 to 0.09 s (mean 0.083 ± 0.0033 s, median 0.075 s). The mean P-R interval is generally 0.21 ± 0.016 s (median 0.21 s; range from 0.16 to 0.24 s). "QS" or "S"

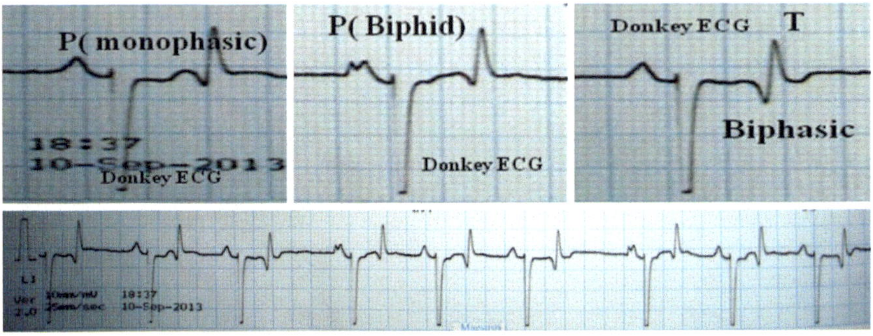

Fig. 19.32 Electrocardiogram (base-apex lead system lead I, sensitivity 1.0, speed 25 mm/s) of an adult male donkey showing biphasic P at two places,'rS' type QRS pattern and biphasic 'T' wave

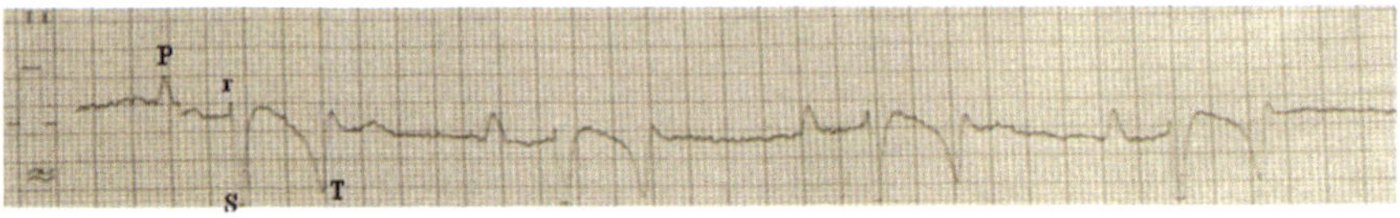

Fig. 19.33 Electrocardiogram (base-apex lead system lead I, sensitivity 1.0, speed 25 mm/s) of an adult male horse showing 'rS' type QRS pattern and – ve 'T' wave
(This ECG strip of the horse is with the courtesy of Dr. Neetu Saini, Department of Medicine, Guru Angad Dev Veterinary and Animal Science University, Ludhiana, Punjab).

Table 19.3 Electrocardiographic parameters of healthy horses (lead I)

Electrocardiographic parameters	Range values
Heart rate	30.0–45.0 bpm
P amplitude	0.30–0.5 mV
P duration	0.08–0.12 s biphid P wave are also seen
P-R interval	0.24–0.30 s
R amplitude	0.50–0.9 mV
S amplitude	1.20–1.50 mV
QRS duration	0.09–0.14 s
S-T segment	0.20–0. 24 s
T amplitude	0.70–1.0 mV
T duration	0.12–0.20 s biphasic T is also common
Q-T interval	0.50–0.58 s

amplitude ranges from 0.5 to 1.55 mV (mean 0.825 ± 0.157 mV, median 0.7 mV); and its duration varies from 0.08 to 0.09 s (mean of 0.081 ± 0.053 s, median 0.085 s). Duration of S-T segment is of 0.193 ± 0.071 s (range 0.16–0.24 s). "T" wave amplitude ranges from 0.4 to 0.8 mV (mean 0.56 ± 0.067, median 0.55 mV) and duration from 0.08 to 0.12 s (mean 0.093 ± 0.0068 s, median 0.09 s). The Q-T interval is large and variable (0.32–0.43 s). The mean value of R-R interval is 1.528 ± 0.11 s (range 1.03–1.8 s, median 1.57 s) and seems related to heart rate. The values of electrocardiographic parameters of healthy horses are given in Table 19.3.

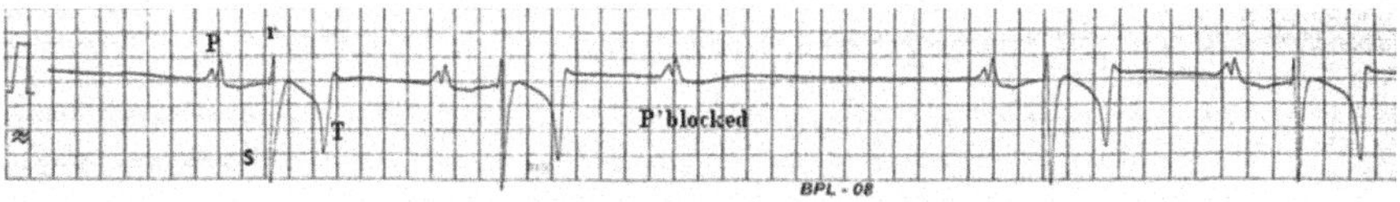

Fig. 19.34 Electrocardiogram (base-apex lead system lead I, sensitivity 1.0, speed 25 mm/s) of an adult male horse showing second-degree AV block. P′ wave, after two complexes from the left, is unconducted. This ECG strip of the horse is with the courtesy of Dr. Neetu Saini, Department of Medicine, Guru Angad Dev Veterinary and Animal Science University, Ludhiana, Punjab)

19.4.3 Characteristics of Equine Electrocardiogram

1. "QRS" is predominantly negative.
2. "r" is very small.
3. "P" monophasic or biphid.
4. "P" broad as compared to canines.
5. "T" monophasic or biphasic.
6. Artifacts due to movement of wire on the animal more common.
7. Muscle tremors are very common due to respiratory movements.

19.4.4 Arrhythmias in Equines

Incidence of arrhythmias in horses is high at rest because of high vagal tone, and these are considered physiological as they abolish on exercise, excitement, or increase in sympathetic tone. Heart rate in resting horses varies from 30 to 45. On the other hand some arrhythmias are exaggerated on exercise and warrants proper investigation. Common arrhythmias in horses are:

1. Sinus bradycardia (less than normal range).
2. Sinus tachycardia (high than normal range).
3. Sinus arrhythmia.
4. Sinus blocks.
5. Atrial premature complexes.
6. Atrial fibrillation.
7. Heart blocks (Fig.19.34).

19.4.5 Association of Arrhythmias with Diseases in Equines

Arrhythmias	Associated conditions
Sinus tachycardia	Fever
	Anemia
	Shock
	Hemorrhages
	Heart failure
Sinus arrhythmias	During rest after exercise
	Mostly physiological

Arrhythmias	Associated conditions
Sinus block or sinus arrest	Generally physiological
	Vagal induced
	May lead to syncope
Atrial premature complexes	Atrial disease
Atrial tachycardia	Atrial myocardial disease
	Systemic diseases
	Electrolyte imbalances
Atrial fibrillation	Common in horses
	Paroxysmal in thoroughbreds during racing
	Cardiac disease
Second-degree AV block	Physiological due to high vagal tone.
	Advance blocks are pathological.
Third-degree AV block	Degenerative or inflammatory AV node disease
Ventricular tachycardia	Cardiac or systemic diseases
Ventricular fibrillation	Terminal event

19.4.6 Electrocardiographic Features of Arrhythmias and Conduction Disturbances in Equines

Type of arrhythmias/conduction disturbance	Electrocardiographic features
Normal sinus rhythm	Resting heart rate is in between 30 and 45 bpm
	R-R intervals are almost same
	There is P wave for every QRS complex
	P-R interval is constant
Sinus arrhythmia	Heart rate is within range
	Pauses are shorter than twice the normal R-R interval
	There is a P wave for every QRS complex
	P-R interval is constant
	P-P and R-R intervals are varying
Sinus arrest	Pauses are equal to or greater than twice the normal R-R interval
	Heart rate is within range
	Morphology of P, QRS, and T is normal.
Sinoatrial block	Pauses are exactly twice or < the normal R-R.
Wandering pace maker	"P" waves vary in amplitude
Sinus tachycardia	Resting heart rate is higher than normal limit
	Heart rhythm is regular
	R-R intervals are normal or varying slightly
	There is a "P" wave for every QRS complex
	P-R interval is constant
	"P" and "QRS" are of normal configuration
Sinus bradycardia	Heart rate is lower than the normal range
	R-R intervals are regular
	There is a P wave for every QRS
Atrial premature contractions	Premature P′ wave has different configuration
	QRS-T complexes are normal

Type of arrhythmias/conduction disturbance	Electrocardiographic features
	There is no QRS-T complex for premature P′
	P-R interval of APC may be prolonged
	Extra P′ appears between two normal P waves
Atrial tachycardia	Four or more APCs occur in succession
	Atrial rate is high
	P′ wave (pre mature P wave) is repeated many times
	Rhythm may be regular or irregular
	Morphology of P′ (premature P wave) may be normal or abnormal
Atrial fibrillation (AF)	Atrium contraction is uncoordinated
	Heart rhythm is irregular
	P wave is replaced by fine baseline undulations ("f" wave)
	QRS is normal without P wave
	R-R interval is irregularly irregular
	Paroxysmal AF is seen in thoroughbreds during racing
Ventricular premature complex (VPC)	Heart rate may remain normal
	Normal rhythm is broken by premature beat
	There is no P wave for premature beats
	In normal beat QRS is related to P wave
	Premature complex is large and bizarre
	T wave direction is opposite to QRS in VPC
	Premature complex is followed by pause.
Ventricular fibrillation	Heartbeat is uncoordinated
	Specific P waves and QRS complexes are absent
	There is a series of baseline undulations
Ventricular tachycardia	In normal complex P is of normal configuration
	Four or more VPCs are in row
	QRS complex is wide and bizarre and independent of P
	Abnormal QRS having same morphology denotes monomorphic ventricular tachycardia
	Abnormal QRS with different morphology denotes polymorphic ventricular tachycardia
	VT is serous and fatal arrhythmia
First-degree atrioventricular block (AV block)	Heart rate is usually normal
	Heart rhythm is sinus rhythm
	There is a P wave for every QRS complex
	P-R interval is consistent and prolonged (0.44 s)
	Morphology of P, QRS, and T is normal
Second-degree atrioventricular block (AV block)	Rhythm is broken by the absence of one or several QRS complexes
	There is a P wave for every QRS complexes, but some P waves are not conducted
	Normal complexes have consistent P-R interval
	P wave and QRS complexes are of normal configuration

Type of arrhythmias/conduction disturbance	Electrocardiographic features
Third-degree atrioventricular block (AV block)	Number of P wave is more than QRS complexes
	P-R interval is varying
	Relationship between atrial and ventricular beats is not consistent
	P waves are of normal configuration
	QRS complexes are near normal or bizarre
	P-P and R-R intervals are relatively constant

19.5 Rabbits

Rabbits are used as an experimental animals in biomedical research and are also being kept as pet animals in urban areas. Nowadays their number in outpatient departments is increasing. For diagnosis of ailments, a complete clinical examination of all organs is required. Since heart rate is fast in rabbits and their body size is small, it becomes difficult to evaluate the heart merely by auscultation and pulse examination. This lacuna can be overcome by electrocardiography that is a noninvasive and handy tool for evaluating cardiac rhythm, rate, and conduction disturbances. Not much work has been done on electrocardiography of rabbits.

19.5.1 Positioning of Rabbits for Electrocardiography

The rabbits are positioned on a table covered with a foam mattress or rubber sheet in either right lateral recumbency (Fig. 19.35), sternal recumbency (Fig. 19.36), or dorsal recumbency(Fig. 19.37).

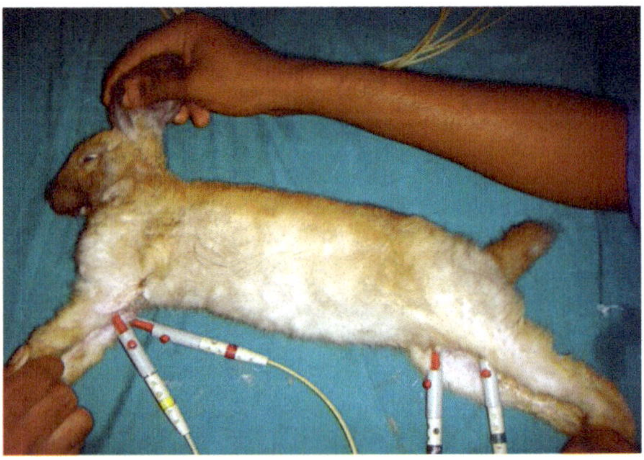

Fig. 19.35 Placement of electrodes in a rabbit in right lateral recumbency position

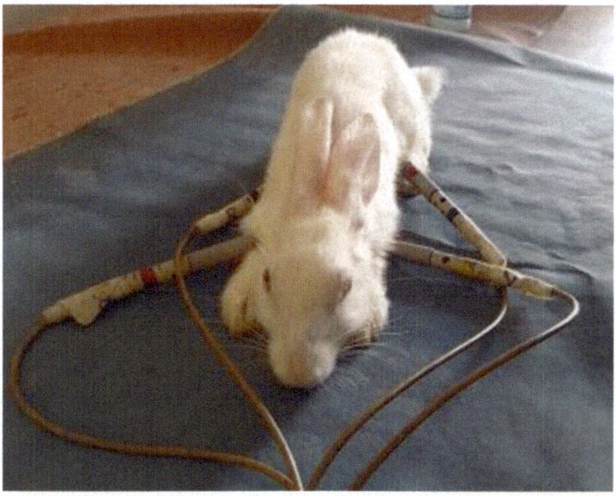

Fig. 19.36 Placement of electrodes in a rabbit in sternal recumbency position

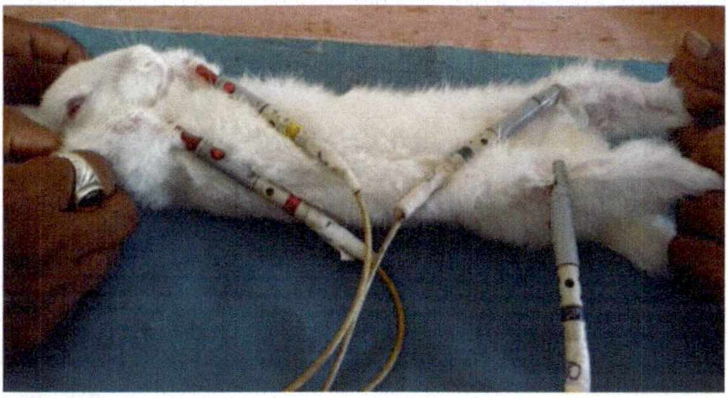

Fig. 19.37 Placement of electrodes in a rabbit in dorsal recumbency position

19.5.2 Placement of the Electrodes in Rabbits

Before putting the electrodes both electrode and skin are moistened with electrocardiographic gel, paste, or alcohol. Alcohol works well. Electrodes are attached directly to the skin. In case nothing is available, water can be used to increase contact between electrode and skin in the unfavorable situation. Placement of electrode is shown in Figs. 19.35, 19.36, and 19.37.

RA	Right forelimb clip (R) electrode is attached proximal to the olecranon on the caudal aspect of the right forelimb
LA	Left forelimb clip electrode (Y) is attached proximal to the olecranon on the caudal aspect of the left forelimb
RL	Right hind limb clip electrode (B)is attached over patellar ligament on the anterior aspect of the right hind limb
LL	Left hind limb clip electrode (G) is attached over patellar ligament on the anterior aspect of the left hind limb

Recording of the electrocardiogram can be done at paper speed of 25 mm/s or 50 mm/s and sensitivity of 1 or 2. At sensitivity of 2 complexes are larger but their numerical value is same as at sensitivity of 1.

19.5.3 Electrocardiogram and Electrocardiographic Indices in Rabbits

Electrocardiogram of rabbits is having "P," "QRS," and "T" complexes and intervals (Figs. 19.38 and 19.39). QRS of rabbits is positive and complexes (P, R and T) are comparatively small. Electrocardiograms taken in dorsal, sternal, or right lateral recumbency do not differ significantly (Fig. 19.40).

Heart rate in rabbits varies from 150 to 250 bpm. The "P" wave is positive in lead I, II, III, and avF and negative polarities in a VR and a VL. QRS in rabbits is predominantly positive in lead I, II, III, and aVF and negative in lead aVR and aVL. "T" polarity follows pattern similar to P and QRS. Amplitude and duration of "P" wave varies from 0.05 to 0.1 mV and 0.03 to 0.04 s, respectively. P-R interval ranges between 0.05 and 0.08 s. In some rabbits "Q" wave is also seen and is small to the tune of 0.05–0.1 mV. "R" amplitude is also small as compared to dogs and cats, and its value ranges in between 0.05 and 0.2 mV. QRS duration varies from 0.03 to 0.04 s. S-T segment is short varying from 0.02 to 0.04 s. Amplitude and duration of "T" wave varies from 0.05 to 0.20 mV and 0.04 to 0.06 s. R-R interval depends on the heart rate.

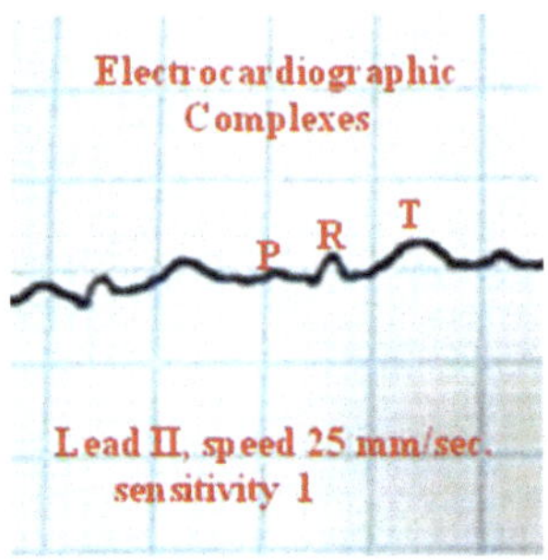

Fig. 19.38 Electrocardiographic complexes (P, R, and T) of an adult male rabbit

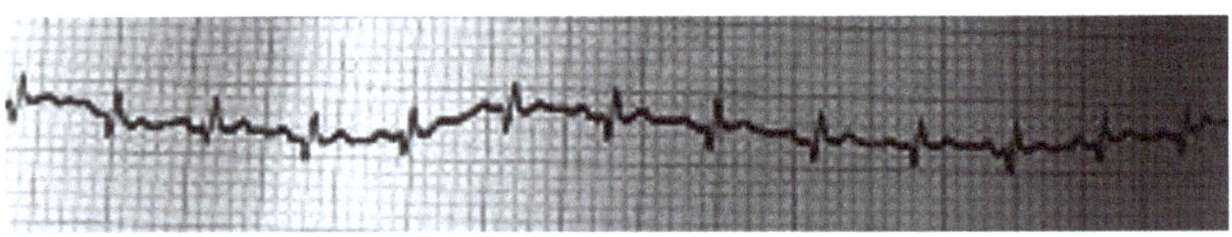

Fig. 19.39 Electrocardiogram (lead II, sensitivity 1, speed 25 mm/s) of a healthy 400 g male rabbit showing heart rate as 220 bpm with sinus rhythm; P 0.05 mV, 0.04 s; P-R 0.08 s: Q 0.1 mV, R 0.2 mV; QRS 0.04 s; S-T 0.02 s: T 0.10 mV, 0.06 s; Q-T interval 0.12 s; and R-R 0.27 s

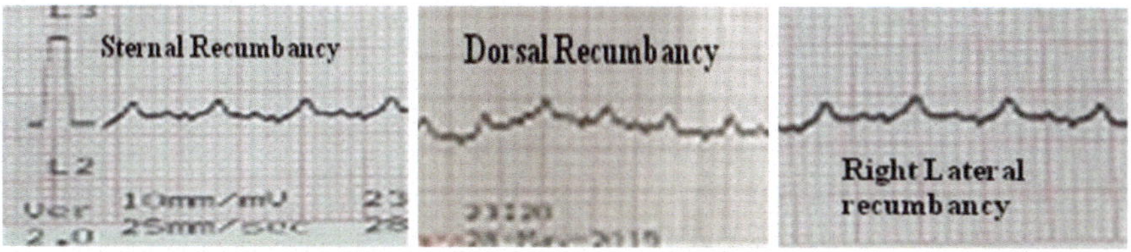

Fig. 19.40 Electrocardiographic complexes (lead II, sensitivity 1, speed 25 mm/s) of the same rabbit in sternal, dorsal and right lateral recumbency. Practically there is no significant difference in the size of the complexes

19.5.4 Abnormal Electrocardiograms in Rabbits (Figs. 19.41 and 19.42)

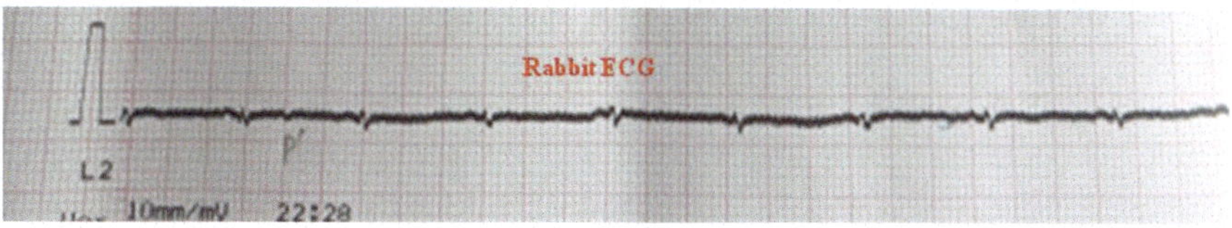

Fig. 19.41 Electrocardiogram (lead II, sensitivity 1, speed 25 mm/s) of a 1.5 kg male rabbit with unconsciousness showing bradycardia (heart rate 96 bpm) and electrical disturbances. The rabbit collapsed within 30 min

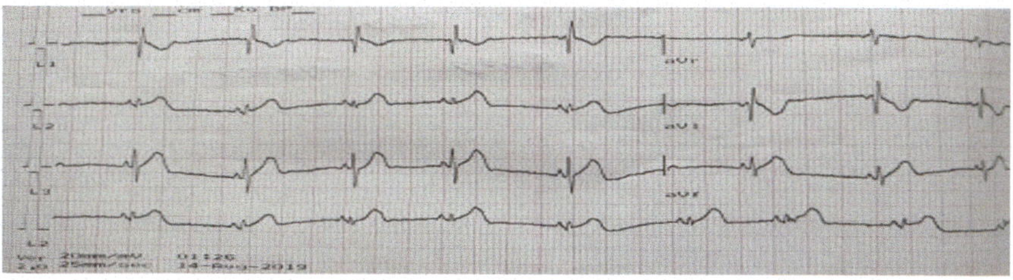

Fig. 19.42 Electrocardiogram (lead I, II, III, aVR, aVL, and aVF sensitivity 2, speed 25 mm/s) of a 650 g male 5–6-month-old rabbit with syncope showing severe bradycardia (heart rate 84 bpm) and large complexes in lead I and III as compared to lead II. In lead II, normal P wave (0.05 mV, 0.04 s), small q(0.025), small r (0.05 mV), normal QRS (0.03 s), ST segment 0.08 s, large and broad T (0.2 mV and 0.08 s), and sinus rhythm (R-R interval 0.66 s) were suggestive of sinus bradycardia with myocardial hypoxia

19.6 Indian Grey Mongoose

Indian grey mongoose (*Herpestes edwardsii*), popularly known as nevlaa in India, often is kept as a pet to keep the dwellings free from rats and other pests. It preys on rats, snakes, birds' eggs, and a variety of invertebrates. It is placed together with cat in suborder Feliformia.

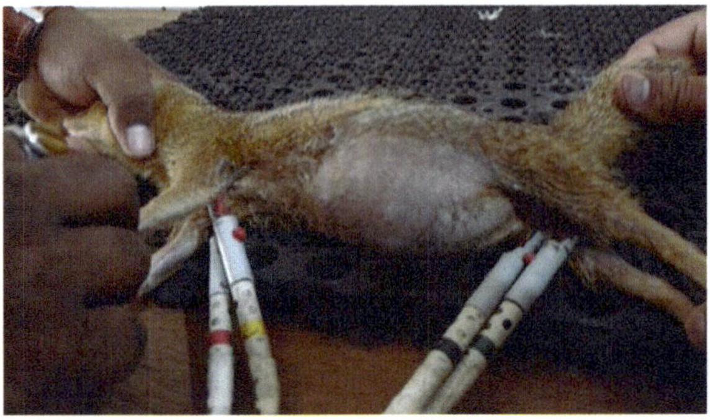

Fig. 19.43 Placement of the electrode in Indian grey mongoose

19.6.1 Positioning of Mongoose and Placement of Electrodes

Positioning of mongoose for electrocardiography is in right lateral recumbency on a table a wooden table or a table covered with a foam mattress and rubber sheet. Electrodes are placed (Fig. 19.43) in similar manner as in case of cats and dogs. Recording of the electrocardiogram is done at paper speed of 25 mm/s and sensitivity of 1.

19.6.2 Electrocardiogram of the Indian Grey Mongoose

Electrocardiogram of an Indian grey mongoose is also having "P," "QRS," and "T" complexes and intervals (Figs. 19.44 and 19.45) similar to cats. P, QRS, and T complexes in lead II, III, and aVF are positive. P and T waves are comparatively small and R is conspicuous. Complexes are small and negative in lead I, aVR, and aVL as seen in canines and felines. No data on normal electrocardiographic values of Indian grey mongoose is available. This electrocardiogram is of a sick Indian grey mongoose.

19.7 Leopard

Importance of electrocardiography in the diagnosis of cardiac diseases of man and animals is well documented. Electrocardiographic studies in various species of domestic and pet animals have been reported in the literature in India and abroad. Nevertheless, information on electrocardiographic indices of leopard is very scanty. In India no report could be traced about electrocardiogram of leopard.

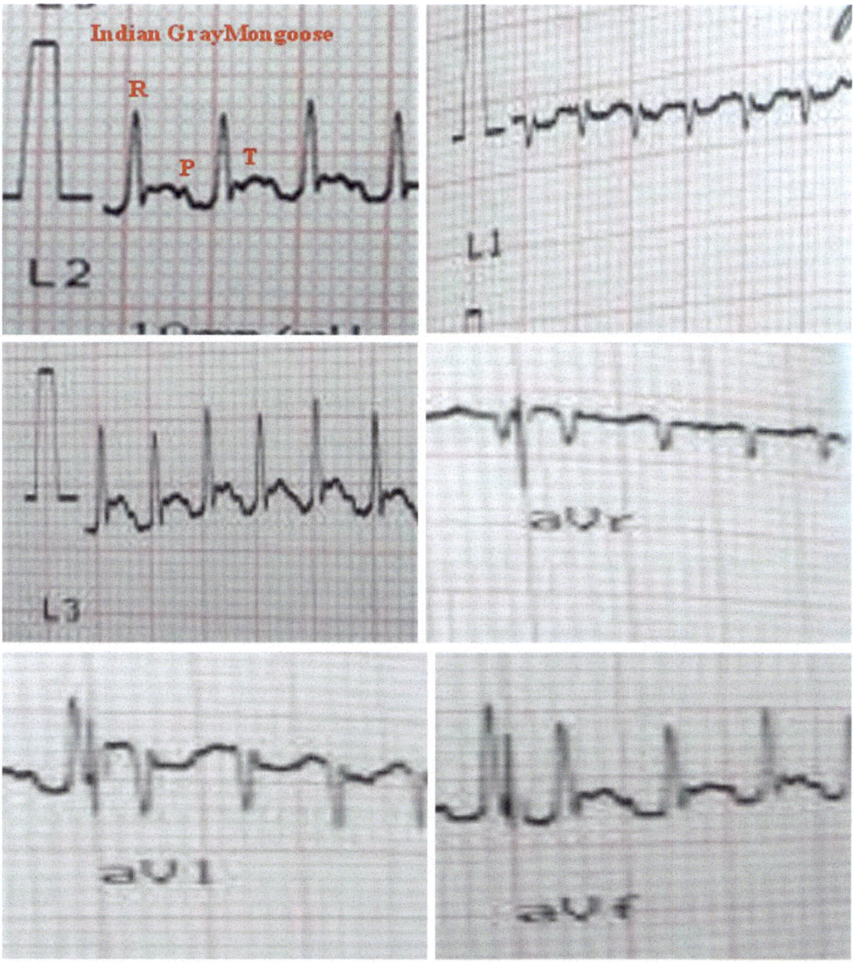

Fig. 19.44 Electrocardiographic complexes of Indian grey mongoose in different leads (sensitivity 1, speed 25 mm/s)

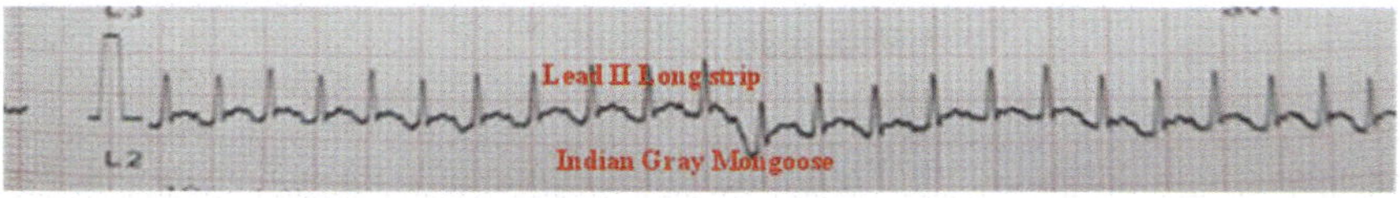

Fig. 19.45 Long strip of electrocardiogram (lead II, sensitivity 1, speed 25 mm/min) of an Indian grey mongoose, weighing 550 g, with ascites owing to liver cirrhosis (hyperechoic liver with serrated margins, fluid in peritoneal cavity and in between liver lobes) showing heart rate as 320 bpm, sinus rhythm; P 0.05 mV, 0.02 s; P-R interval 0.07 s: R 0.6 mV, QRS 0.02 s; S-T 0.03 s; T 0.075 mV, 0.03 s; R-R interval 0.18 s; Q-T 0.08 s suggesting sinus tachycardia as per feline standard (no basic data on Indian grey mongoose is available)

19.7.1 Control of Animal

Wild animals are anesthetized with xylazine (1.0 mg/kg) and ketamine (10 mg/kg) so that they can cooperate during examinations including electrocardiography, echocardiography, or radiography.

19.7.2 Positioning of Leopard and Placement of Electrodes for Electrocardiography

Positioning of leopard for electrocardiography is in right lateral recumbency on a wooden table (Fig. 19.46) or a table covered with a foam mattress and rubber sheet. Electrodes are placed (Fig. 19.46) in similar manner as in case of cats and dogs. Recording of the electrocardiogram is done at paper speed of 25 mm/s and sensitivity of 1.

19.7.3 Electrocardiogram of Leopard

Electrocardiogram of a male 6-month-old leopard cub (Figs. 19.47 and 19.48) is having complexes ("P," "QRS," and "T"), segments, and intervals (P-R interval, S-T segment, Q-T interval, and R-R interval) similar to cats. P, QRS, and T complexes in lead II, III, and aVF are positive. P and T waves are comparatively small and R is conspicuous. Complexes are small and negative in aVR and aVL as seen in canines and felines. Electrocardiogram of the leopard cub revealed heart rate as 140 bpm with sinus rhythm; P 0.1 mV, 0.04 s; P-R interval 0.07 s; R 0.25 mV, QRS 0.04 s; S-T 0.12 s; T 0.2 mV, 0.08 s; R-R interval 0.46 s; and Q-T interval 0.24 s (Fig. 19.49).

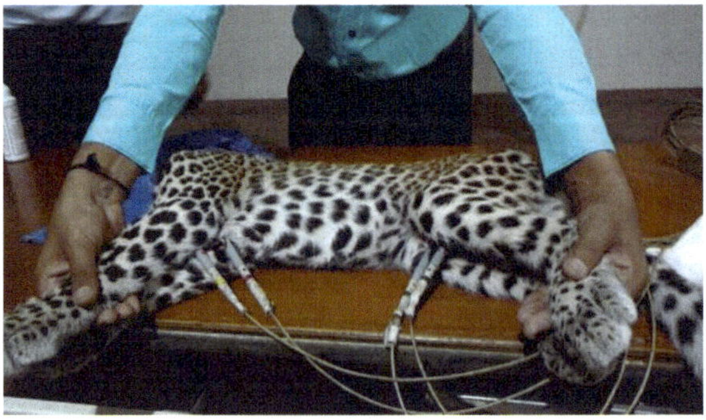

Fig. 19.46 Placement of the electrode in leopard cub

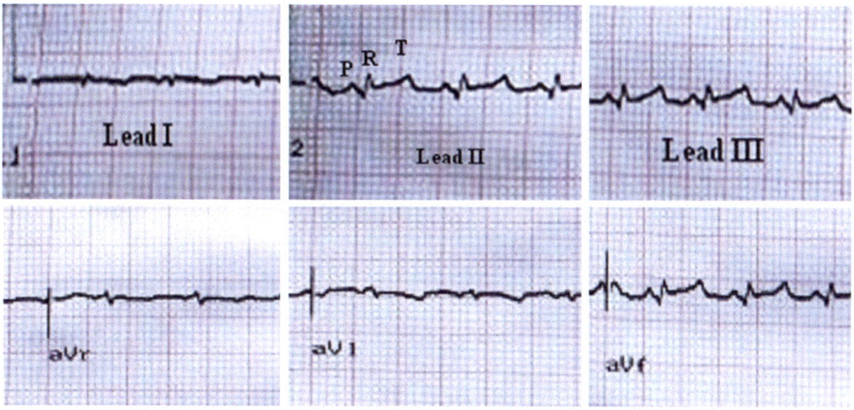

Fig. 19.47 Electrocardiographic complexes of leopard cub in different leads (sensitivity 1, speed 25 mm/s)

Fig. 19.48 Magnified view of electrocardiogram of a leopard cub (lead II, sensitivity 1,speed 25 mm/s) showing different complexes and intervals

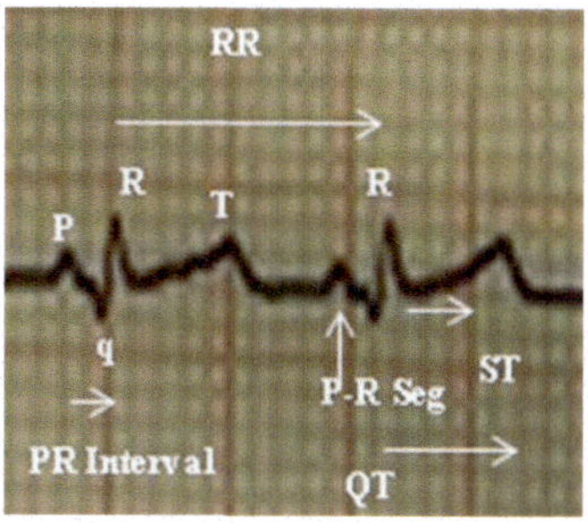

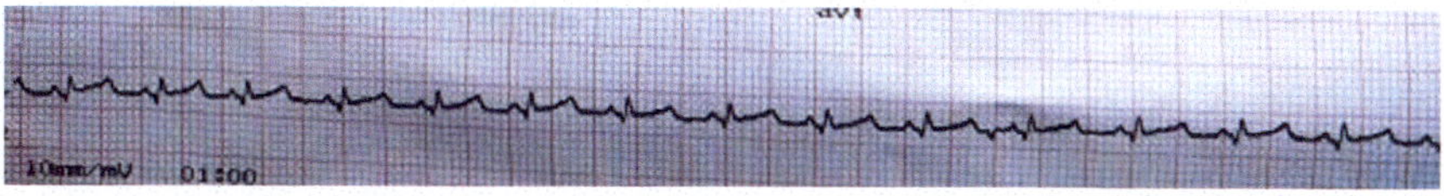

Fig. 19.49 Electrocardiogram (lead II, sensitivity 1, speed 25 mm/min) of a male 6-month-old leopard cub weighing 8.6 kg, with unconsciousness due to first cervical vertebra injury showing heart rate as 140 bpm, sinus rhythm; P 0.1 mV, 0.04 s; P-R interval 0.07 s: R 0.25 mV, QRS 0.04 s; S-T 0.12 s; T 0.2 mV, 0,08 s; R-R interval 0.46 s; and Q-T interval 0.24 s suggesting sinus rhythm with elevation of S-T segment

19.8 Electrocardiographic Complexes of Different Species of Animals at a Glance (Fig. 19.50)

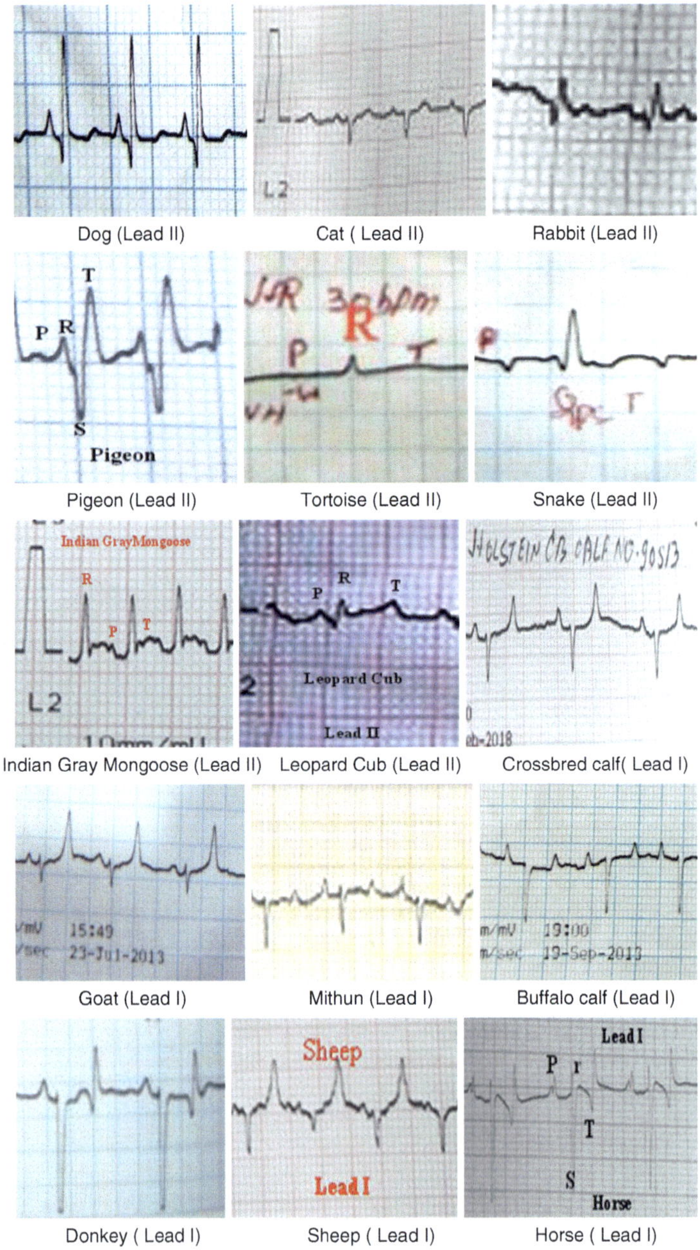

Fig. 19.50 Electrocardiographic complexes of dog, cat, rabbit, pigeon, tortoise, snake, Indian grey mongoose, leopard cub, crossbred calf, goat, mithun, buffalo calf, donkey, sheep, and horse are at a glance

References

Varshney JP (2014) Cardiomyopathy in a parrot. Blue cross book 30:61–62

Varshney JP, Chaudhary PS (2014) Cardiomyopathy in a cobra snake: A clinical case report. Ind J Vet Med 34:70–71

Varshney JP (2016) Diagnosis and management of anasarca in a turtle. Intas Polivet 17:209–210

Varshney JP (2017a) Electrocardiographic studies in domestic pigeons (*Columba livia domestica*). Intas Polivet 18:226–230

Varshney JP (2017b) Electrocardiographic studies in healthy conscious turtle. J Anim Res 7:741–744

Varshney JP, Monapara HD (2019) Radiographic diagnoses in chelonians. Intas Polivet 20:201–206

Further Reading

Fregin GF (1992) Medical evaluation of the cardiovascular system. Vet Clin N Am Equine Pract 8:329–346

Lopez-Murca MM, Bernal LJ, Montes AM, Garcia-Martiniz JD, Ayala I (2005) The normal electrocardiogram of the unanaesthised 'Spanish Pouler' pigeon (Columba livia gutturosa). J Vet Med A 52:347–349

Lumeij JT, Stokhof AA (1985) Electrocardiogram of the racing pigeon (Columba livia domestica). Res Vet Sci 38:275–278

Lumeij JT, Ritchie BW (1994) Cardiovascular system. In: Ritchie BW, Harrsion GJ, Harrsion LR (eds) Avian medicine, principles and application. Brentwood, TN, HBD Int'l Inc, pp 694–722

Marja JLK, Mark AM (2000) Reptile cardiology: a review of anatomy and physiology, diagnostic approaches and clinical disease. Seminars in Avian and Exotic Pet Medicine 14:52–60

Norton TM (2005) Chelonian emergency and critical care. Topics in medicine and surgery. Sem Avian Exot Pet Med 14:106–130

Oda SG, Yamato RJ, Fedullo JD, Leomil NM, Larsson MH (2009) Standardization of some electrocardiographic parameters of captive leopard cats (*Leopardus tigrinus*). J Zoo Wildl Med 40:414–420

Papahn AA, Naddaf H, Rezakhani A, Mayahi M (2006) Electrocardiogram of homing pigeons. J Appl Anim Res 30:129–132

Penn A, Lu M-X, Parkes JL (1990) Ingestion of chlorinated water has no effect upon indicators of cardiovascular diseases in pigeons. Toxicology 63:301–313

Verheyen T, Decloedt A, DeClercq D, Deprez P, Sys SU, van Loon G (2010) Electrocardiography in horses. Part 2: How to read equine ECG. Vlaams Diergeneeskundig Tijdschrift 79:337–344

The manufacturer's authorised representative in the EU is Springer Nature Customer Service Centre GmbH, Europaplatz 3, 69115 Heidelberg, Germany. If you have any concerns regarding our products, please contact ProductSafety@springernature.com

Printed and bound by CPI Group (UK) Ltd, Croydon, CR0 4YY
30/09/2026
02225241-0004